THE SOCIETY OF MIND

Marvin Minsky

Illustrations by Juliana Lee

A TOUCHSTONE BOOK
Published by Simon & Schuster Inc.
NEW YORK • LONDON • TORONTO • SYDNEY • TOKYO

Copyright © 1985, 1986 by Marvin Minsky

First Touchstone Edition, 1988

Published by Simon & Schuster, Inc.
Simon & Schuster Building
Rockefeller Center
1230 Avenue of the Americas
New York, NY 10020

TOUCHSTONE and colophon are registered trademarks
of Simon & Schuster, Inc.

Designed by Irving Perkins Associates
Manufactured in the United States of America

3 5 7 9 10 8 6 4 2
7 9 10 8 6   Pbk.

Library of Congress Cataloging in Publication Data

Minsky, Marvin Lee, date.
The society of mind.

Includes index.
1. Intellect. 2. Human information processing.
3. Science—Philosophy. I. Title.
BF431.M553   1986   153   86-20322
ISBN 0-671-60740-5
ISBN 0-671-65713-5 Pbk.

The author is grateful for permission to reprint the following material:

Excerpts by Theodore Melnechuk. Reprinted by permission of the author.
Excerpt by Manfred Clynes. Reprinted by permission of the author.
From "Burnt Norton" in *Four Quartets* by T. S. Eliot, © 1943 by T. S. Eliot; renewed 1971 by Esme Valerie
Eliot. Reprinted by permission of Harcourt Brace Jovanovich, Inc.
From "Choruses from 'The Rock'" in *Collected Poems 1909–1962* by T. S. Eliot, © 1936 by Harcourt Brace
Jovanovich, Inc.

CONTENTS

CONTENTS

CONTENTS   15

CONTENTS

CHAPTER 1

PROLOGUE

Everything should be made as simple as possible,
but not simpler.

—ALBERT EINSTEIN

This book tries to explain how minds work. How can intelligence emerge from nonintelligence? To answer that, we'll show that you can build a mind from many little parts, each mindless by itself.

I'll call "Society of Mind" this scheme in which each mind is made of many smaller processes. These we'll call *agents*. Each mental agent by itself can only do some simple thing that needs no mind or thought at all. Yet when we join these agents in societies—in certain very special ways—this leads to true intelligence.

There's nothing very technical in this book. It, too, is a society—of many small ideas. Each by itself is only common sense, yet when we join enough of *them* we can explain the strangest mysteries of mind.

One trouble is that these ideas have lots of cross-connections. My explanations rarely go in neat, straight lines from start to end. I wish I could have lined them up so that you could climb straight to the top, by mental stair-steps, one by one. Instead they're tied in tangled webs.

Perhaps the fault is actually mine, for failing to find a tidy base of neatly ordered principles. But I'm inclined to lay the blame upon the nature of the mind: much of its power seems to stem from just the messy ways its agents cross-connect. If so, that complication can't be helped; it's only what we must expect from evolution's countless tricks.

What can we do when things are hard to describe? We start by sketching out the roughest shapes to serve as scaffolds for the rest; it doesn't matter very much if some of those forms turn out partially wrong. Next, draw details to give these skeletons more lifelike flesh. Last, in the final filling-in, discard whichever first ideas no longer fit.

That's what we do in real life, with puzzles that seem very hard. It's much the same for shattered pots as for the cogs of great machines. Until you've seen some of the rest, you can't make sense of any part.

1.1 THE AGENTS OF THE MIND

Good theories of the mind must span at least three different scales of time: slow, for the billion years in which our brains have evolved; fast, for the fleeting weeks and months of infancy and childhood; and in between, the centuries of growth of our ideas through history.

To explain the mind, we have to show how minds are built from mindless stuff, from parts that are much smaller and simpler than anything we'd consider smart. Unless we can explain the mind in terms of things that have no thoughts or feelings of their own, we'll only have gone around in a circle. But what could those simpler particles be—the "agents" that compose our minds? This is the subject of our book, and knowing this, let's see our task. There are many questions to answer.

Function:	*How do agents work?*
Embodiment:	*What are they made of?*
Interaction:	*How do they communicate?*
Origins:	*Where do the first agents come from?*
Heredity:	*Are we all born with the same agents?*
Learning:	*How do we make new agents and change old ones?*
Character:	*What are the most important kinds of agents?*
Authority:	*What happens when agents disagree?*
Intention:	*How could such networks want or wish?*
Competence:	*How can groups of agents do what separate agents cannot do?*
Selfness:	*What gives them unity or personality?*
Meaning:	*How could they understand anything?*
Sensibility:	*How could they have feelings and emotions?*
Awareness:	*How could they be conscious or self-aware?*

How could a theory of the mind explain so many things, when every separate question seems too hard to answer by itself? These questions all seem difficult, indeed, when we sever each one's connections to the other ones. But once we see the mind as a society of agents, each answer will illuminate the rest.

1.2 THE MIND AND THE BRAIN

It was never supposed [the poet Imlac said] *that cogitation is
inherent in matter, or that every particle is a thinking being. Yet if
any part of matter be devoid of thought, what part can we suppose
to think? Matter can differ from matter only in form, bulk,
density, motion and direction of motion: to which of these,
however varied or combined, can consciousness be annexed? To be
round or square, to be solid or fluid, to be great or little, to be
moved slowly or swiftly one way or another, are modes of material
existence, all equally alien from the nature of cogitation. If matter
be once without thought, it can only be made to think by some new
modification, but all the modifications which it can admit are
equally unconnected with cogitative powers.*
—SAMUEL JOHNSON

How could solid-seeming brains support such ghostly things as thoughts? This question troubled many thinkers of the past. The world of thoughts and the world of things appeared to be too far apart to interact in any way. So long as thoughts seemed so utterly different from everything else, there seemed to be no place to start.

A few centuries ago it seemed equally impossible to explain Life, because living things appeared to be so different from anything else. Plants seemed to grow from nothing. Animals could move and learn. Both could reproduce themselves—while nothing else could do such things. But then that awesome gap began to close. Every living thing was found to be composed of smaller cells, and cells turned out to be composed of complex but comprehensible chemicals. Soon it was found that plants did not create any substance at all but simply extracted most of their material from gases in the air. Mysteriously pulsing hearts turned out to be no more than mechanical pumps, composed of networks of muscle cells. But it was not until the present century that John von Neumann showed theoretically how cell-machines could reproduce while, almost independently, James Watson and Francis Crick discovered how each cell actually makes copies of its own hereditary code. No longer does an educated person have to seek any special, vital force to animate each living thing.

Similarly, a century ago, we had essentially no way to start to explain how thinking works. Then psychologists like Sigmund Freud and Jean Piaget produced their theories about child development. Somewhat later, on the mechanical side, mathematicians like Kurt Gödel and Alan Turing began to reveal the hitherto unknown range of what machines could be made to do. These two streams of thought began to merge only in the 1940s, when Warren McCulloch and Walter Pitts began to show how machines might be made to see, reason, and remember. Research in the modern science of Artificial Intelligence started only in the 1950s, stimulated by the invention of modern computers. This inspired a flood of new ideas about how machines could do what only minds had done previously.

Most people still believe that no machine could ever be conscious, or feel ambition, jealousy, humor, or have any other mental life-experience. To be sure, we are still far from being able to create machines that do all the things people do. But this only means that we need better theories about how thinking works. This book will show how the tiny machines that we'll call "agents of the mind" could be the long sought "particles" that those theories need.

1.3 THE SOCIETY OF MIND

You know that everything you think and do is thought and done by you. But what's a "you"? What kinds of smaller entities cooperate inside your mind to do your work? To start to see how minds are like societies, try this: *pick up a cup of tea!*

> *Your GRASPING agents want to keep hold of the cup.*
> *Your BALANCING agents want to keep the tea from spilling out.*
> *Your THIRST agents want you to drink the tea.*
> *Your MOVING agents want to get the cup to your lips.*

Yet none of these consume your mind as you roam about the room talking to your friends. You scarcely think at all about *Balance*; *Balance* has no concern with *Grasp*; *Grasp* has no interest in *Thirst*; and *Thirst* is not involved with your social problems. Why not? Because they can depend on one another. If each does its own little job, the really big job will get done by all of them together: drinking tea.

How many processes are going on, to keep that teacup level in your grasp? There must be at least a hundred of them, just to shape your wrist and palm and hand. Another thousand muscle systems must work to manage all the moving bones and joints that make your body walk around. And to keep everything in balance, each of those processes has to communicate with some of the others. What if you stumble and start to fall? Then many other processes quickly try to get things straight. Some of them are concerned with how you lean and where you place your feet. Others are occupied with what to do about the tea: you wouldn't want to burn your own hand, but neither would you want to scald someone else. You need ways to make quick decisions.

All this happens while you talk, and none of it appears to need much thought. But when you come to think of it, neither does your talk itself. What kinds of agents choose your words so that you can express the things you mean? How do those words get arranged into phrases and sentences, each connected to the next? What agencies inside your mind keep track of all the things you've said—and, also, whom you've said them to? How foolish it can make you feel when you repeat—unless you're sure your audience is new.

We're always doing several things at once, like planning and walking and talking, and this all seems so natural that we take it for granted. But these processes actually involve more machinery than anyone can understand all at once. So, in the next few sections of this book, we'll focus on just one ordinary activity—*making things with children's building-blocks*. First we'll break this process into smaller parts, and then we'll see how each of them relates to all the other parts.

In doing this, we'll try to imitate how Galileo and Newton learned so much by studying the simplest kinds of pendulums and weights, mirrors and prisms. Our study of how to build with blocks will be like focusing a microscope on the simplest objects we can find, to open up a great and unexpected universe. It is the same reason why so many biologists today devote more attention to tiny germs and viruses than to magnificent lions and tigers. For me and a whole generation of students, the world of work with children's blocks has been the prism and the pendulum for studying intelligence.

In science, one can learn the most by studying what seems the least.

1.4 THE WORLD OF BLOCKS

Imagine a child playing with blocks, and imagine that this child's mind contains a host of smaller minds. Call them mental agents. Right now, an agent called *Builder* is in control. *Builder*'s specialty is making towers from blocks.

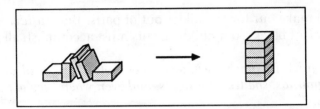

Our child likes to watch a tower grow as each new block is placed on top. But building a tower is too complicated a job for any single, simple agent, so *Builder* has to ask for help from several other agents:

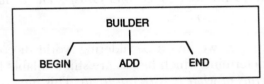

Choose a place to start the tower.
Add a new block to the tower.
Decide whether it is high enough.

In fact, even to find another block and place it on the tower top is too big for a job for any single agent. So *Add*, in turn, must call for other agents' help. Before we're done, we'll need more agents than would fit in any diagram.

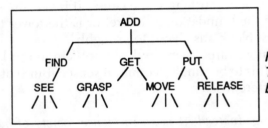

First ADD must FIND a new block.
Then the hand must GET that
block and PUT it on the tower top.

Why break things into such small parts? Because minds, like towers, are made that way—except that they're composed of processes instead of blocks. And if making stacks of blocks seems insignificant—remember that you didn't always feel that way. When first you found some building toys in early childhood, you probably spent joyful weeks of learning what to do with them. If such toys now seem relatively dull, then you must ask yourself how *you* have changed. Before you turned to more ambitious things, it once seemed strange and wonderful to be able to build a tower or a house of blocks. Yet, though all grown-up persons know how to do such things, *no one understands how we learn to do them!* And *that* is what will concern us here. To pile up blocks into heaps and rows: these are skills each of us learned so long ago that we can't remember learning them at all. Now they seem mere common sense—and that's what makes psychology hard. This forgetfulness, the amnesia of infancy, makes us assume that all our wonderful abilities were always there inside our minds, and we never stop to ask ourselves how they began and grew.

1.5 COMMON SENSE

> *You cannot think about thinking, without thinking about thinking about something.*
> —SEYMOUR PAPERT

We found a way to make our tower builder out of parts. But *Builder* is really far from done. To build a simple stack of blocks, our child's agents must accomplish all these other things.

> ***See*** *must recognize its blocks, whatever their color, size, and place—in spite of different backgrounds, shades, and lights, and even when they're partially obscured by other things.*

> *Then, once that's done,* ***Move*** *has to guide the arm and hand through complicated paths in space, yet never strike the tower's top or hit the child's face.*

> *And think how foolish it would seem, if* ***Find*** *were to see, and* ***Grasp*** *were to grasp, a block supporting the tower top!*

When we look closely at these requirements, we find a bewildering world of complicated questions. For example, how could *Find* determine which blocks are still available for use? *It would have to "understand" the scene in terms of what it is trying to do.* This means that we'll need theories both about what it means to understand and about how a machine could have a goal. Consider all the *practical* judgments that an actual *Builder* would have to make. It would have to decide whether there are enough blocks to accomplish its goal and whether they are strong and wide enough to support the others that will be placed on them.

What if the tower starts to sway? A real builder must guess the cause. It is because some joint inside the column isn't square enough? Is the foundation insecure, or is the tower too tall for its width? Perhaps it is only because the last block was placed too roughly.

All children learn about such things, but we rarely ever think about them in our later years. By the time we are adults we regard all of this to be simple "common sense." But that deceptive pair of words conceals almost countless different skills.

> *Common sense is not a simple thing. Instead, it is an immense society of hard-earned practical ideas—of multitudes of life-learned rules and exceptions, dispositions and tendencies, balances and checks.*

If common sense is so diverse and intricate, what makes it seem so obvious and natural? This illusion of simplicity comes from losing touch with what happened during infancy, when we formed our first abilities. As each new group of skills matures, we build more layers on top of them. As time goes on, the layers below become increasingly remote until, when we try to speak of them in later life, we find ourselves with little more to say than *"I don't know."*

1.6 AGENTS AND AGENCIES

We want to explain intelligence as a combination of simpler things. This means that we must be sure to check, at every step, that none of our agents is, itself, intelligent. Otherwise, our theory would end up resembling the nineteenth-century "chessplaying machine" that was exposed by Edgar Allan Poe to actually conceal a human dwarf inside. Accordingly, whenever we find that an agent has to do anything complicated, we'll replace it with a subsociety of agents that do simpler things. Because of this, the reader must be prepared to feel a certain sense of loss. When we break things down to their smallest parts, they'll each seem dry as dust at first, as though some essence has been lost.

For example, we've seen how to construct a tower-building skill by making *Builder* from little parts like *Find* and *Get*. Now, where does its "knowing-how-to-build" reside when, clearly, it is not in any part—and yet those parts are all that *Builder* is? The answer: It is not enough to explain only what each separate agent does. We must also understand how those parts are interrelated—that is, how *groups* of agents can accomplish things.

Accordingly, each step in this book uses two different ways to think about agents. If you were to watch *Builder* work, from the outside, with no idea of how it works inside, you'd have the impression that it knows how to build towers. But if you could see *Builder* from the inside, you'd surely find no knowledge there. You would see nothing more than a few switches, arranged in various ways to turn each other on and off. *Does **Builder** "really know" how to build towers?* The answer depends on how you look at it. Let's use two different words, "*agent*" and "*agency*," to say why *Builder* seems to lead a double life. As agency, it seems to know its job. As agent, it cannot know anything at all.

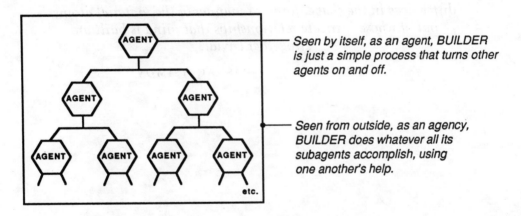

Seen by itself, as an agent, BUILDER is just a simple process that turns other agents on and off.

Seen from outside, as an agency, BUILDER does whatever all its subagents accomplish, using one another's help.

When you drive a car, you regard the steering wheel as an agency that you can use to change the car's direction. You don't care how it works. But when something goes wrong with the steering, and you want to understand what's happening, it's better to regard the steering wheel as just one agent in a larger agency: it turns a shaft that turns a gear to pull a rod that shifts the axle of a wheel. Of course, one doesn't always want to take this microscopic view; if you kept all those details in mind while driving, you might crash because it took too long to figure out which way to turn the wheel. Knowing how is not the same as knowing why. In this book, we'll always be switching between agents and agencies because, depending on our purposes, we'll have to use different viewpoints and kinds of descriptions.

CHAPTER 2

WHOLES AND PARTS

It is the nature of the mind that makes individuals kin, and the differences in the shape, form, or manner of the material atoms out of whose intricate relationships that mind is built are altogether trivial.

—ISAAC ASIMOV

2.1 COMPONENTS AND CONNECTIONS

We saw that *Builder*'s skill could be reduced to the simpler skills of *Get* and *Put*. Then we saw how these, in turn, could be made of even simpler ones. *Get* merely needs to *Move* the hand to *Grasp* the block that *Find* just found. *Put* only has to *Move* the hand so that it puts that block upon the tower top. So it might appear that all of *Builder*'s functions have been "reduced" to things that simpler parts can do.

But something important has been left out. *Builder* is not merely a collection of parts like *Find*, *Get*, *Put*, and all the rest. For *Builder* would not work at all unless those agents were linked to one another by a suitable network of interconnections.

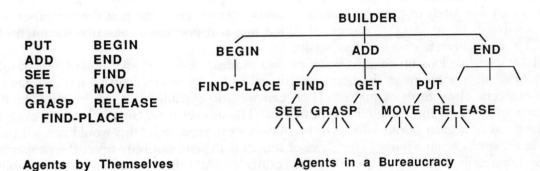

Agents by Themselves   **Agents in a Bureaucracy**

Could you predict what *Builder* does from knowing just that left-hand list? Of course not: you must also know which agents work for which. Similarly, you couldn't predict what would happen in a human community from knowing only what each separate individual can do; you must also know how they are organized—that is, who talks to whom. And it's the same for understanding any large and complex thing. First, we must know how each separate part works. Second, we must know how each part interacts with those to which it is connected. And third, we have to understand how all these local interactions combine to accomplish what that system *does*—as seen from the outside.

In the case of the human brain, it will take a long time to solve these three kinds of problems. First we will have to understand how brain cells work. This will be difficult because there are hundreds of different types of brain cells. Then we'll have to understand how the cells of each type interact with the other types of cells to which they connect. There could be thousands of these different kinds of interactions. Then, finally, comes the hardest part: we'll also have to understand how our billions of brain cells are organized into societies. To do this, we'll need to develop many new theories and organizational concepts. The more we can find out about how our brains evolved from those of simpler animals, the easier that task will be.

2.2 NOVELISTS AND REDUCTIONISTS

It's always best when mysteries can be explained in terms of things we know. But when we find this hard to do, we must decide whether to keep trying to make old theories work or to discard them and try new ones. I think this is partly a matter of personality. Let's call "Reductionists" those people who prefer to build on old ideas, and "Novelists" the ones who like to champion new hypotheses. Reductionists are usually right—at least at science's cautious core, where novelties rarely survive for long. Outside that realm, though, novelists reign, since older ideas have had more time to show their flaws.

It really is amazing how certain sciences depend upon so few kinds of explanations. The science of physics can now explain virtually everything we see, *at least in principle*, in terms of how a very few kinds of particles and force-fields interact. Over the past few centuries reductionism has been remarkably successful. What makes it possible to describe so much of the world in terms of so few basic rules? No one knows.

Many scientists look on chemistry and physics as ideal models of what psychology should be like. After all, the atoms in the brain are subject to the same all-inclusive physical laws that govern every other form of matter. Then can we also explain what our brains actually do entirely in terms of those same basic principles? The answer is no, simply because even if we understood how each of our billions of brain cells work separately, this would not tell us how the brain works as an agency. The "laws of thought" depend not only upon the properties of those brain cells, but also on how they are connected. And these connections are established not by the basic, "general" laws of physics, but by the particular arrangements of the millions of bits of information in our inherited genes. To be sure, "general" laws apply to everything. But, for that very reason, they can rarely explain anything in particular.

Does this mean that psychology must reject the laws of physics and find its own? Of course not. It is not a matter of *different* laws, but of *additional* kinds of theories and principles that operate at higher levels of organization. Our ideas of how *Builder* works as an agency need not, and must not, conflict with our knowledge of how *Builder*'s lower-level agents work. Each higher level of description must *add* to our knowledge about lower levels, rather than replace it. We'll return to the idea of "level" at many places in this book.

Will psychology ever resemble any of the sciences that have successfully reduced their subjects to only a very few principles? That depends on what you mean by "few." In physics, we're used to explanations in terms of perhaps a dozen basic principles. For psychology, our explanations will have to combine hundreds of smaller theories. To physicists, that number may seem too large. To humanists, it may seem too small.

2.3 PARTS AND WHOLES

We're often told that certain wholes are "more than the sum of their parts." We hear this expressed with reverent words like "holistic" and "gestalt," whose academic tones suggest that they refer to clear and definite ideas. But I suspect the actual function of such terms is to anesthetize a sense of ignorance. We say "gestalt" when things combine to act in ways we can't explain, "holistic" when we're caught off guard by unexpected happenings and realize we understand less than we thought we did. For example, consider the two sets of questions below, the first "subjective" and the second "objective":

What makes a drawing more than just its separate lines?
How is a personality more than a set of traits?
In what way is a culture more than a mere collection of customs?

What makes a tower more than separate blocks?
Why is a chain more than its various links?
How is a wall more than a set of many bricks?

Why do the "objective" questions seem less mysterious? Because we have good ways to answer them—in terms of how things interact. To explain how walls and towers work, we just point out how every block is held in place by its neighbors and by gravity. To explain why chain-links cannot come apart, we can demonstrate how each would get in its neighbors' way. These explanations seem almost self-evident to adults. However, they did not seem so simple when we were children, and it took each of us several years to learn how real-world objects interact—for example, to prevent any two objects from ever being in the same place. We regard such knowledge as "obvious" only because we cannot remember how hard it was to learn.

Why does it seem so much harder to explain our reactions to drawings, personalities, and cultural traditions? Many people assume that those "subjective" kinds of questions are impossible to answer because they involve our minds. But that doesn't mean they can't be answered. It only means that we must first know more about our minds.

"Subjective" reactions are also based on how things interact. The difference is that here we are not concerned with objects in the world outside, but with processes inside our brains.

In other words, those questions about arts, traits, and styles of life are actually quite technical. They ask us to explain what happens among the agents in our minds. But this is a subject about which we have never learned very much—and neither have our sciences. Such questions will be answered in time. But it will just prolong the wait if we keep using pseudo–explanation words like "holistic" and "gestalt." True, sometimes giving names to things can help by leading us to focus on some mystery. It's harmful, though, when naming leads the mind to think that names alone bring meaning close.

2.4 HOLES AND PARTS

> *It has been the persuasion of an immense majority of human beings that sensibility and thought [as distinguished from matter] are, in their own nature, less susceptible of division and decay, and that, when the body is resolved into its elements, the principle which animated it will remain perpetual and unchanged. However, it is probable that what we call thought is not an actual being, but no more than the relation between certain parts of that infinitely varied mass, of which the rest of the universe is composed, and which ceases to exist as soon as those parts change their position with respect to each other.*
>
> —PERCY BYSSHE SHELLEY

What is Life? One dissects a body but finds no life inside. What is Mind? One dissects a brain but finds no mind therein. Are life and mind so much more than the "sum of their parts" that it is useless to search for them? To answer that, consider this parody of a conversation between a Holist and an ordinary Citizen.

Holist: *"I'll prove no box can hold a mouse. A box is made by nailing six boards together. But it's obvious that no box can hold a mouse unless it has some 'mouse-tightness' or 'containment.' Now, no single board contains any containment, since the mouse can just walk away from it. And if there is no containment in one board, there can't be any in six boards. So the box can have no mousetightness at all. Theoretically, then, the mouse can escape!"*

Citizen: *"Amazing. Then what **does** keep a mouse in a box?"*

Holist: *"Oh, simple. Even though it has no real mousetightness, a good box can 'simulate' it so well that the mouse is fooled and can't figure out how to escape."*

What, then, keeps the mouse confined? Of course, it is the way a box prevents motion in all directions, because each board bars escape in a certain direction. The left side keeps the mouse from going left, the right from going right, the top keeps it from leaping out, and so on. The secret of a box is simply in how the boards are arranged to prevent motion in *all* directions! That's what *containing* means. So it's silly to expect any separate board by itself to contain any *containment*, even though each contributes to the containing. It is like the cards of a straight flush in poker: only the full hand has any value at all.

The same applies to words like *life* and *mind*. It is foolish to use these words for describing the smallest components of living things because these words were invented to describe how larger assemblies interact. Like *boxing-in*, words like *living* and *thinking* are useful for describing phenomena that result from certain combinations of relationships. The reason *box* seems nonmysterious is that everyone understands how the boards of a well-made box interact to prevent motion in any direction. In fact, the word *life* has already lost most of its mystery—at least for modern biologists, because they understand so many of the important interactions among the chemicals in cells. But *mind* still holds its mystery—because we still know so little about how mental agents interact to accomplish all the things they do.

2.5 EASY THINGS ARE HARD

In the late 1960s *Builder* was embodied in the form of a computer program at the MIT Artificial Intelligence Laboratory. Both my collaborator, Seymour Papert, and I had long desired to combine a mechanical hand, a television eye, and a computer into a robot that could build with children's building-blocks. It took several years for us and our students to develop *Move*, *See*, *Grasp*, and hundreds of other little programs we needed to make a working *Builder*-agency. I like to think that this project gave us glimpses of what happens inside certain parts of children's minds when they learn to "play" with simple toys. The project left us wondering if even a thousand microskills would be enough to enable a child to fill a pail with sand. It was this body of experience, more than anything we'd learned about psychology, that led us to many ideas about societies of mind.

To do those first experiments, we had to build a mechanical Hand, equipped with sensors for pressure and touch at its fingertips. Then we had to interface a television camera with our computer and write programs with which that Eye could discern the edges of the building-blocks. It also had to recognize the Hand itself. When those programs didn't work so well, we added more programs that used the fingers' feeling-sense to verify that things were where they visually seemed to be. Yet other programs were needed to enable the computer to move the Hand from place to place while using the Eye to see that there was nothing in its way. We also had to write higher-level programs that the robot could use for planning what to do—and still more programs to make sure that those plans were actually carried out. To make this all work reliably, we needed programs to verify at every step (again by using Eye and Hand) that what had been planned inside the mind did actually take place outside—or else to correct the mistakes that occurred.

In attempting to make our robot work, we found that many everyday problems were much more complicated than the sorts of problems, puzzles, and games adults consider hard. At every point, in that world of blocks, when we were forced to look more carefully than usual, we found an unexpected universe of complications. Consider just the seemingly simple problem of not reusing blocks already built into the tower. To a person, this seems simple common sense: *"Don't use an object to satisfy a new goal if that object is already involved in accomplishing a prior goal."* No one knows exactly how human minds do this. Clearly we learn from experience to recognize the situations in which difficulties are likely to occur, and when we're older we learn to plan ahead to avoid such conflicts. But since we cannot be sure what will work, we must learn policies for dealing with uncertainty. Which strategies are best to try, and which will avoid the worst mistakes? Thousands and, perhaps, millions of little processes must be involved in how we anticipate, imagine, plan, predict, and prevent—and yet all this proceeds so automatically that we regard it as "ordinary common sense." But if thinking is so complicated, what makes it seem so simple? At first it may seem incredible that our minds could use such intricate machinery and yet be unaware of it.

In general, we're least aware of what our minds do best.

It's mainly when our other systems start to fail that we engage the special agencies involved with what we call "consciousness." Accordingly, we're more aware of simple processes that don't work well than of complex ones that work flawlessly. This means that we cannot trust our offhand judgments about which of the things we do are simple, and which require complicated machinery. Most times, each portion of the mind can only sense how quietly the other portions do their jobs.

2.6 ARE PEOPLE MACHINES?

Many people feel offended when their minds are likened to computer programs or machines. We've seen how a simple tower-building skill can be composed of smaller parts. But could anything like a real mind be made of stuff so trivial?

"Ridiculous," most people say. *"I certainly don't feel like a machine!"*

But if you're not a machine, what makes you an authority on what it feels like to be a machine? A person might reply, *"I think, therefore I know how the mind works."* But that would be suspiciously like saying, *"I drive my car, therefore I know how its engine works."* Knowing how to use something is not the same as knowing how it works.

"But everyone knows that machines can behave only in lifeless, mechanical ways."

This objection seems more reasonable: indeed, a person *ought* to feel offended at being likened to any *trivial* machine. But it seems to me that the word "machine" is getting to be out of date. For centuries, words like "mechanical" made us think of simple devices like pulleys, levers, locomotives, and typewriters. (The word "computerlike" inherited a similar sense of pettiness, of doing dull arithmetic by little steps.) But we ought to recognize that we're still in an early era of machines, with virtually no idea of what they may become. What if some visitor from Mars had come a billion years ago to judge the fate of earthly life from watching clumps of cells that hadn't even learned to crawl? In the same way, we cannot grasp the range of what machines may do in the future from seeing what's on view right now.

Our first intuitions about computers came from experiences with machines of the 1940s, which contained only thousands of parts. But a human brain contains billions of cells, each one complicated by itself and connected to many thousands of others. Present-day computers represent an intermediate degree of complexity; they now have millions of parts, and people already are building billion-part computers for research on Artificial Intelligence. And yet, in spite of what is happening, we continue to use old words as though there had been no change at all. We need to adapt our attitudes to phenomena that work on scales never before conceived. The term "machine" no longer takes us far enough.

But rhetoric won't settle anything. Let's put these arguments aside and try instead to understand what the vast, unknown mechanisms of the brain may do. Then we'll find more self-respect in knowing what wonderful machines we are.

CONFLICT AND COMPROMISE

3.1 CONFLICT

Most children not only like to build, they also like to knock things down. So let's imagine another agent called *Wrecker*, whose specialty is knocking-down. Our child loves to hear the complicated noises and watch so many things move all at once.

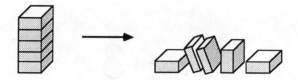

Suppose *Wrecker* gets aroused, but there's nothing in sight to smash. Then *Wrecker* will have to get some help—by putting *Builder* to work, for example. But what if, at some later time, *Wrecker* considers the tower to be high enough to smash, while *Builder* wants to make it taller still? Who could settle that dispute?

The simplest policy would be to leave that decision to *Wrecker*, who was responsible for activating *Builder* in the first place. But in a more realistic picture of a child's mind, such choices would depend on many other agencies. For example, let's assume that both *Builder* and *Wrecker* were originally activated by a higher-level agent, *Play-with-Blocks*. Then, a conflict might arise if *Builder* and *Wrecker* disagree about whether the tower is high enough.

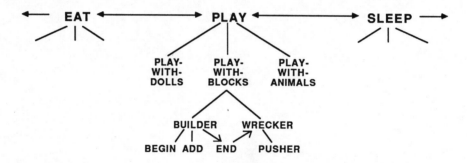

What aroused *Play-with-Blocks* in the first place? Perhaps some even higher-level agent, *Play*, was active first. Then, inside *Play*, the agent *Play-with-Blocks* achieved control, in spite of two competitors, *Play-with-Dolls* and *Play-with-Animals*. But even *Play* itself, their mutual superior in chief, must have had to compete with other higher-level agencies like *Eat* and *Sleep*. For, after all, a child's play is not an isolated thing but always happens in the context of other real-life concerns. Whatever we may choose to do, there are always other things we'd also like to do.

In several sections of this book, I will assume that conflicts between agents tend to migrate upward to higher levels. For example, any prolonged conflict between *Builder* and *Wrecker* will tend to weaken their mutual superior, *Play-with-Blocks*. In turn, this will reduce *Play-with-Blocks'* ability to suppress *its* rivals, *Play-with-Dolls* and *Play-with-Animals*. Next, if *that* conflict isn't settled soon, it will weaken the agent *Play* at the next-higher level. Then *Eat* or *Sleep* might seize control.

3.2 NONCOMPROMISE

To settle arguments, nations develop legal systems, corporations establish policies, and individuals may argue, fight, or compromise—or turn for help to mediators that lie outside themselves. What happens when there are conflicts inside minds?

Whenever several agents have to compete for the same resources, they are likely to get into conflicts. If those agents were left to themselves, the conflicts might persist indefinitely, and this would leave those agents paralyzed, unable to accomplish any goal. What happens then? We'll assume that those agents' supervisors, too, are under competitive pressure and likely to grow weak themselves whenever their subordinates are slow in achieving their goals, no matter whether because of conflicts between them or because of individual incompetence.

> **The Principle of Noncompromise:** *The longer an internal conflict persists among an agent's subordinates, the weaker becomes that agent's status among its own competitors. If such internal problems aren't settled soon, other agents will take control and the agents formerly involved will be "dismissed."*

So long as playing with blocks goes well, *Play* can maintain its strength and keep control. In the meantime, though, the child may also be growing hungry and sleepy, because other processes are arousing the agents *Eat* and *Sleep*. So long as *Eat* and *Sleep* are not yet strongly activated, *Play* can hold them both at bay. However, any conflict inside *Play* will weaken it and make it easier for *Eat* or *Sleep* to take over. Of course, *Eat* or *Sleep* must conquer in the end, since the longer they wait, the stronger they get.

We see this in our own experience. We all know how easy it is to fight off small distractions when things are going well. But once some trouble starts inside our work, we become increasingly impatient and irritable. Eventually we find it so hard to concentrate that the least disturbance can allow another, different, interest to take control. Now, when any of our agencies loses the power to control what other systems do, that doesn't mean it has to cease its own internal activity. An agency that has lost control can continue to work inside itself—and thus become prepared to seize a later opportunity. However, we're normally unaware of all those other activities proceeding deep inside our minds.

Where does it stop, this process of yielding control to other agencies? Must every mind contain some topmost center of control? Not necessarily. We sometimes settle conflicts by appealing to superiors, but other conflicts never end and never cease to trouble us.

At first, our principle of noncompromise may seem too extreme. After all, good human supervisors plan ahead to avoid conflicts in the first place, and—when they can't—they try to settle quarrels locally before appealing to superiors. But we should not try to find a close analogy between the low-level agents of a single mind and the members of a human community. Those tiny mental agents simply cannot know enough to be able to negotiate with one another or to find effective ways to adjust to each other's interference. Only larger agencies could be resourceful enough to do such things. Inside an actual child, the agencies responsible for *Building* and *Wrecking* might indeed become versatile enough to negotiate by offering support for one another's goals. *"Please, **Wrecker,** wait a moment more till **Builder** adds just one more block: it's worth it for a louder crash!"*

3.3 HIERARCHIES

> bu•reauc´ra•cy *n.* the administration of government
> through departments and subdivisions managed by
> sets of officials following an inflexible routine.
> —*Webster's Unabridged Dictionary*

As an agent, *Builder* does no physical work but merely turns on *Begin*, *Add*, and *End*. Similarly, *Add* just orders *Find*, *Put*, and *Get* to do their jobs. Then these divide into agents like *Move* and *Grasp*. It seems that it will never stop—this breaking-down to smaller things. Eventually, it all must end with agents that do actual work, but there are many steps before we get to all the little muscle-motor agents that actually move the arms and hands and finger joints. Thus *Builder* is like a high-level executive, far removed from those subordinates who actually produce the final product.

Does this mean that *Builder*'s administrative work is unimportant? Not at all. Those lower-level agents need to be controlled. It's much the same in human affairs. When any enterprise becomes too complex and large for one person to do, we construct organizations in which certain agents are concerned, not with the final result, but only with what some other agents do. Designing any society, be it human or mechanical, involves decisions like these:

> *Which agents choose which others to do what jobs?*
> *Who will decide which jobs are done at all?*
> *Who decides what efforts to expend?*
> *How will conflicts be settled?*

How much of ordinary human thought has *Builder*'s character? The *Builder* we described is not much like a human supervisor. It doesn't decide which agents to assign to which jobs, because that has already been arranged. It doesn't plan its future work but simply carries out fixed steps until *End* says the job is done. Nor has it any repertoire of ways to deal with unexpected accidents.

Because our little mental agents are so limited, we should not try to extend very far the analogy between them and human supervisors and workers. Furthermore, as we'll shortly see, the relations between mental agents are not always strictly hierarchical. And in any case, such roles are always relative. To *Builder*, *Add* is a subordinate, but to *Find*, *Add* is a boss. As for yourself, it all depends on how you live. Which sorts of thoughts concern you most—the orders you are made to take or those you're being forced to give?

3.4 HETERARCHIES

A hierarchical society is like a tree in which the agent at each branch is exclusively responsible for the agents on the twigs that branch from it. This pattern is found in every field, because dividing work into parts like that is usually the easiest way to start solving a problem. It is easy to construct and understand such organizations because each agent has only a single job to do: it needs only to "look up" for instructions from its supervisor, then "look down" to get help from its subordinates.

But hierarchies do not always work. Consider that when two agents need to use each other's skills, then neither one can be "on top." Notice what happens, for example, when you ask your vision-system to decide whether the left-side scene below depicts three blocks—or only two.

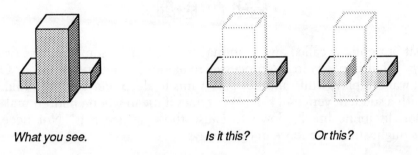

What you see.   Is it this?   Or this?

The agent *See* could answer that if it could *Move* the front block out of the line of view. But, in the course of doing that, *Move* might have to *See* if there were any obstacles that might interfere with the arm's trajectory. At such a moment, *Move* would be working for *See*, and *See* would be working for *Move*, both at the same time. This would be impossible inside a simple hierarchy.

Most of the diagrams in the early parts of this book depict simple hierarchies. Later, we'll see more cross-connected rings and loops—when we are forced to consider the need for memory, which will become a constant subject of concern in this book. People often think of memory in terms of keeping records of the past, for recollecting things that happened in earlier times. But agencies also need other kinds of memory as well. *See*, for example, requires some sort of temporary memory in order to keep track of what next to do, when it starts one job before its previous job is done. If each of *See*'s agents could do only one thing at a time, it would soon run out of resources and be unable to solve complicated problems. But if we have enough memory, we can arrange our agents into circular loops and thus use the same agents over and over again to do parts of several different jobs at the same time.

3.5 DESTRUCTIVENESS

In any actual child's mind, the urge to *Play* competes with other demanding urges, such as *Eat* and *Sleep*. What happens if another agent wrests control from *Play*, and what happens to the agents *Play* controlled?

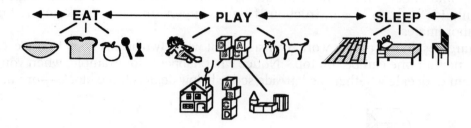

Suppose that our child is called away, no matter whether by someone else or by an internal urge like *Sleep*. What happens to the processes remaining active in the mind? One part of the child may still want to play, while another part wants to sleep. Perhaps the child will knock the tower down with a sudden, vengeful kick. What does it mean when children make such scenes? Is it that inner discipline breaks down to cause those savage acts? Not necessarily. Those "childish" acts might still make sense in other ways.

> *Smashing takes so little time that **Wrecker,** freed from **Play**'s constraint, need persist for only one more kick to gain the satisfaction of a final crash.*

> *Though childish violence might seem senseless by itself, it serves to communicate frustration at the loss of goal. Even if the parent scolds, that just confirms how well the message was transmitted and received.*

> *Destructive acts can serve constructive goals by leaving fewer problems to be solved. That kick may leave a mess outside, yet tidy up the child's mind.*

When children smash their treasured toys, we shouldn't ask for *the* reason why—since no such act has a single cause. Besides, it isn't true, in a human mind, that when *Sleep* starts, then *Play* must quit and all its agents have to cease. A real child can go to bed—yet still build towers in its head.

3.6 PAIN AND PLEASURE SIMPLIFIED

When you're in pain, it's hard to keep your interest in other things. You feel that nothing's more important than finding some way to stop the pain. That's why pain is so powerful: it makes it hard to think of anything else. Pain simplifies your point of view.

When something gives you pleasure, then, too, it's hard to think of other things. You feel that nothing's more important than finding a way to make that pleasure last. That's why pleasure is so powerful. It also simplifies your point of view.

Pain's power to distract us from our other goals is not an accident; that's how it helps us to survive. Our bodies are endowed with special nerves that detect impending injuries, and the signals from these nerves for pain make us react in special ways. Somehow, they disrupt our concerns with long-term goals—thus forcing us to focus on immediate problems, perhaps by transferring control to our lowest-level agencies. Of course, this can do more harm than good, especially when, in order to remove the source of pain, one has to make a complex plan. Unfortunately, pain interferes with making plans by undermining interest in anything that's not immediate. Too much suffering diminishes us by restricting the complexities that constitute our very selves. It must be the same for pleasure as well.

We think of pleasure and pain as opposites, since pleasure makes us draw its object near while pain impels us to reject its object. We also think of them as similar, since both make rival goals seem small by turning us from other interests. They both distract. Why do we find such similarities between antagonistic things? Sometimes two seeming opposites are merely two extremes along a single scale, or one of them is nothing but the absence of the other—as in the case of sound and silence, light and darkness, interest and unconcern. But what of opposites that are genuinely different, like pain and pleasure, fear and courage, hate and love?

> *In order to appear opposed, two things must serve related goals—or otherwise engage the selfsame agencies.*

Thus, affection and abhorrence both involve our attitudes toward relationships; and pleasure and pain both engage constraints that simplify our mental scenes. The same goes for courage and cowardice: each does best by knowing both. When on attack, you have to press against whatever weakness you can find in your opponent's strategy. When on defense, it's much the same: you still must guess the other's plan.

CHAPTER 4

THE SELF

We are what we pretend to be, so we must be careful about what we pretend to be.

—KURT VONNEGUT

4.1   THE SELF

> **self** *n. 1. the identity, character, or essential qualities of any person or thing. 2. the identity, personality, individuality, etc. of a given person; one's own person as distinct from all others.*
> —*Webster's Unabridged Dictionary*

We all believe that human minds contain those special entities we call selves. But no one agrees about what they are. To keep things straight, I shall write "self" when speaking in a general sense about an entire person and reserve "Self" for talking about that more mysterious sense of personal identity. Here are some of the things people say about the Self:

> *Self is the part of mind that's really me, or rather, it's the part of me—that is, part of my mind—that actually does the thinking and wanting and deciding and enjoying and suffering. It's the part that's most important to me because it's that which stays the same through all experience—the **identity** which ties everything together. And whether you can treat it scientifically or not, I know it's there, because it's me. Perhaps it's the sort of thing that Science can't explain.*

This isn't much of a definition, but I don't think it is a good idea to try to find a better one. It often does more harm than good to force definitions on things we don't understand. Besides, only in logic and mathematics do definitions ever capture concepts perfectly. The things we deal with in practical life are usually too complicated to be represented by neat, compact expressions. Especially when it comes to understanding minds, we still know so little that we can't be sure our ideas about psychology are even aimed in the right directions. In any case, one must not mistake defining things for knowing what they are. You can know what a tiger is without defining it. You may define a tiger, yet know scarcely anything about it.

Even if our old ideas about the mind are wrong, we can learn a lot by trying to understand why we believe them. Instead of asking, *"What are Selves?"* we can ask, instead, *"What are our ideas about Selves?"*—and then we can ask, *"What psychological functions do those ideas serve?"* When we do this, it shows us that we do not have one such idea, but many.

Our ideas about our Selves include beliefs about what we *are*. These include beliefs both about what we are capable of doing and about what we may be disposed to do. We exploit these beliefs whenever we solve problems or make plans. I'll refer to them, rather vaguely, as a person's *self-images*. In addition to our self-images, our ideas about ourselves also include ideas about what we'd *like* to be and ideas about what we *ought* to be. These, which I'll call a person's *self-ideals*, influence each person's growth from infancy, but we usually find them hard to express because they're inaccessible to consciousness.

4.2 ONE SELF OR MANY?

One common image of the Self suggests that every mind contains some sort of Voyeur-Puppeteer inside—to feel and want and choose for us the things we feel, want, and choose. But if we had *those* kinds of Selves, what would be the use of having Minds? And, on the other hand, if Minds could do such things themselves, why have Selves? Is this concept of a Self of any real use at all? It is indeed—provided that we think of it not as a centralized and all-powerful entity, but as a society of ideas that include both our images of what the mind is and our ideals about what it ought to be.

Besides, we're often of two minds about ourselves. Sometimes we regard ourselves as single, self-coherent entities. Other times we feel decentralized or dispersed, as though we were made of many different parts with different tendencies. Contrast these views:

> **SINGLE-SELF VIEW.** *"I think, I want, I feel. It's me, myself, who thinks my thoughts. It's not some nameless crowd or cloud of selfless parts."*

> **MULTIPLE-SELF VIEW.** *"One part of me wants this, another part wants that. I must get better control of myself."*

We're never wholly satisfied with either view. We all sense feelings of disunity, conflicting motives, compulsions, internal tensions, and dissensions. We carry on negotiations in our head. We hear scary tales in which some person's mind becomes enslaved by compulsions and commands that seem to come from somewhere else. And the times we feel most reasonably unified can be just the times that others see us as the most confused.

But if there is no single, central, ruling Self inside the mind, what makes us feel so sure that one exists? What gives that myth its force and strength? A paradox: perhaps it's *because* there are no persons in our heads to make us do the things we want—nor even ones to make us *want to want*—that we construct the myth that *we're* inside ourselves.

4.3   THE SOUL

And we thank Thee that darkness reminds us of light.
—T. S. Eliot

A common concept of the soul is that the essence of a self lies in some spark of invisible light, a thing that cowers out of body, out of mind, and out of sight. But what might such a symbol mean? It carries a sense of anti-self-respect: that there is no significance in anyone's accomplishments.

People ask if machines can have souls. And I ask back whether souls can learn. It does not seem a fair exchange—if souls can live for endless time and yet not use that time to learn—to trade all change for changelessness. And that's exactly what we get with inborn souls that cannot grow: a destiny the same as death, an ending in a permanence incapable of any change and, hence, devoid of intellect.

Why try to frame the value of a Self in such a singularly frozen form? The art of a great painting is not in any one idea, nor in a multitude of separate tricks for placing all those pigment spots, but in the great network of relationships among its parts. Similarly, the agents, raw, that make our minds are by themselves as valueless as aimless, scattered daubs of paint. What counts is what we make of them.

We all know how an ugly husk can hide an unexpected gift, like a treasure buried in the dust or a graceless oyster bearing a pearl. But minds are just the opposite. We start as little embryos, which then build great and wondrous selves—whose merit lies entirely within their own coherency. The value of a human self lies not in some small, precious core, but in its vast, constructed crust.

What are those old and fierce beliefs in spirits, souls, and essences? *They're all insinuations that we're helpless to improve ourselves.* To look for our virtues in such thoughts seems just as wrongly aimed a search as seeking art in canvas cloths by scraping off the painter's works.

4.4   THE CONSERVATIVE SELF

How do we control our minds? Ideally, we first choose what we want to do, then make ourselves do it. But that's harder than it sounds: we spend our lives in search of schemes for self-control. We celebrate when we succeed, and when we fail, we're angry with ourselves for not behaving as we wanted to—and then we try to scold or shame or bribe ourselves to change our ways. But wait! How could a self be angry with itself? Who would be mad at whom? Consider an example from everyday life.

> *I was trying to concentrate on a certain problem but was getting bored and sleepy. Then I imagined that one of my competitors, Professor Challenger, was about to solve the same problem. An angry wish to frustrate Challenger then kept me working on the problem for a while. The strange thing was, this problem was not of the sort that ever interested Challenger.*

What makes us use such roundabout techniques to influence ourselves? Why be so indirect, inventing misrepresentations, fantasies, and outright lies? Why can't we simply tell ourselves to do the things we want to do?

To understand how something works, one has to know its purposes. Once, no one understood the heart. But as soon as it was seen that hearts move blood, a lot of other things made sense: those things that looked like pipes and valves were really pipes and valves indeed—and anxious, pounding, pulsing hearts were recognized as simple pumps. New speculations could then be formed: was this to give our tissues drink or food? Was it to keep our bodies warm or cool? For sending messages from place to place? In fact, all those hypotheses were correct, and when that surge of functional ideas led to the guess that blood can carry air as well, more puzzle parts fell into place.

To understand what we call the Self, we first must see what Selves are for. *One function of the Self is to keep us from changing too rapidly.* Each person must make some long-range plans in order to balance single-purposeness against attempts to do everything at once. But it is not enough simply to instruct an agency to start to carry out our plans. We also have to find some ways to constrain the changes we might later make—to prevent ourselves from turning those plan-agents off again! If we changed our minds too recklessly, we could never know what we might want next. We'd never get much done because we could never depend on ourselves.

Those ordinary views are wrong that hold that Selves are magic, self-indulgent luxuries that enable our minds to break the bonds of natural cause and law. Instead, those Selves are practical necessities. The myths that say that Selves embody special kinds of liberty are merely masquerades. Part of their function is to hide from us the nature of our self-ideals—the chains we forge to keep ourselves from wrecking all the plans we make.

4.5 EXPLOITATION

Let's look more closely at that episode of Professor Challenger. Apparently, what happened was that my agency for *Work* exploited *Anger* to stop *Sleep*. But why should *Work* use such a devious trick?

To see why we have to be so indirect, consider some alternatives. If *Work* could simply turn off *Sleep*, we'd quickly wear our bodies out. If *Work* could simply switch *Anger* on, we'd be fighting all the time. Directness is too dangerous. We'd die.

Extinction would be swift indeed for species that could simply switch off hunger or pain. Instead, there must be checks and balances. We'd never get through one full day if any agency could seize and hold control over all the rest. This must be why our agencies, in order to exploit each other's skills, have to discover such roundabout pathways. All direct connections must have been removed in the course of our evolution.

This must be one reason why we use fantasies: to provide the missing paths. You may not be able to make yourself angry simply by deciding to be angry, but you can still imagine objects or situations that *make* you angry. In the scenario about Professor Challenger, my agency *Work* exploited a particular memory to arouse my *Anger*'s tendency to counter *Sleep*. This is typical of the tricks we use for self-control.

Most of our self-control methods proceed unconsciously, but we sometimes resort to conscious schemes in which we offer rewards to ourselves: *"If I can get this project done, I'll have more time for other things."* However, it is not such a simple thing to be able to bribe yourself. To do it successfully, you have to discover which mental incentives will actually work on yourself. This means that you—or rather, your agencies—have to learn something about one another's dispositions. In this respect the schemes we use to influence ourselves don't seem to differ much from those we use to exploit other people—and, similarly, they often fail. When we try to induce ourselves to work by offering ourselves rewards, we don't always keep our bargains; we then proceed to raise the price or even to deceive ourselves, much as one person may try to conceal an unattractive aspect of a bargain from another person.

Human self-control is no simple skill, but an ever-growing world of expertise that reaches into everything we do. Why is it that, in the end, so few of our self-incentive tricks work well? Because, as we have seen, directness is too dangerous. If self-control were easy to obtain, we'd end up accomplishing nothing at all.

4.6 SELF-CONTROL

*Those who really seek the path to Enlightenment dictate terms to
their mind. Then they proceed with strong determination.*
—BUDDHA

The episode of Professor Challenger showed just one way we can control ourselves: by exploiting an emotional aversion in order to accomplish an intellectual purpose. Consider all the other kinds of tricks we use to try to force ourselves to work when we're tired or distracted.

WILLPOWER: *Tell yourself, "Don't give in to that," or, "Keep on trying."*

Such self-injunctions can work at first—but finally they always fail, as though some engine in the mind runs out of fuel. Another style of self-control involves more physical activity:

ACTIVITY: *Move around. Exercise. Inhale. Shout.*

Certain physical acts are peculiarly effective, especially the facial expressions involved in social communication: they affect the sender as much as the recipient.

EXPRESSION: *Set jaw. Stiffen upper lip. Furrow brow.*

Another kind of stimulating act is moving to a stimulating place. And we often perform actions that directly change the brain's chemical environment.

CHEMISTRY: *Take coffee, amphetamines, or other brain-affecting drugs.*

Then there are actions in the mind with which we set up thoughts and fantasies that move our own emotions, arousing hopes and fears through self-directed offers, bribes, and even threats.

EMOTION: *"If I win, there's much to gain, but more to lose if I fail!"*

Perhaps most powerful of all are those actions that promise gain or loss of the regard of certain special persons.

ATTACHMENT: *Imagine admiration if you succeed—or disapproval if you fail—
especially from those to whom you are attached.*

So many schemes for self-control! How do we choose which ones to use? There isn't any easy way. Self-discipline takes years to learn; it grows inside us stage by stage.

4.7 LONG-RANGE PLANS

In the search for truth there are certain questions that are not important. Of what material is the universe constructed? Is the universe eternal? Are there limits or not to the universe? What is the ideal form of organization for human society? If a man were to postpone his search and practice for Enlightenment until such questions were solved, he would die before he found the path.
—BUDDHA

We often become involved in projects that we can't complete. It is easy to solve small problems because we can treat them as though they were detached from all our other goals. But it is different for projects that span larger portions of our lives, like learning a trade, raising a child, or writing a book. We cannot simply "decide" or "choose" to accomplish an enterprise that makes a large demand for time, because it will inevitably conflict with other interests and ambitions. Then we'll be forced to ask questions like these:

What must I give up for this?
What will I learn from it?
Will it bring power and influence?
Will I remain interested in it?
Will other people help me with it?
Will they still like me?

Perhaps the most difficult question of all is, *"How will adopting this goal change me?"* Just wanting to own a large, expensive house, for instance, can lead to elaborate thoughts like these:

"That means I'd have to save for years and not get other things I'd like. I doubt that I could bear it. True, I could reform myself, and try to be more thrifty and deliberate. But that's just not the sort of person I am."

Until such doubts are set aside, all the plans we make will be subject to the danger that we may "change our mind." So how can any long-range plan succeed? The easiest path to "self-control" is doing only what one is already disposed to do.

Many of the schemes we use for self-control are the same as those we learn to use for influencing other people. We make ourselves behave by exploiting our own fears and desires, offering ourselves rewards, or threatening the loss of what we love. But when short-range tricks won't keep us to our projects for long enough, we may need some way to make changes that won't let us change ourselves back again. I suspect that, in order to commit ourselves to our largest, most ambitious plans, we learn to exploit agencies that operate on larger spans of time.

Which are our slowest-changing agencies of all? Later we'll see that these must include the silent, hidden agencies that shape what we call *character.* These are the systems that are concerned not merely with the things we *want,* but with what we *want ourselves to be*—that is, the ideals we set for ourselves.

4.8 IDEALS

We usually reserve the word "ideals" to refer to how we think we ought to conduct our ethical affairs. But I'll use the term in a broader sense, to include the standards we maintain—consciously or otherwise—for how we ought to think about ordinary matters.

We're always involved with goals of varying spans and scales. What happens when a transient inclination clashes with a long-term self-ideal? What happens, for that matter, when our ideals disagree among themselves, as when there is an inconsistency between the things we want to do and those we feel we ought to do? These disparities give rise to feelings of discomfort, guilt, and shame. To lessen such disturbances, we must either change the things we do—or change the ways we feel. Which should we try to modify—our immediate wants or our ideals? Such conflicts must be settled by the multilayered agencies that are formed in the early years of the growth of our personalities.

In childhood, our agencies acquire various types of goals. Then we grow in overlapping waves, in which our older agencies affect the making of the new. This way, the older agencies can influence how our later ones will behave. Outside the individual, similar processes go on in every human community; we find children "taking after" persons other than themselves by absorbing values from their parents, families, and peers, even from the heroes and villains of mythology.

Without enduring self-ideals, our lives would lack coherence. As individuals, we'd never be able to trust ourselves to carry out our personal plans. In a social group, no one person would be able to trust the others. A working society must evolve mechanisms that stabilize ideals— and many of the social principles that each of us regards as personal are really "long-term memories" in which our cultures store what they have learned across the centuries.

CHAPTER 5

INDIVIDUALITY

PUNCH AND JUDY, TO THEIR AUDIENCE

Our puppet strings are hard to see,
So we perceive ourselves as free,
Convinced that no mere objects could
Behave in terms of bad and good.

To you, we mannikins seem less
than live, because our consciousness
is that of dummies, made to sit
on laps of gods and mouth their wit;

Are you, our transcendental gods,
likewise dangled from your rods,
and need, to show spontaneous charm,
some higher god's inserted arm?

We seem to form a nested set,
with each the next one's marionette,
who, if you asked him, would insist
that he's the last ventriloquist.

—THEODORE MELNECHUK

5.1 CIRCULAR CAUSALITY

Whenever we can, we like to explain things in terms of simple cause and effect. We explained the case of Professor Challenger by assuming that my wish to *Work* came first, then *Work* exploited *Anger*'s aptitude for fighting *Sleep*. But in real life the causal relations between feelings and thoughts are rarely so simple. My desire to work and my annoyance with Challenger were probably so intermingled, all along, that it is inappropriate to ask which came first, *Anger* or *Work*. Most likely, *both* agencies exploited one another simultaneously, thus combining both into a single fiendish synthesis that accomplished two goals at once; *Work* thus got to do its work —and, thereby, injured Challenger! (In an academic rivalry, a technical accomplishment can hurt more than a fist.) Two goals can support each other.

A causes B   *"John wanted to go home because he felt tired of work."*
B causes A   *"John felt tired of work because he wanted to go home."*

There need be no "first cause" since John could start out with both distaste for work and inclination to go home. Then a loop of circular causality ensues, in which each goal gains support from the other until their combined urge becomes irresistible. We're always enmeshed in causal loops. Suppose you had borrowed past your means and later had to borrow more in order to pay the interest on your loan. If you were asked what the difficulty was, it would not be enough to say simply, *"Because I have to pay the interest,"* or to say only, *"Because I have to pay the principal."* Neither alone is the actual cause, and you'd have to explain that you're caught in a loop.

We often speak of "straightening things out" when we're involved in situations that seem too complicated. It seems to me that this metaphor reflects how hard it is to find one's way through a maze that has complicated loops in it. In such a situation, we always try to find a "path" through it by seeking "causal" explanations that go in only one direction. There's a good reason for doing this.

> *There are countless different types of networks that contain loops. But all networks that contain no loops are basically the same: each has the form of a simple chain.*

Because of this, we can apply the very same types of reasoning to *everything* we can represent in terms of chains of causes and effects. Whenever we accomplish that, we can proceed from start to end without any need for a novel thought; that's what we mean by "straightening out." But frequently, to construct such a path, we have to ignore important interactions and dependencies that run in other directions.

5.2 UNANSWERABLE QUESTIONS

And while it shall please thee to continue me in this world, where there is much to be done and little to be known, teach me, by thy Holy Spirit, to withdraw my mind from unprofitable and dangerous enquiries, from difficulties vainly curious, and doubts impossible to be solved.
—SAMUEL JOHNSON

When we reflect on anything for long enough, we're likely to end up with what we sometimes call "basic" questions—ones we can see no way at all to answer. For we have no perfect way to answer even this question: *How can one tell when a question has been properly answered?*

What caused the universe, and why?   *What is the purpose of life?*
How can you tell which beliefs are true?   *How can you tell what is good?*

These questions seem different on the surface, but all of them share one quality that makes them impossible to answer: *all of them are circular!* You can never find a final cause, since you must always ask one question more: *"What caused that cause?"* You can never find any ultimate goal, since you're always obliged to ask, *"Then what purpose does **that** serve?"* Whenever you find out why something is good—or is true—you still have to ask what makes *that* reason good and true. No matter what you discover, at every step, these kinds of questions will always remain, because you have to challenge every answer with, *"Why should I accept **that** answer?"* Such circularities can only waste our time by forcing us to repeat, over and over and over again, *"What good is Good?"* and, *"What god made God?"*

When children keep on asking, *"Why?"* we adults learn to deal with this by simply saying, *"Just because!"* This may seem obstinate, but it's also a form of self-control. What stops adults from dwelling on such questions endlessly? The answer is that every culture finds special ways to deal with these questions. One way is to brand them with shame and taboo; another way is to cloak them in awe or mystery; both methods make those questions undiscussable. Consensus is the simplest way—as with those social styles and trends wherein we each accept as true whatever all the others do. I think I once heard W. H. Auden say, *"We are all here on earth to help others. What I can't figure out is what the others are here for."*

All human cultures evolve institutions of law, religion, and philosophy, and these institutions both adopt specific answers to circular questions and establish authority-schemes to indoctrinate people with those beliefs. One might complain that such establishments substitute dogma for reason and truth. But in exchange, they spare whole populations from wasting time in fruitless reason loops. Minds can lead more productive lives when working on problems that can be solved.

But when thinking keeps returning to its source, it doesn't always mean something's wrong. For circular thinking can lead to growth when it results, at each return, in deeper and more powerful ideas. Then, because we can communicate, such systems of ideas may even find the means to cross the boundaries of selfish selves—and thus take root in other minds. This way, a language, science, or philosophy can transcend the limitation of each single mind's mortality. Now, we cannot know that any individual is destined for some paradise. Yet certain religions are oddly right; they manage to achieve their goal of offering an afterlife—if only to their own strange souls.

5.3 THE REMOTE-CONTROL SELF

When people have no answers to important questions, they often give some anyway.

> *What controls the brain?* The Mind.
> *What controls the mind?* The Self.
> *What controls the Self?* Itself.

To help us think about how our minds are connected to the outer world, our culture teaches schemes like this:

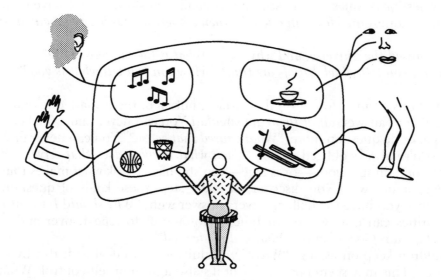

This diagram depicts our sensory machinery as sending information to the brain, wherein it is projected on some inner mental movie screen. Then, inside that ghostly theater, a lurking Self observes the scene and then considers what to do. Finally, that Self may act—somehow reversing all those steps—to influence the real world by sending various signals back through yet another family of remote-control accessories.

This concept simply doesn't work. It cannot help for you to think that inside yourself lies someone else who does your work. This notion of "homunculus"—a little person inside each self—leads only to a paradox since, then, *that inner Self requires yet another movie screen inside itself, on which to project what it has seen!* And then, to watch that play-within-a-play, we'd need yet another Self-inside-a-Self—to do the thinking for the last. And then this would all repeat again, as each new Self requires yet another one to do its job!

> *The idea of a single, central Self doesn't explain anything. This is because a thing with no parts provides nothing that we can use as pieces of explanation!*

Then why do we so often embrace the strange idea that what we do is done by Someone Else —that is, our Self? Because so much of what our minds do is hidden from the parts of us that are involved with verbal consciousness.

5.4 PERSONAL IDENTITY

Whate'er the passion—knowledge, fame, or pelf,
Not one will change his neighbor with himself.
—ALEXANDER POPE

Why do we accept that paradoxical image of a central Self inside the self? Because it serves us well in many spheres of practical life. Here are some reasons to regard a person as a single thing.

> **The Physical World:** *Our bodies act like other objects that take up space. Because of that, we must base our plans and decisions on having a single body. Two people cannot fit where there is room for only one—nor can a person walk through walls or stay aloft without support.*

> **Personal Privacy:** *When Mary tells Jack something, she must remember to "whom" it was told, and she must not assume that every other person knows it, too. Also, without the concept of an individual, we could have no sense of responsibility.*

> **Mental Activity:** *We often find it hard to think two different thoughts at once, particularly when they're similar, because we get "confused" when the same agencies are asked to do different jobs at the same time.*

Why do our mental processes so often seem to us to flow in "streams of consciousness"? Perhaps because, in order to keep control, we have to simplify how we represent what's happening. Then, when that complicated mental scene is "straightened out," it seems as though a single pipeline of ideas were flowing through the mind.

These are all compelling reasons why it helps to see ourselves as singletons. Still, each of us must also learn not only that different people have their own identities, but that the same person can entertain different beliefs, plans, and dispositions at the same time. For finding good ideas about psychology, the single-agent image has become a grave impediment. To comprehend the human mind is surely one of the hardest tasks any mind can face. The legend of the single Self can only divert us from the target of that inquiry.

5.5 FASHION AND STYLE

*The notes I handle no better than many pianists. But the pauses
between the notes—ah, that is where the art resides!*
—ARTUR SCHNABEL

Why do we like so many things that seem to us to have no earthly use? We often speak of this with mixtures of defensiveness and pride.

"Art for Art's sake."
"I find it aesthetically pleasing."
"I just like it."
"There's no accounting for it."

Why do we take refuge in such vague, defiant declarations? "There's no accounting for it" sounds like a guilty child who's been told to keep accounts. And "I just like it" sounds like a person who is hiding reasons too unworthy to admit. However, we often do have sound practical reasons for making choices that have no reasons by themselves but have effects on larger scales.

Recognizability: *The legs of a chair work equally well if made square or round. Then why do we tend to choose our furniture according to systematic styles or fashions? Because familiar styles make it easier for us to recognize and classify the things we see.*

Uniformity: *If every object in a room were interesting in itself, our furniture might occupy our minds too much. By adopting uniform styles, we protect ourselves from distractions.*

Predictability: *It makes no difference whether a single car drives on the left or on the right. But it makes all the difference when there are many cars! Societies need rules that make no sense for individuals.*

It can save a lot of mental work if one makes each arbitrary choice the way one did before. The more difficult the decision, the more this policy can save. The following observation by my associate, Edward Fredkin, seems important enough to deserve a name:

Fredkin's Paradox: *The more equally attractive two alternatives seem, the harder it can be to choose between them—no matter that, to the same degree, the choice can only matter less.*

No wonder we often can't account for "taste"—if it depends on hidden rules that we use when ordinary reasons cancel out! I do not mean to say that fashion, style, and art are all the same—only that they often share this strategy of using forms that lie beneath the surface of our thoughts. When should we quit reasoning and take recourse in rules of style? Only when we're fairly sure that further thought will just waste time. Perhaps that's why we often feel such a sense of being free from practicality when we make "aesthetic" choices. Such decisions might seem more constrained if we were aware of how they're made. And what about those fleeting hints of guilt we sometimes feel for "just liking" art? Perhaps they're how our minds remind themselves not to abandon thought too recklessly.

5.6 TRAITS

Isn't it remarkable that words can portray human individuals? You might suppose this should be impossible, considering how much there is to say. Then what permits a writer to depict such seemingly real personalities? It is because we all agree on so many things that are left unsaid. For example, we assume that all the characters are possessed of what we call "commonsense knowledge," and we also agree on many generalities about what we call "human nature."

Hostility evokes defensiveness. Frustration arouses aggression.

We also recognize that individuals have particular qualities and traits of character.

Jane is tidy. Mary's timid. Grace is smart.
That's not the sort of thing Charles does. It's not his style.

Why should traits like these exist? Humanists are prone to boast about how hard it is to grasp the measure of a mind. But let's ask instead, "*What makes personalities so easy to portray?*" Why, for example, should any person tend toward a general quality of being neat, rather than simply being tidy about some things and messy about others? Why should our personalities show such coherencies? How could it be that a system assembled from a million agencies can be described by short and simple strings of words? Here are some possible reasons.

Selectivity: *First we should face the fact that our images of other minds are often falsely clear. We tend to think of another person's "personality" in terms of that which we can describe—and tend to set aside the rest as though it simply weren't there.*

Style: *To escape the effort of making decisions we consider unimportant, we tend to develop policies that become so systematic that they can be discerned from the outside and characterized as personal traits.*

Predictability: *Because it is hard to maintain friendship without trust, we try to conform to the expectations of our friends. Then, to the extent that we frame our images of our associates in terms of traits, we find ourselves teaching ourselves to behave in accord with those same descriptions.*

Self-Reliance: *Thus, over time, imagined traits can make themselves actual! For even to carry out our own plans, we must be able to predict what we ourselves are likely to do—and that will become easier the more we simplify ourselves.*

It's nice to be able to trust our friends, but we need to be able to trust ourselves. How can that be possible when we can't be sure what's in our own heads? One way to accomplish this is by thinking of ourselves in terms of traits—and then proceeding to train ourselves to behave according to those self-images. Still, a personality is merely the surface of a person. What we call traits are only the regularities we manage to perceive. We never really know ourselves because there are so many other processes and policies that never show themselves directly in our behavior but work behind the scenes.

5.7 PERMANENT IDENTITY

*There are causes for all human suffering, and there is a way by
which they may be ended, because everything in the world is the
result of a vast concurrence of causes and conditions, and
everything disappears as these causes and conditions
change and pass away.*
—BUDDHA

What do we signify by words like "me," "myself," and "I"? What does a story mean that starts with "In *my* childhood"? What is that strange possession "you," which stays the same throughout your life? Are you the same person you were before you learned to read? You scarcely can imagine, now, how words looked then. Just try to look at these words without reading them:

So far as consciousness is concerned, we find it almost impossible to separate the appearances of things from what they've come to mean to us. But if we cannot recollect how things appeared to us before we learned to link new meanings to those things, what makes us think we can recollect how we ourselves appeared to us in previous times? What would you say if someone asked questions like these:

> **"Are you the same person now that you once were, before you learned to talk?"**
> "Of course I am. Why, who else could I be?"
> **"Do you mean that you haven't changed at all?"**
> "Of course not. I only mean I'm the same person—the same in some ways,
> different in others—but still the same me."
> **"But how can you be the same as the person you were before you had even
> learned to remember things? Can you even imagine what that was like?"**
> "Perhaps I can't—yet still there must have been some continuity. Even if I can't
> remember it, I surely was that person, too."

We all experience that sense of changelessness in spite of change, not only for the past but also for the future, too! Consider how you are generous to future self at present self's expense. Today, you put some money in the bank in order that sometime later you can take it out. Whenever did that future self do anything so good for you? Is "you" the body of those memories whose meanings change only slowly? Is it the never-ending side effects of all your previous experience? Or is it just whichever of your agents change the least as time and life proceed?

CHAPTER 6

INSIGHT AND INTROSPECTION

MIND. A mysterious form of matter secreted by the brain. Its chief activity consists in the endeavor to ascertain its own nature, the futility of the attempt being due to the fact that it has nothing but itself to know itself with.

—AMBROSE BIERCE

6.1 CONSCIOUSNESS

> **con•scious** *a. 1. having a feeling or knowledge (of one's sensations, feelings, etc., or of external things); knowing or feeling (that something is or was happening or existing); . . . 3. aware of oneself as a thinking being; knowing what one is doing and why.*
> —Webster's Unabridged Dictionary

In real life, you often have to deal with things you don't completely understand. You drive a car, not knowing how its engine works. You ride as passenger in someone else's car, not knowing how that driver works. Most strange of all, you drive your body and your mind, not knowing how your own self works. Isn't it amazing that we can think, not knowing what it means to think? Isn't it remarkable that we can get ideas, yet not explain what ideas are?

In every normal person's mind there seem to be some processes that we call consciousness. We usually regard them as enabling us to know what's happening inside our minds. But this reputation of self-awareness is not so well deserved, because our conscious thoughts reveal to us so little of what gives rise to them.

Consider how a driver guides the immense momentum of a motorcar, not knowing how its engine works or how its steering wheel directs it to the left or right. Yet when one comes to think of it, we drive our bodies in much the same way. So far as conscious thought is concerned, you turn yourself to walk in a certain direction in much the way you steer a car; you are aware only of some general intention, and all the rest takes care of itself. To change your direction of motion is actually quite complicated. If you simply took a larger or smaller step on one side, the way you would turn a rowboat, you would fall toward the outside of the turn. Instead, you start to turn by making yourself fall toward the *inside*—and then use centrifugal force to right yourself on the next step. This incredible process involves a huge society of muscles, bones, and joints, all controlled by hundreds of interacting programs that even specialists don't yet understand. Yet all you think is, *Turn that way*, and your wish is automatically fulfilled.

We give the name "signals" to acts whose consequences are not inherent in their own character but have merely been assigned to them. When you accelerate your car by pressing on the gas pedal, this is not what does the work; it is merely a signal to make the engine push the car. Similarly, rotating the steering wheel is merely a signal that makes the steering mechanism turn the car. The car's designer could easily have assigned the pedal to steer the car or made the steering wheel control its speed. But practical designers try to exploit the use of signals that already have acquired some significance.

Our conscious thoughts use signal-signs to steer the engines in our minds, controlling countless processes of which we're never much aware. Not understanding how it's done, we learn to gain our ends by sending signals to those great machines, much as the sorcerers of older times used rituals to cast their spells.

6.2 SIGNALS AND SIGNS

How do we ever understand anything? Almost always, I think, by using one or another kind of analogy—that is, by representing each new thing as though it resembles something we already know. Whenever a new thing's internal workings are too strange or complicated to deal with directly, we represent whatever parts of it we can in terms of more familiar signs. This way, we make each novelty seem similar to some more ordinary thing. It really is a great discovery, the use of signals, symbols, words, and names. They let our minds transform the strange into the commonplace.

Suppose an alien architect has invented a radically new way to go from one room to another. This invention serves the normal functions of a door, but it has a form and mechanism so far outside our experience that to see it, we would never recognize it as a door, nor guess how to use it. All its physical details are wrong. It is not what we normally expect a door to be—a hinged, swinging, wooden slab set into a wall. No matter: just superimpose on its exterior some decoration, symbol, icon, token, word, or sign that can remind us of its use. Clothe it in a rectangular shape, or add to it a push-plate lettered *EXIT* in red and white, and every visitor from the planet Earth will know, without a conscious thought, just what that pseudoportal's purpose is, and use it as though it were a door.

At first it may seem mere trickery, to assign the symbol for a door to an invention that is not really a door. But we're always in that same predicament. There are no doors inside our minds, only connections among our signs. To overstate the case a bit, what we call "consciousness" consists of little more than menu lists that flash, from time to time, on mental screen displays that other systems use. It is very much like the way the players of computer games use symbols to invoke the processes inside their complicated game machines without the slightest understanding of how they work.

And when you come to think about it, it scarcely could be otherwise! Consider what would happen if we actually could confront the trillion-wire networks in our brains. Scientists have peered at tiny fragments of those structures for many years, yet failed to comprehend what they do. Fortunately, for the purposes of everyday life, it is enough for our words or signals to evoke some useful happenings within the mind. Who cares how they work, so long as they work! Consider how you can scarcely see a hammer except as something to hit with, or see a ball except as something to throw and catch. Why do we see things, less as they are, and more in view of how they can be used? It is because our minds did not evolve to serve as instruments for science or philosophy, but to solve practical problems of nutrition, defense, procreation, and the like. We tend to think of knowledge as good in itself, but knowledge is useful only when we can exploit it to help us reach our goals.

6.3 THOUGHT-EXPERIMENTS

How do you discover things about the world? Just look and see! It seems simple—but it's not. Each casual glance employs a billion brain cells to represent the present scene and to summarize its differences from records of other experiences. Your agencies formulate little bits of theories about what happens in the world and then make you do small experiments to confirm or reformulate those conjectures. It only seems simple because you're unaware of what is happening.

How do you discover things about your mind? You use a similar technique. You make up little bits of theories about how you think, then test them with tiny experiments. The trouble is that thought-experiments don't often lead to the sorts of clear, crisp findings that scientists seek. Ask yourself what happens when you try to imagine a round square—or when you try to be happy and sad at the same time. Why is it so hard to describe the results of such experiments or draw useful conclusions from them? It is because we get confused. Our thoughts about our mind-experiments are mind-experiments themselves—and therefore interfere with one another.

Thinking affects our thoughts.

People who program computers encounter similar problems when new programs malfunction because of unexpected interactions among their parts. To find out what's happening, programmers have developed special programs for "debugging" other programs. But just as in thought-experiments, there is a danger that the program being watched might change the one that's watching it. To prevent this, all modern computers are equipped with special "interruption" machinery that detects any other program's attempt to alter a debugging program; when this happens, the culprit is "frozen" in its tracks so that the debugging program can examine it. To do this, the interruption machinery must be supplied with a private memory bank that can store enough information to make it possible, later, to restart the frozen program as though nothing had happened.

Are brains equipped to do similar things? It was easy to build self-examination systems into computers that did only one thing at a time, but it would be much harder to do in a system that, like the brain, engages many processes at once. The problem is that if you were to freeze only one process without stopping the others, it would change the situation you're trying to examine. However, if you were to stop all those processes all at once, you couldn't experiment on how they interact.

Later, we'll see that consciousness is connected with our most immediate memories. This means that there are limits on what consciousness can tell us about itself—because it can't do perfect self-experiments. That would require keeping perfect records of what happens inside one's memory machinery. But any such machinery must get confused by self-experiments that try to find out how it works—since such experiments must change the very records they are trying to inspect! We cannot handle interruptions perfectly. This doesn't mean that consciousness cannot be understood, in principle. It only means that to study it, we'll have to use the less direct methods of science, because we cannot simply "look and see."

6.4 *B*-BRAINS

There *is* one way for a mind to watch itself and still keep track of what's happening. Divide the brain into two parts, A and B. Connect the A-brain's inputs and outputs to the real world—so it can sense what happens there. But don't connect the B-brain to the outer world at all; instead, connect it so that the A-brain is the B-brain's world!

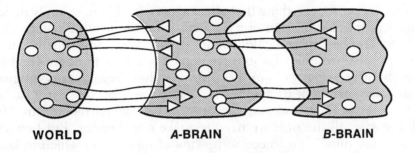

WORLD   *A*-BRAIN   *B*-BRAIN

Now A can see and act upon what happens in the outside world—while B can "see" and influence what happens inside A. What uses could there be for such a B? Here are some A-activities that B might learn to recognize and influence.

A seems disordered and confused.	*Inhibit that activity.*
A appears to be repeating itself.	*Make A stop. Do something else.*
*A does something **B** considers good.*	*Make A remember this.*
A is occupied with too much detail.	*Make A take a higher-level view.*
A is not being specific enough.	*Focus A on lower-level details.*

This two-part arrangement could be a step toward having a more "reflective" mind-society. The B-brain could do experiments with the A-brain, just as the A-brain can experiment with the body or with the objects and people surrounding it. And just as A can attempt to predict and control what happens in the outer world, B can try to predict and control what A will do. For example, the B-brain could supervise how the A-brain learns, either by making changes in A directly or by influencing A's own learning processes.

Even though B may have no concept of what A's activities mean in relation to the outer world, it is still possible for B to be useful to A. This is because a B-brain could learn to play a role somewhat like that of a counselor, psychologist, or management consultant, who can assess a client's mental strategy without having to understand all the details of that client's profession. Without having any idea of what A's goals are, B might be able to learn to tell when A is not accomplishing them but only going around in circles or wandering, confused because certain A-agents are repeating the same things over and over again. Then B might try some simple remedies, like suppressing some of those A-agents. To be sure, this could also result in B's activities becoming nuisances to A. For example, if A had the goal of adding up a long column of numbers, B might start to interfere with this because, from B's point of view, A appears to have become trapped in a repetitive loop. This could cause a person accustomed to more variety to find it difficult to concentrate on such a task and complain of being bored.

To the extent that the B-brain knows what is happening in A, the entire system could be considered to be partly "self-aware." However, if we connect A and B to "watch" each other too closely, then anything could happen, and the entire system might become unstable. In any case, there is no reason to stop with only two levels; we could connect a C-brain to watch the B-brain, and so on.

6.5 FROZEN REFLECTION

Time present and time past
Are both perhaps present in time future,
And time future contained in time past.
— T. S. ELIOT

No supervisor can know everything that all its agents do. There's simply never time enough. Each bureaucrat sees but a fraction of what happens underneath its place in the pyramid of information flow. The best subordinates are those that work most quietly. Indeed, that's *why* we build administrative pyramids for jobs we don't know how to do or don't have time to do ourselves. It is also why so many of our thoughts must hide beyond our consciousness.

Good scientists never try to learn too much at once. Instead, they select particular aspects of a situation, observe carefully, and make records. Experimental records are "frozen phenomena." They let us take all the time we need to make our theories. But how could we do the same thing inside the mind? We'd need some kind of memory in which to keep such records safe.

We'll see how this could work when we come to the chapters on memory. We'll conjecture that your brain contains a host of agents called "K-lines," which you can use to make records of what some of your brain-agents are doing at a certain moment. Later, when you activate the same K-lines, this restores those agents to their previous states. This makes you "remember" part of your previous mental state, by making those parts of your mind do just what they did before. Then, the other parts of your mind will react as though the same events were happening again! Of course, such memories will always be incomplete, since nothing could have capacity enough to record every detail of its own state. (Otherwise, it would have to be larger than itself.) Since we can't remember everything, each individual mind faces the same problem that scientists always face: they have no foolproof way to know, before the fact, what are the most important things to notice and record.

Using the mind to examine itself is like science in another way. Just as physicists cannot see the atoms they talk about, psychologists can't watch the processes they try to examine. We only "know" such things through their effects. But the problem is worse where the mind is concerned, since scientists can read each other's notes, but different parts of the mind can't read each other's memories.

We've now seen several reasons why we cannot simply watch our minds by sitting still and waiting till our vision clears. The only course left for us is to study the mind the way scientists do when something is too large or small to see—by building theories based on evidence. Make a guess; test it with a shrewd experiment; collect one's thoughts and guess again. When introspection seems to work, it's not because we've found a magic way to see inside ourselves. Instead, it means that we've done some well-designed experiment.

6.6 MOMENTARY MENTAL TIME

What do you think you're thinking now? You might reply, "Why, just the thoughts I'm thinking now!" And that makes sense, in ordinary life, where "now" means "at this moment in time." But the meaning of "now" is far less clear for an agent inside a society.

It takes some time for changes in one part of a mind to affect the other parts. There's always some delay.

For example, suppose you meet your friend Jack. Your agencies for *Voices* and *Faces* may recognize Jack's voice and face, and both send messages to an agency *Names*, which may recall Jack's name. But *Voices* may also send a "word-message" to *Quotes*, a language-based agency that has a way to remember phrases Jack has said before, while *Faces* may also send a message to *Places*, an agency concerned with space, which might recall some earlier place in which Jack's face was seen.

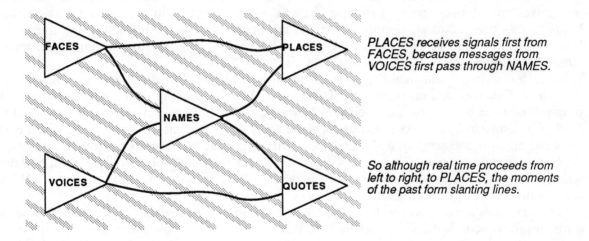

PLACES receives signals first from FACES, because messages from VOICES first pass through NAMES.

So although real time proceeds from left to right, to PLACES, the moments of the past form slanting lines.

Now suppose we could ask both *Places* and *Quotes* which had happened first, seeing Jack or hearing his voice? We'd get two different answers! *Places* will first detect the face—while *Quotes* will first detect the voice. The seeming order of events depends upon which message reached each agent first—so the seeming sequence of events differs from one agent to another. Each agent will react in its own, slightly different way—because it has been affected by a slightly different "causal history," which spreads like a wave into the past.

It is simply impossible, in general, for any agent **P** to know for certain what another agent **Q** is doing at precisely the same time. The best that **P** can do is send a query straight to **Q** and hope that **Q** can get a truthful message back before other agents change **Q**'s state—or change its message along the way. No portion of a mind can ever know everything that is happening at the same time in all the other agencies. Because of this, each agency must have at least a slightly different sense both of what has happened in the past—and of what is happening "now." Each different agent of the mind lives in a slightly different world of time.

6.7 THE CAUSAL NOW

To know anything [said the poet] *we must know its effects; to see men we must know their works, that we may know what reason has dictated, or passion has incited, and find what are the most powerful motives of action. To judge rightly of the present, we must oppose it to the past; for all judgment is comparative, and of the future nothing can be known. The truth is, that no mind is much employed upon the present: recollection and anticipation fill up almost all our moments. Our passions are joy and grief, love and hatred, hope and fear; even love and hatred respect the past, for the cause must have been before the effect.*
— SAMUEL JOHNSON

Our everyday ideas about the progression of mental time are wrong: they leave no room for the fact that every agent has a different causal history. To be sure, those different pasts are intermixed over longer spans of time, and every agent is eventually influenced by what has happened in the common, remote history of its society. But that's not what one means by "now." The problem is with the connections between the moment-to-moment activities of largely separate agencies.

When a pin drops, you might say, *"I just heard a pin drop."* But no one says, *"I hear a pin dropping."* Our speaking agencies know from experience that the physical episode of pin dropping will be over before you can even start to speak. But you would say, *"I am in love,"* rather than *"I was just in love,"* because your speaking agencies know that the agencies involved with personal attachments work at a slower pace, with states that may persist for months or years. And, in between, when someone asks, *"What sorts of feelings have you now?"* we often find our half-formed answers wrong before they can be expressed, as other feelings intervene. What seems only a moment to one agency may seem like an era to another.

Our memories are only indirectly linked to physical time. We have no absolute sense of when a memorable event "actually" happened. At best, we can only know some temporal relations between it and certain other events. You might be able to recall that X and Y occurred on different days but be unable to determine which of those days came earlier. And many memories seem not to be linked to intervals of time at all—like knowing that four comes after three, or that "I am myself."

The slower an agency operates—that is, the longer the intervals between each change of state—the more external signals can arrive inside those intervals. Does this mean that the outside world will appear to move faster to a slow agency than to a faster agency? Does life seem swift to tortoises, but tedious to hummingbirds?

6.8 THINKING WITHOUT THINKING

Just as we walk without thinking, we think without thinking! We don't know how our muscles make us walk—nor do we know much more about the agencies that do our mental work. When you have a hard problem to solve, you think about it for a time. Then, perhaps, the answer seems to come all at once, and you say, *"Aha, I've got it. I'll do such and such."* But if someone were to ask how you found the solution, you could rarely say more than things like the following:

> *"I suddenly realized . . ."*
> *"I just got this idea . . ."*
> *"It occurred to me that . . ."*

If we could really sense the workings of our minds, we wouldn't act so often in accord with motives we don't suspect. We wouldn't have such varied and conflicting theories for psychology. And when we're asked how people get their good ideas, we wouldn't be reduced to metaphors about "ruminating," and "digesting," "conceiving" and "giving birth" to concepts—as though our thoughts were anywhere but in the head. If we could see inside our minds, we'd surely have more useful things to say.

Many people seem absolutely certain that no computer could ever be sentient, conscious, self-willed, or in any other way "aware" of itself. But what makes everyone so sure that they themselves possess those admirable qualities? It's true that if we're sure of anything at all, it is that *"I'm aware—hence I'm aware."* Yet what do such convictions really mean? If self-awareness means to know what's happening inside one's mind, no realist could maintain for long that people have much insight, in the literal sense of seeing-in. Indeed, the evidence that we are self-aware—that is, that we have any special aptitude for finding out what's happening inside ourselves—is very weak indeed. It is true that certain people have a special excellence at assessing the attitudes and motivations of other persons (and, more rarely, of themselves). But this does not justify the belief that how we learn things about people, including ourselves, is fundamentally different from how we learn about other things. Most of the understandings we call "insights" are merely variants of our other ways to "figure out" what's happening.

6.9 HEADS IN THE CLOUDS

What we call a mind is nothing but a heap or collection of different perceptions, united together by certain relations and suppos'd, tho' falsely, to be endow'd with a perfect simplicity and identity.
—DAVID HUME

We'll take the view that nothing can have meaning by itself, but only in relation to whatever other meanings we already know. One might complain that this has the quality of the old question, "Which came first, the chicken or the egg?" If each thing one knows depends on other things one knows, isn't that like castles built on air? What keeps them from all falling down, if none are tied to solid ground?

Well, first, there's nothing basically wrong with the idea of a society in which each part lends meaning to the other parts. Some sets of thoughts are much like twisted ropes or woven cloths in which each strand holds others both together and apart. Consider all the music tunes you know. Among them you can surely find two tunes of which you like each one the more because of how it's similar to or different from the other one. Besides, no human mind remains entirely afloat. Later we'll see how our conceptions of space and time can be based entirely on networks of relationships, yet can still reflect the structure of reality.

If every mind builds somewhat different things inside itself, how can any mind communicate with a different mind? In the end, surely, communication is a matter of degree but it is not always lamentable when different minds don't understand each other perfectly. For then, provided *some* communication remains, we can share the richness of each other's thoughts. What good would other people be if we were all identical? In any case, the situation is the same *inside* your mind—since even you yourself can never know precisely what *you* mean! How useless any thought would be if, afterward, your mind returned to the selfsame state. But that never happens, because every time we think about a certain thing, our thoughts go off in different ways.

> *The secret of what anything means to us depends on how we've connected it to all the other things we know. That's why it's almost always wrong to seek the "real meaning" of anything. A thing with just one meaning has scarcely any meaning at all.*

An idea with a single sense can lead you along only one track. Then, if anything goes wrong, it just gets stuck—a thought that sits there in your mind with nowhere to go. That's why, when someone learns something "by rote"—that is, with no sensible connections—we say that they "don't really understand." Rich meaning-networks, however, give you many different ways to go: if you can't solve a problem one way, you can try another. True, too many indiscriminate connections will turn a mind to mush. But well-connected meaning-structures let you turn ideas around in your mind, to consider alternatives and envision things from many perspectives until you find one that works. And that's what we mean by thinking!

6.10 WORLDS OUT OF MIND

All these beautiful, evolutionary qualities spontaneously blossom
in individual and collective life . . . where consciousness is found
identified with the unified field of all the laws of nature.
—BULLETIN, MAHARISHI
INTERNATIONAL
UNIVERSITY, 1984

There is no singularly real world of thought; each mind evolves its own internal universe. The worlds of thought that we appear to like the best are those where goals and actions seem to mesh in regions large enough to spend our lives in—and thus become a Buddhist, or Republican, or poet, or topologist. Some mental starting points grow into great, coherent continents. In certain parts of mathematics, science, and philosophy, a relatively few but clear ideas may lead into an endless realm of complex yet consistent new structures. Yet even in mathematics, a handful of seemingly innocent rules can lead to complications far beyond our grasp. Thus we feel we understand perfectly the rules of addition and multiplication—yet when we mix them together, we encounter problems about prime numbers that have remained unsolved for centuries.

Minds also make up pleasant worlds of practical affairs—which work because we make them work, by putting things in order there. In the physical realm, we keep our books and clothes in self-made shelves and cabinets—thus building artificial boundaries to keep our things from interacting very much. Similarly, in mental realms, we make up countless artificial schemes to force things to seem orderly, by specifying legal codes, grammar rules and traffic laws. When growing up in such a world, it all seems right and natural—and only scholars and historians recall the mass of precedents and failed experiments it took to make it work so well. These "natural" worlds are actually more complex than the technical worlds of philosophy. They're far too vast to comprehend—except where we impose on them the rules we make.

There is also a different and more sinister way to make the world seem orderly, in which the mind has merely found a way to simplify itself. This is what we must suspect whenever some idea seems to explain too much. Perhaps no problem was actually solved at all; instead, the mind has merely found some secondary pathway in the brain, through which one can mechanically dislodge each doubt and difference from its rightful place! This may be what happens in some of those experiences that leave a person with a sense of revelation—in a state in which no doubts remain, or with a vision of astounding clarity—yet unable to recount any details. Some accident of mental stress has temporarily suppressed the capacity to question, doubt, or probe. One remembers that no questions went unanswered but forgets that none were asked! One can acquire certainty only by amputating inquiry.

When victims of these incidents become compelled to recapture them, their lives and personalities are sometimes permanently changed. Then others, seeing the radiance in their eyes and hearing of the glory to be found, are drawn to follow them. But to offer hospitality to paradox is like leaning toward a precipice. You can find out what it is like by falling in, but you may not be able to fall out again. Once contradiction finds a home, few minds can spurn the sense-destroying force of slogans such as "all is one."

Suppose that while you walked and talked, you could watch the signals that traverse your brain. Would they make any sense to you? Many people have done experiments to make such signals audible and visible, by using biofeedback devices. This often helps a person to learn to control various muscles and glands that are not usually under conscious control. But it never leads to comprehending how their hidden circuits work.

Scientists encounter similar problems when they use electronic instruments to tap into brain signals. This has led to a good deal of knowledge about how nervous systems work—but those insights and understandings never came from observation by itself. One cannot use data without having at least the beginnings of some theory or hypothesis. Even if we could directly sense *all* the interior details of mental life, it wouldn't tell us how to understand them. It might even make that enterprise more difficult, by overwhelming our capacity to interpret what we see. The causes and functions of what we observe are not themselves things we can observe.

Where *do* we get the ideas we need? Most of our concepts come from the communities in which we're raised. Even the ideas we "get" for ourselves come from communities—this time, the ones inside our heads. Brains don't manufacture thoughts in the direct ways that muscles exert forces or ovaries make estrogens; instead, to get a good idea, one must engage huge organizations of submachines that do a vast variety of jobs. Each human cranium contains hundreds of kinds of computers, developed over hundreds of millions of years of evolution, each with a somewhat different architecture. Each specialized agency must learn to call on other specialists that can serve its purposes. Certain sections of the brain distinguish the sounds of voices from other sorts of sounds; other specialized agencies distinguish the sights of faces from other types of objects. No one knows how many different such organs lie in our brains. But it is almost certain that they all employ somewhat different types of programming and forms of representation; they share no common language code.

If a mind whose parts use different languages and modes of thought attempted to look inside itself, few of those agencies would be able to comprehend one another. It is hard enough for people who speak different human languages to communicate, and the signals used by different portions of the mind are surely even less similar. If agent **P** asked any question of an unrelated agent **Q**, how could **Q** sense what was asked, or **P** understand its reply? Most pairs of agents can't communicate at all.

6.12 INTERNAL COMMUNICATION

If agents can't communicate, how is it that people can—in spite of having such different backgrounds, thoughts, and purposes? The answer is that we overestimate how much we actually communicate. Instead, despite those seemingly important differences, much of what we do is based on common knowledge and experience. So even though we can scarcely speak at all about what happens in our lower-level mental processes, we can exploit their common heritage. Although we can't express what we mean, we can often cite various examples to indicate how to connect structures we're sure must already exist inside the listener's mind. In short, we can often indicate which sorts of thoughts to think, even though we can't express how they operate.

The words and symbols we use to summarize our higher-level goals and plans are not the same as the signals used to control lower-level ones. So when our higher-level agencies attempt to probe into the fine details of the lower-level submachines that they exploit, they cannot understand what's happening. This must be why our language-agencies cannot express such things as how we balance on our bicycles, distinguish pictures from real things, or fetch our facts from memory. We find it particularly hard to use our language skills to talk about the parts of the mind that learned such skills as balancing, seeing, and remembering, before we started to learn to speak.

"Meaning" itself is relative to size and scale: it makes sense to talk about a meaning only in a system large enough to have many meanings. For smaller systems, that concept seems vacant and superfluous. For example, *Builder*'s agents require no sense of meaning to do their work; *Add* merely has to turn on *Get* and *Put*. Then *Get* and *Put* do not need any subtle sense of what those turn-on signals "mean"—because they're wired up to do only what they're wired up to do. In general, the smaller an agency is, the harder it will be for other agencies to comprehend its tiny "language."

> *The smaller two languages are, the harder it will be to translate between them. This is not because there are too many meanings, but because there are too few. The fewer things an agent does, the less likely that what another agent does will correspond to any of those things. And if two agents have nothing in common, no translation is conceivable.*

In the more familiar difficulty of translating between human languages, each word has many meanings, and the main problem is to narrow them down to something they share. But in the case of communication between unrelated agents, narrowing down cannot help if the agents have nothing in common from the start.

6.13 SELF-KNOWLEDGE IS DANGEROUS

To "know oneself" more perfectly might seem to promise something powerful and good. But there are fallacies concealed behind that happy thought. No doubt, a mind that wants to change itself could benefit from knowing how it works. But such knowledge might as easily encourage us to wreck ourselves—if we had ways to poke our clumsy mental fingers into the tricky circuits of the mind's machinery. Could this be why our brains force us to play those games of mental hide and seek?

Just see how prone we are to risk experiments that change ourselves; how fatally we're drawn to drugs, to meditation, music, even conversation—all powerful addictions that can change our very personalities. Just see how everyone is entranced by any promise to transgress the bounds of normal pleasure and reward.

In ordinary life, our pleasure systems help us learn—and, therefore, to behave ourselves—by forcing checks and balances on us. Why, for example, do we become bored when doing the same thing over and over, even if that activity was pleasant at first? This appears to be one property of our pleasure systems; without enough variety, they tend to satiate. Every learning machine must have some such protective scheme, since otherwise it could get trapped into endlessly repeating the same activity. We are fortunate to be equipped with mechanisms that keep us from wasting too much time, and it is fortunate, too, that we find it hard to suppress such mechanisms.

> *If we could deliberately seize control of our pleasure systems, we could reproduce the pleasure of success without the need for any actual accomplishment. And that would be the end of everything.*

What prevents such meddling? Our minds are bound by many self-constraints. For example, we find it hard to determine what's happening inside the mind. Later, when we talk about infant development, we'll see that even if our inner eyes could see what's there, we'd find it singularly hard to change the agents we might want most to change—the ones that, in our infancy, helped shape our longest-lasting self-ideals.

These agents are hard to change because of their special evolutionary origin. The long-term stability of many other mental agencies depends on how slowly we change our images of what we ought to be like. Few of us would survive if, left to random chance, our most adventurous impulses could tamper freely with the basis of our personalities. Why would that be such a bad thing to do? Because an ordinary "change of mind" can be reversed if it leads to a bad result. But when you change your self-ideals—then nothing is left to turn you back.

Sigmund Freud theorized that each person's growth is governed by unconscious needs to please, placate, oppose, or terminate our images of parental authority. If we recognized the influence of those old images, however, we might consider them too infantile or too unworthy to tolerate and seek to replace them with something better. But then what would we substitute for them—once we divested ourselves of all those ties to instinct and society? We'd each end up as instruments of even more capricious sorts of self-invented goals.

6.14 CONFUSION

It's mainly when our systems fail that consciousness becomes engaged. For example, we walk and talk without much sense of how we actually do those things. But a person with an injured leg may, for the first time, begin to formulate theories about how walking works ("*To turn to the left, I'll have to push myself that way*") and then perhaps consider which muscles might accomplish that goal. When we recognize that we're confused, we begin to reflect on how our minds solve problems and engage the little we know about our strategies of thought. Then we find ourselves saying things like this:

> "*Now I must get organized. Why can't I concentrate on the important questions and not get distracted by those other nonessential details?*"

Paradoxically, it is smart to realize that one is confused—as opposed to being confused without knowing it. For that stimulates us to apply our intellect to altering or repairing the defective process. Yet we dislike and disparage the sense of confusion, not appreciating the quality of this recognition.

However, once your B-brains make you start to ask yourself "*What was I really attempting to do?*" you can exploit that as an opportunity to change your goals or change how you describe your situation. That way, you can escape the distress of feeling trapped because there seem to be no adequate alternatives. The conscious experience of confusion can resemble pain; perhaps this is because of how they both impel us to discover ways to escape from a predicament. The difference is that confusion is directed against a person's own failing state of mind, whereas pain reflects exterior disturbances. In either case, internal processes must be demolished and rebuilt.

Both confusion and pain have injurious effects when they lead us to abandon goals on larger scales than appropriate: "*The entire subject makes me feel ill. Perhaps I should abandon the whole project, occupation, or relationship.*" But even such dispiriting thoughts can serve as probes for finding other agencies that might be engaged for help.

CHAPTER 7

PROBLEMS AND GOALS

7.1 INTELLIGENCE

Many people insist on having some definition of "intelligence."

> CRITIC: *How can we be sure that things like plants and stones, or storms and streams, are not intelligent in ways that we have not yet conceived?*

It doesn't seem a good idea to use the same word for different things, unless one has in mind important ways in which they are the same. Plants and streams don't seem very good at solving the kinds of problems we regard as needing intelligence.

> CRITIC: *What's so special about solving problems? And why don't you define "intelligence" precisely, so that we can agree on what we're discussing?*

That isn't a good idea, either. An author's job is using words the ways other people do, not telling others how to use them. In the few places the word "intelligence" appears in this book, it merely means what people usually mean—the ability to solve hard problems.

> CRITIC: *Then you should define what you mean by a "hard" problem. We know it took a lot of human intelligence to build the pyramids—yet little coral reef animals build impressive structures on even larger scales. So don't you have to consider them intelligent? Isn't it hard to build gigantic coral reefs?*

Yes, but it is only an illusion that animals can "solve" those problems! No individual bird *discovers* a way to fly. Instead, each bird exploits a solution that evolved from countless reptile years of evolution. Similarly, although a person might find it very hard to design an oriole's nest or a beaver's dam, no oriole or beaver ever figures out such things at all. Those animals don't "solve" such problems themselves; they only exploit procedures available within their complicated gene-built brains.

> CRITIC: *Then wouldn't you be forced to say that evolution itself must be intelligent, since it solved those problems of flying and building reefs and nests?*

No, because people also use the word "intelligence" to emphasize swiftness and efficiency. Evolution's time rate is so slow that we don't see it as intelligent, even though it finally produces wonderful things we ourselves cannot yet make. Anyway, it isn't wise to treat an old, vague word like "intelligence" as though it must define any definite thing. Instead of trying to say what such a word "means," it is better simply to try to explain how we use it.

> *Our minds contain processes that enable us to solve problems we consider difficult. "Intelligence" is our name for whichever of those processes we don't yet understand.*

Some people dislike this "definition" because its meaning is doomed to keep changing as we learn more about psychology. But in my view that's exactly how it ought to be, because the very concept of intelligence is like a stage magician's trick. Like the concept of "*the unexplored regions of Africa,*" it disappears as soon as we discover it.

7.2 UNCOMMON SENSE

We've all heard jokes about how stupid present-day computers are. They send us bills and checks for zero dollars and zero cents. They don't mind working in endless loops, repeating the same thing a billion times. Their total lack of common sense is another reason people think that no machine could have a mind.

It is interesting to note that some of the earliest computer programs excelled at what people consider to be "expert" skills. A 1956 program solved hard problems in mathematical logic, and a 1961 program solved college-level problems in calculus. Yet not till the 1970s could we construct robot programs that could see and move well enough to arrange children's building-blocks into simple towers and playhouses. Why could we make programs do grown-up things before we could make them do childish things? The answer may seem paradoxical: much of "expert" adult thinking is actually simpler than what is involved when ordinary children play! Why is it easier to program what experts do than what children do?

What people vaguely call common sense is actually more intricate than most of the technical expertise we admire. Neither that "expert" program for logic nor the one for calculus embodied more than a hundred or so "facts"—and most of them were rather similar to one another. Yet these were enough to solve college-level problems. In contrast, think of all the different *kinds* of things a child must know merely to build a house of blocks—a process that involves knowledge of shapes and colors, space and time, support and balance, and an ability to keep track of what one is doing.

> *To be considered an "expert," one needs a large amount of knowledge of only a relatively few varieties. In contrast, an ordinary person's "common sense" involves a much larger variety of different **types** of knowledge—and this requires more complicated management systems.*

There is a simple reason why it is easier to acquire specialized knowledge than commonsense knowledge. Each type of knowledge needs some form of "representation" and a body of skills adapted to using that style of representation. Once that investment has been made, it is relatively easy for a specialist to accumulate further knowledge, provided the additional expertise is uniform enough to suit the same style of representation. A lawyer, doctor, architect, or composer who has learned to deal with a range of cases in some particular field finds it relatively easy to acquire more knowledge of a similar character. Think how much longer it would take a single person to learn to deal competently with a few diseases *and* several kinds of law cases *and* a small variety of architectural blueprints *and* a few orchestral scores. The greater variety of representations would make it much harder to acquire the "same amount" of knowledge. For each new domain, our novice would have to learn another type of representation and new skills for using it. It would be like learning many different languages, each with its own grammar, lexicon, and idioms. When seen this way, what children do seems all the more remarkable, since so many of their actions are based upon their own inventions and discoveries.

7.3 THE PUZZLE PRINCIPLE

Many people reason that machines do only what they're programmed to do—and hence can never be creative or original. The trouble is that this argument presumes what it purports to show: that you can't program a machine to be creative! In fact, it is surprisingly easy to program a computer so that it will proceed to do more different things than any programmer could imagine in advance. This is possible because of what we'll call the "puzzle principle."

> **Puzzle Principle:** *We can program a computer to solve any problem by trial and error, without knowing how to solve it in advance, provided only that we have a way to recognize when the problem is solved.*

By "trial and error" we mean programming the machine systematically to generate all possible structures within some universe of possibilities. For example, suppose you wished to have a robot machine that could build a bridge across a stream. The most efficient program for this would simply execute a specific procedure, planned out in advance, to precisely place some boards and nails. Of course, you couldn't write such a program unless you already knew how to build a bridge. But consider the alternative below, which is sometimes called the *generate and test* method. It consists of writing a two-part program.

> **Generate.** *The first process simply produces, one after another, every possible arrangement of the boards and nails. At first, you might expect such a program to be hard to write. But it turns out to be surprisingly easy, once you appreciate that there is no requirement for each arrangement to make any sense whatsoever!*

> **Test.** *The second part of the process examines each arrangement to see whether the problem has been solved. If the goal were to build a dam, the test is simply whether it holds back the stream. If the goal were to build a bridge, the test is simply whether it spans the stream.*

This possibility makes us reexamine all our old ideas about intelligence and creativity, since it means that, in principle, at least, we can make machines solve any problems whose solutions we can recognize. This is rarely practical, however. Consider that there must be a thousand ways to attach two boards, a million ways to connect three of them, and a billion ways to nail four boards together. It would take inconceivably long before the puzzle principle produced a workable bridge. But it does help, philosophically, to replace our feeling of mystery about creativity by more specific and concrete questions about the efficiency of processes. The main problem with our bridge-building machine is the lack of connection between its generator and its test. Without some notion of progress toward a goal, it is hard to do better than mindless chance.

7.4   PROBLEM SOLVING

In principle, we can use the "generate and test" method—that is, trial and error—to solve any problem whose solution we can recognize. But in practice, it can take too long for even the most powerful computer to test enough possible solutions. Merely assembling a simple house from a dozen wooden blocks would require searching through more possibilities than a child could try in a lifetime. Here is one way to improve upon blind trial-and-error search.

> **The Progress Principle:** *Any process of exhaustive search can be greatly reduced if we possess some way to detect when "progress" has been made. Then we can trace a path toward a solution, just as a person can climb an unfamiliar hill in the dark— by feeling around, at every step, to find the direction of steepest ascent.*

Many easy problems can be solved this way, but for a hard problem, it may be almost as difficult to recognize "progress" as to solve the problem itself. Without a larger overview, that "hill climber" may get stuck forever on some minor peak and never find the mountaintop. There is no foolproof way to avoid this.

> **Goals and Subgoals.** *The most powerful way we know for discovering how to solve a hard problem is to find a method that splits it into several simpler ones, each of which can be solved separately.*

Much research in the field called Artificial Intelligence has been concerned with finding methods machines can use for splitting a problem into smaller subproblems and then, if necessary, dividing these into yet smaller ones. In the next few sections we'll see how this can be done by formulating our problems in terms of "goals."

> **Using Knowledge.** *The most efficient way to solve a problem is to already know how to solve it. Then one can avoid search entirely.*

Accordingly, another branch of Artificial Intelligence research has sought to find ways to embody knowledge in machines. But this problem itself has several parts: we must discover how to acquire the knowledge we need, we must learn how to represent it, and, finally, we must develop processes that can exploit our knowledge effectively. To accomplish all that, our memories must represent, in preference to vast amounts of small details, only those relationships that may help us reach our goals. This research has led to many practical "knowledge-based" problem-solving systems. Some of these are often called "expert systems" because they're based on imitating the methods of particular human practitioners.

A curious phenomenon emerged from this research. It often turned out easier to program machines to solve specialized problems that educated people considered hard—such as playing chess or proving theorems about logic or geometry—than to make machines do things that most people considered easy—such as building toy houses with children's blocks. This is why I've emphasized so many "easy" problems in this book.

7.5 LEARNING AND MEMORY

There is an old and popular idea that we learn only what we are rewarded for. Some psychologists have claimed that human learning is based entirely on "reinforcement" by reward: that even when we train ourselves with no external inducements, we are still learning from reward —only now in the form of signals from inside ourselves. But we cannot trust an argument that assumes what it purports to prove, and in any case, when we try to use this idea to explain how people learn to solve hard problems, we encounter a deadly circularity. *You first must be able to do something before you can be rewarded for doing it!*

This circularity was no great problem when Ivan Pavlov studied conditioned reflexes nearly a century ago, because in his experiments the animals never needed to produce new kinds of behavior; they only had to link new stimuli to old behaviors. Decades later, Pavlov's research was extended by the Harvard psychologist B. F. Skinner, who recognized that higher animals did indeed sometimes exhibit new forms of behavior, which he called "operants." Skinner's experiments confirmed that when a certain operant is followed by a reward, it is likely to reappear more frequently on later occasions. He also discovered that this kind of learning has much larger effects *if the animal cannot predict when it will be rewarded.* Under names like "operant conditioning" and "behavior modification," Skinner's discoveries had a wide influence in psychology and education, but never led to explaining how brains produce new operants. Furthermore, few of these animal experiments shed much light on how humans learn to form and carry out their complex plans; the trouble is that other animals can scarcely learn such things at all. Those twin ideas—*reward/success* and *punish/failure*—do not explain enough about how people learn to produce the new ideas that enable them to solve difficult problems that could not otherwise be solved without many lifetimes of ineffectual trial and error.

The answer must lie in learning better ways to learn. In order to discuss these things, we'll have to start by using many ordinary words like *goal, reward, learning, thinking, recognizing, liking, wanting, imagining,* and *remembering*—all based on old and vague ideas. We'll find that most such words must be replaced by new distinctions and ideas. Still, there's something common to them all: in order to solve any hard problem, we must use various kinds of memories. At each moment, we must keep track of what we've just done—or else we might repeat the same steps over and over again. Also, we must somehow maintain our goals—or we'll end up doing pointless things. Finally, once our problem is solved, we need access to records of how it was done, for use when similar problems arise in the future.

Much of this book will be concerned with memory—that is, with records of the mental past. Why, how, and when should such records be made? When the human brain solves a hard problem, many millions of agents and processes are involved. Which agents could be wise enough to guess what changes should then be made? The high-level agents can't know such things; they scarcely know which lower-level processes exist. Nor can lower-level agents know which of their actions helped us to reach our high-level goals; they scarcely know that higher-level goals exist. The agencies that move our legs aren't concerned with whether we are walking toward home or toward work—nor do the agents involved with such destinations know anything of controlling individual muscle units. Where in the mind are judgments made about which agents merit praise or blame?

7.6 REINFORCEMENT AND REWARD

*For learning to take place, each play of the game must yield much
more information. This is achieved by breaking the problem into
components. The unit of success is the goal. If a goal is achieved,
its subgoals are reinforced; if not, they are inhibited.*
—ALLEN NEWELL

One thing is sure: we always find it easier to do things we've done before. What happens in our minds to make that possible? Here's one idea: In the course of solving some problem, certain agents must have aroused certain other agents. So let's take "reward" to mean that if agent **A** has been involved in arousing agent **B**, the effect of reward is, somehow, to make it easier for **A** to arouse **B** in the future and also, perhaps, to make it harder for **A** to arouse other agents. At one time, I was so taken with this idea that I designed a machine called the *Snarc*, which learned according to this principle; it was composed of forty agents, each connected to several others, more or less at random, through a "reward" system that, when activated after each success, made each agent more likely to rearouse the same recipients at later times.

We presented this machine with problems like learning to find a path through a maze while avoiding a hostile predator. It quickly learned to solve easy problems but never could learn to solve hard problems like building towers or playing chess. It became clear that, in order to solve complicated problems, any machine of limited size must be able to reuse its agents in different ways in different contexts—as *See* must do when involved in two concurrent tasks. But when the *Snarc* tried to learn its way through a complicated maze, a typical agent might suggest a good direction to move in at one moment, then suggest a bad direction at another moment. Later, when we rewarded it for doing something we liked, *both* those decisions became more likely—and all those "goods" and "bads" tended to cancel one another out!

This poses a dilemma in designing machines that learn by "reinforcing" the connections between agents. In the course of solving a hard problem, one will usually try several bad moves before finding a good one—for this is virtually what we *mean* by calling a problem "hard." To avoid learning those bad moves, we could design a machine to reinforce only what happened in the last few moments before success. But such a machine would be able to learn only to solve problems whose solutions require just a few steps. Alternatively, we could design the reward to work over longer spans of time; however, that would not only reward the bad decisions along with the good but would also erase other things that it had previously learned to do. We cannot learn to solve hard problems by indiscriminately reinforcing agents or their connections. Why is it that among all the animals, only the great-brained relatives of man can learn to solve problems that require many steps or involve using the same agencies for different purposes? We'll seek the answer in the policies our agencies use for accomplishing goals.

You might argue that a beaver goes through many steps to build a dam, as does a colony of termites when it builds its complex castle nest. However, these wonderful animals do not learn such accomplishments as individuals but use the procedures that have become encoded in their species' genes over millions of years of evolution. You cannot train a beaver to build a termite nest or teach termites to build beaver dams.

7.7 LOCAL RESPONSIBILITY

Suppose that Alice, who owns a wholesale store, asks the manager, Bill, to increase sales. Bill instructs the salesman, Charles, to sell more radios. Charles secures a large order, on profitable terms. But then the firm can't actually deliver those radios, because they are in short supply. Who is to blame? Alice would be justified in punishing Bill, whose job it was to verify the inventory. The question is, should Charles be rewarded? From Alice's viewpoint, Charles's actions have only disgraced the firm. But from Bill's viewpoint, Charles succeeded in his mission to get sales, and it wasn't *his* fault that this failed to accomplish his supervisor's goal. Consider this example from two perspectives—call them "local reward" and "global reward."

> The **Local** scheme rewards each agent that helps accomplish its supervisor's goal. So Bill rewards Charles, even though Charles's action served no higher-level goal.

> The **Global** scheme rewards only agents that help accomplish top-level goals. So Charles gets no reward at all.

It is easy to invent machinery to embody local learning policies, since each assignment of credit depends only on the relation between an agent and its supervisor. It is harder to implement a global learning scheme because this requires machinery to find out which agents are connected all the way to the original goal by unbroken chains of accomplished subgoals. The local scheme is relatively generous to Charles by rewarding him whenever he accomplishes what is asked of him. The global scheme is much more parsimonious. It dispenses no credit whatever to Charles, even though he does as his supervisor requests, unless his action also contributes to the top-level enterprise. In such a scheme, agents will often learn nothing at all from their experiences. Accordingly, global policies lead to learning more slowly.

Both schemes have various advantages. The cautiousness of the global policy is appropriate when mistakes are very dangerous or when the system has plenty of time. This can lead to more "responsible" behavior—since it could make Charles learn, in time, to check the inventory for himself instead of slavishly obeying Bill. The global policy does not permit one to justify a bad action with "*I was only obeying the orders of my superior.*" On the other side, the local policy can lead to learning many more different things at once, since each agent can constantly improve its ability to achieve its local goals, regardless of how they relate to those of other portions of the mind. Surely our agencies have several such options. Which ones they use may depend, from moment to moment, upon the states of other agencies whose job it is to learn, themselves, *which learning strategies to use, depending on the circumstances.*

The global scheme requires some way to distinguish not only which agents' activities have helped to solve a problem, but also which agents helped with which subproblems. For example, in the course of building a tower, you might find it useful to push a certain block aside to make room for another one. Then you'd want to remember that pushing can help in building a tower —but if you were to conclude that pushing is a generally useful thing to do, you'd never get another tower built. When we solve a hard problem, it usually is not enough to say that what a certain agent did was "good" or "bad" for the entire enterprise; one must make such judgments depend, to some extent, on the local circumstances—that is, on how the work of each agent helped or hindered the work of related agents. The effect of rewarding an agent must be to make that agent react in ways that help to accomplish some specific goal—without too much interference with other, more important goals. All this is simple common sense, but in order to pursue it further, we'll have to clarify our language. We have all experienced the pursuit of goals, but experience is not the same as understanding. What is a goal, and how can a machine have one?

7.8 DIFFERENCE-ENGINES

Whenever we talk about a "goal," we mix a thousand meanings in one word. Goals are linked to all the unknown agencies that are engaged whenever we try to change ourselves or the outside world. If "goal" involves so many things, why tie them all to a single word? Here's some of what we usually expect when we think that someone has a goal:

> A "goal-driven" system does not seem to react directly to the stimuli or situations it encounters. Instead, it treats the things it finds as objects to exploit, avoid, or ignore, as though it were concerned with something else that doesn't yet exist. When any disturbance or obstacle diverts a goal-directed system from its course, that system seems to try to remove the interference, go around it, or turn it to some advantage.

What kind of process inside a machine could give the impression of having a goal—of purpose, persistence, and directedness? There is indeed a certain particular type of machine that appears to have those qualities; it is built according to the principles below, which were first studied in the late 1950s by Allen Newell, C. J. Shaw, and Herbert A. Simon. Originally, these systems were called *general problem solvers*, but I'll simply call them *difference-engines*.

> A *difference-engine must contain a description of a "desired" situation.*
>
> *It must have subagents that are aroused by various differences between the desired situation and the actual situation.*
>
> *Each subagent must act in a way that tends to diminish the difference that aroused it.*

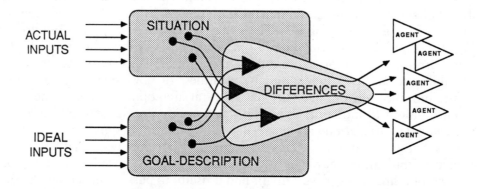

At first, this may seem both too simple and too complicated. Psychologically, a difference-engine might appear to be too primitive to represent the complex of ambitions, frustrations, satisfactions, and disappointments involved in the pursuit of a human goal. But these aren't really aspects of our goals themselves but emerge from the interactions among the many agencies that become engaged in pursuit of those goals. On the other side, one might wonder whether the notion of a goal really needs to engage such a complicated four-way relationship among agents, situations, descriptions, and differences. Presently we'll see that this is actually simpler than it seems, because most agents are already concerned with differences.

7.9 INTENTIONS

When we watch a ball roll down a slope, we notice it seems to try to get around obstacles that lie in its path. If we didn't know about gravity, we might be tempted to think that the ball has the goal of moving down. But we know that the ball isn't "trying" to do anything; the impression of intention is only in the watcher's mind.

When we experiment with *Builder* we also get the sense that it has a goal. Whenever you take its blocks away, it reaches out and takes them back. Whenever you knock its tower down, it rebuilds it. It seems to *want* a tower there, and it perseveres until the tower is done. Certainly *Builder* seems smarter than the rolling ball because it overcomes more complicated obstacles. But once we know how *Builder* works, we see that it's not so different from that ball: all it does is keep on finding blocks and putting them on top of other blocks. Does *Builder* really have a goal?

One ingredient of having a goal is persistence. We wouldn't say that *Builder* wants a tower, if it didn't keep persisting in attempts to build one. But persistence alone is not enough—and neither *Builder* nor that rolling ball have any sense of *where* they want to go. The other critical ingredient of goal is to have some image or description of a wanted or desired state. Before we'd agree that *Builder* wants a tower, we'd have to make sure that it contains something like an image or a description of a tower. The idea of a difference-engine embodies both elements: a representation of some outcome and a mechanism to make it persist until that outcome is achieved.

Do difference-engines "really" want? It is futile to ask that kind of question because it seeks a distinction where none exists—except in some observer's mind. We can think of a ball as a perfectly passive object that merely reacts to external forces. But the eighteenth-century physicist Jean Le Rond d'Alembert showed that one can also perfectly predict the behavior of a rolling ball by describing it as a difference-engine whose goal is to reduce its own energy. We need not force ourselves to decide questions like whether machines can have goals or not. Words should be our servants, not our masters. The notion of goal makes it easy to describe certain aspects of what people and machines can do; it offers us the opportunity to use simple descriptions in terms of active purposes instead of using unmanageably cumbersome descriptions of machinery.

To be sure, this doesn't capture everything that people mean by "having goals." We humans have so many ways of wanting things that no one scheme can embrace them all. Nevertheless, this idea has already led to many important developments both in Artificial Intelligence and in psychology. The difference-engine scheme remains the most useful conception of goal, purpose, or intention yet discovered.

7.10 GENIUS

We naturally admire our Einsteins, Shakespeares, and Beethovens—and we wonder if machines could ever create such wondrous theories, plays, and symphonies. Most people think that accomplishments like these require "talents" or "gifts" that cannot be explained. If so, then it follows that computers can't create such things—since anything machines do can be explained. But why assume that what our greatest artists do is very different from what ordinary people do—when we know so little about what ordinary people do! Surely it is premature to ask how great composers write great symphonies before we know how ordinary people think of ordinary tunes. I don't believe there is much difference between normal and "creative" thought. Right now, if asked which seems the more mysterious, I'd have to say the ordinary kind.

We shouldn't let our envy of distinguished masters of the arts distract us from the wonder of how each of us gets new ideas. Perhaps we hold on to our superstitions about creativity in order to make our own deficiencies seem more excusable. For when we tell ourselves that masterful abilities are simply unexplainable, we're also comforting ourselves by saying that those superheroes come *endowed* with all the qualities we don't possess. Our failures are therefore no fault of our own, nor are those heroes' virtues to their credit, either. *If it isn't learned, it isn't earned.*

When we actually meet the heroes whom our culture views as great, we don't find any singular propensities—only combinations of ingredients quite common in themselves. Most of these heroes are intensely motivated, but so are many other people. They're usually very proficient in some field—but in itself we simply call this craftsmanship or expertise. They often have enough self-confidence to stand up to the scorn of peers—but in itself, we might just call that stubbornness. They surely think of things in some novel ways, but so does everyone from time to time. And as for what we call "intelligence," my view is that each person who can speak coherently already has the better part of what our heroes have. Then what makes genius *appear* to stand apart, if we each have most of what it takes?

I suspect that genius needs one thing more: in order to accumulate outstanding qualities, one needs unusually effective ways to learn. It's not enough to learn a lot; one also has to *manage* what one learns. Those masters have, beneath the surface of their mastery, some special knacks of "higher-order" expertise, which help them organize and apply the things they learn. It is those hidden tricks of mental management that produce the systems that create those works of genius. Why do certain people learn so many more and better skills? These all-important differences could begin with early accidents. One child works out clever ways to arrange some blocks in rows and stacks; a second child plays at rearranging how it thinks. Everyone can praise the first child's castles and towers, but no one can see what the second child has done, and one may even get the false impression of a lack of industry. But if the second child persists in seeking better ways to learn, this can lead to silent growth in which some better ways to learn may lead to better ways to *learn to learn*. Then, later, we'll observe an awesome, qualitative change, with no apparent cause—and give to it some empty name like talent, aptitude, or gift.

Finally, an awful thought: perhaps what we call genius is rare because our evolution works without respect for individuals. Could any tribe or culture endure in which each individual discovered novel ways to think? If not, how sad, since the genes for genius might then lead not to nurturing, but only to frequent weeding-out.

A THEORY OF MEMORY

I compared these various happy impressions with one another and found that they had this in common, namely, that I felt them as though they were occurring simultaneously in the present moment and in some distant past, which the sound of the spoon against the plate, or the unevenness of the flagstones, or the peculiar flavor of the madeleine *even went so far as to make coincide with the present, leaving me uncertain in which period I was. In truth, the person within me who was at this moment enjoying this impression enjoyed in it the qualities which it possessed that were common to both an earlier day and the present moment; and this person came into play only when, by this process of identifying past with present, he could find himself in the only environment in which he could live, that is to say, entirely outside of time.*

—MARCEL PROUST

8.1 K-LINES: A THEORY OF MEMORY

We often talk of memory as though the things we know were stored away in boxes of the mind, like objects we keep in closets in our homes. But this raises many questions.

How is knowledge represented?
How is it stored?
How is it retrieved?
Then, how is it used?

Whenever we try to answer any of these, others seem to get more complicated, because we can't distinguish clearly what we know from how it's used. The next few sections explain a theory of memory that tries to answer all these questions at once by suggesting that *we keep each thing we learn close to the agents that learn it in the first place.* That way, our knowledge becomes easy to reach and easy to use. The theory is based on the idea of a type of agent called a "Knowledge-line," or "K-line" for short.

Whenever you "get a good idea," solve a problem, or have a memorable experience, you activate a K-line to "represent" it. A K-line is a wirelike structure that attaches itself to whichever mental agents are active when you solve a problem or have a good idea.

When you activate that K-line later, the agents attached to it are aroused, putting you into a "mental state" much like the one you were in when you solved that problem or got that idea. This should make it relatively easy for you to solve new, similar problems!

In other words, we "memorize" what we're thinking about by making a list of the agents involved in that activity. Making a K-line is like making a list of the people who came to a successful party. Here is another image of how K-lines work, suggested by Kenneth Haase, a student at the MIT Artificial Intelligence Laboratory who had a great deal of influence on this theory.

"You want to repair a bicycle. Before you start, smear your hands with red paint. Then every tool you need to use will end up with red marks on it. When you're done, just remember that red means 'good for fixing bicycles.' Next time you fix a bicycle, you can save time by taking out all the red-marked tools in advance.

"If you use different colors for different jobs, some tools will end up marked with several colors. That is, each agent can become attached to many different K-lines. Later, when there's a job to do, just activate the proper K-line for that kind of job, and all the tools used in the past for similar jobs will automatically become available."

This is the basic idea of the K-line theory. But suppose you had tried to use a certain wrench, and it didn't fit. It wouldn't be so good to paint *that* tool red. To make our K-lines work efficiently, we'd need more clever policies. Still, the basic idea is simple: for each familiar kind of mental job, your K-lines can refill your mind with fragments of ideas you've used before on similar jobs. In such a moment, you become in those respects more like an earlier version of yourself.

8.2 RE-MEMBERING

Suppose once, long ago, you solved a certain problem P. Some of your agents were active then; others were quiet. Now let's suppose that a certain "learning process" caused the agents that were active then to become attached to a certain agent **kP**, which we'll call a K-line. If you ever activate **kP** afterward, it will turn on just the agents that were active then, when you first solved that problem P!

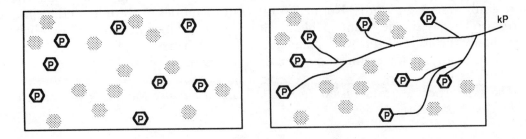

Today you have a different problem. Your mind is in a new state, with agents Q aroused. Something in your mind suspects that Q is similar to P—and activates **kP**.

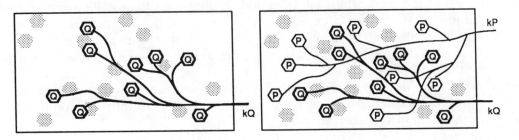

Now two sets of agents are active in your mind at once: the Q-agents of your recent thoughts and the P-agents aroused by that old memory. If everything goes well, perhaps both sets of agents will work together to solve today's problem. And that's our simplest concept of what memories are and how they're formed.

What happens if the now active agents get into conflicts with those the K-line tries to activate? One policy might be to give priority to the K-line's agents. But we wouldn't want our memories to rearouse old states of mind so strongly that they overwhelm our present thoughts—for then we might lose track of what we're thinking now and wipe out all the work we've done. We only want some hints, suggestions, and ideas. Another policy would give the presently active agents priority over the remembered ones, and yet another policy would suppress both, according to the principle of noncompromise. This diagram shows what happens for each of these policies if we assume that neighboring agents tend to get into conflicts:

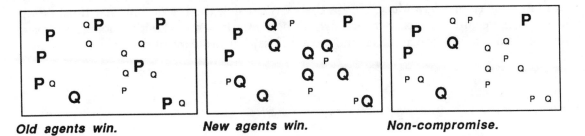

Old agents win.   **New agents win.**   **Non-compromise.**

The ideal scheme would activate exactly those P's that would be most helpful in solving the present problem. But that would be too much to ask of *any* simple strategy.

8.3 MENTAL STATES AND DISPOSITIONS

Many modern scientists think it quaint to talk about "mental states." They feel this idea is too "subjective" to be scientific, and they prefer to base their theories of psychology on ideas about information processing. This has produced many good theories about problem solving, pattern recognition, and other important facets of psychology, but on the whole it hasn't led to useful ways to describe the workings of our dispositions, attitudes, and feelings.

Is this because, as many think, our feelings are inherently more complicated than the things we more easily describe in words? Not necessarily: our memories of attitudes and feelings could come from relatively simple K-line mechanisms—yet still be inexpressible. This is because K-lines can easily record relatively widespread and diffuse activities and, later, reactivate them all at once. This helps explain a familiar psychological phenomenon:

> *The experiences we find easiest to recollect are often just the kinds we find the hardest to describe.*

For example, a novice can remember how it felt to be at a concert. A more proficient amateur can remember more of the music itself—the rhythms and the harmonies and melodies. But only skilled musicians can recall the smaller details of timbre, texture, and arrangement. Why do we find it easier to recollect our attitudes and feelings than to describe what actually took place? That's just what we should expect from memories of the K-line kind. Suppose that a certain sentiment or disposition involved the activities of many different agents. It would be easy to construct a huge K-line with which we could, later, make ourselves approximately reexperience that complicated state—simply by rearousing the same activities. But this would not automatically enable us to *describe* those feelings, which is another matter entirely, because it would require us to summarize that huge, dispersed activity in terms of some much more compact arrangement of verbal expressions.

We cannot always judge the complexity of our mental states by how easily we can express them in words. A certain state of mind might involve a mass of information simply too enormous and diverse to express in any small number of words, yet not be very complicated in any interesting sense. Furthermore, the things we *can* express in words are, to a large extent, constrained by the social process through which we learn to use those words. In order for a word to have a predictable effect on other persons, we must maintain strict, public discipline on how that word is used—whereas each individual's private, internal signals need not be so constrained. The signals that come from our nonverbal agents can have K-line connections that branch out very rapidly to arouse other agents. If each member of such a society were to arouse a mere hundred others, then in only three or four steps the activity of a single one of them could affect a million other agents.

Once we think in terms of K-line memories, it becomes easy to imagine, at least in principle, how a person could recall a *general impression* of a complex previous experience—but it becomes hard to understand how a person can so easily comprehend a *specific statement* like *"John has more candy than Mary."* If this theory is correct, the traditional view must be upside down, which regards it as easy to understand how minds can deal with "facts" and "propositions," but hard to see how minds could have diffuse, hard-to-express dispositions.

8.4 PARTIAL MENTAL STATES

We make our new ideas by merging parts of older ones—and that means keeping more than one idea in mind at once. Let's oversimplify matters for the moment and imagine that the mind is composed of many "divisions," each involved with a different activity, like vision, locomotion, language, and so forth. This pattern repeats on smaller scales, so that even the thought of the simplest ordinary object is made up of smaller thoughts in smaller agencies. Thinking about a small white rubber ball could activate some divisions like these:

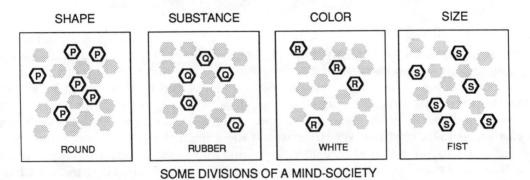

SOME DIVISIONS OF A MIND-SOCIETY

We'll need some way to talk about the states of many agencies at once. So, in this book, I'll use the expression "mental state" or "total mental state" when talking about the states of *all* of one's agents. The new phrase "partial mental state" is for talking about the states of smaller groups of agents. Now in order to be clear, we'll have to simplify our picture of the situation, the way scientists do. *We shall assume that each agent in our society, at each moment, is either in a "quiet state" or an "active state."* Why can't an agent be partially aroused, instead of only "on" or "off"? They could indeed, but there are technical reasons why this would not make any fundamental difference to the issues we are discussing here. In any case, this assumption allows us to be precise:

> A "total state" of mind is a list that specifies which agents are active and which are quiet at a certain moment.

> A "partial state" of mind merely specifies that certain agents are active but does not say which other agents are quiet.

Notice that according to this definition, a mind can have exactly one total state at any moment, but it can be in many partial states at the same time—*because partial states are incomplete descriptions.* The picture above shows a mind-society made up of several separate divisions, so we can think of each division's state as one partial state, and this lets us imagine that the entire system can "think several thoughts at once," just as a crowd of separate people can. When your speech division is being occupied with what your friend is saying while your vision division looks for a door to exit through—then your mind is in two partial states at once.

The situation is more interesting when two K-lines activate agents in the *same* division at the same time: imposing two different partial mental states on the same agency can lead to conflicts. It is easy to think of *a small white ball* because this activates K-lines that connect to unrelated sets of agents. But when you try to imagine *a round square*, your agents for *round* and *square* are forced to compete to control the same set of shape-describing agents. If the conflict is not settled soon, noncompromise may eliminate both—and leave you with the sense of an undefined shape.

8.5 LEVEL-BANDS

kite *n. a toy consisting of a light frame, usually of wood, with paper or other light material stretched upon it; mostly in the form of an isosceles triangle with a circular arc as base, or a quadrilateral symmetrical about the longer diagonal; constructed (usually with a tail of some kind for the purpose of balancing it) to be flown in a strong wind by means of a long string attached.*
—*Oxford English Dictionary*

"Jack is flying his kite." What knowledge do you need to understand this? It helps to know that you can't fly kites without any wind. It helps to know how to fly a kite. You would understand it better if you knew how kites are made, or where they're found, or what they cost. Understanding never ends. It is remarkable how much we can imagine about Jack's activity. Neither you nor I have ever seen Jack's kite, nor do we know its color, shape, or size, and yet our minds supply details from memories of other kites we've seen before. That sentence may have made you think of string, yet string wasn't mentioned. How does your mind arouse so many memories so quickly? And how does your mind know not to arouse *too many* memories —something that could also lead to serious problems? To explain this, I'll introduce what I call the *level-band* theory.

The basic idea is simple: we learn by attaching agents to K-lines, but we don't attach them all with equal firmness. Instead, we make strong connections at a certain level of detail, but we make weaker connections at higher and lower levels. A K-line for a kite might include some properties like these:

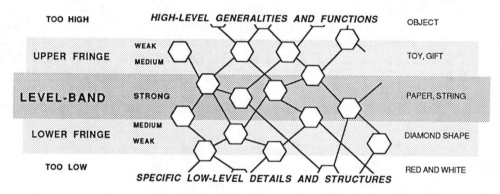

Whenever we turn on this K-line, it tries to activate all these agents, but those near the fringes are attached as though by twice used tape and tend to retreat when other agents challenge them. If most of the kites you've seen before were red and diamond-shaped, then when you hear about Jack's kite, those weak connections will lead you to assume that Jack's kite, too, is red and diamond-shaped. But if you should hear that Jack's kite is green, your weakly activated red-color agent memories will be suppressed by your strongly activated green-color agents. Let's call these kinds of weakly activated memories *assumptions by default.* Default assumptions, once aroused, stay active only when there are no conflicts. In psychological terms, they are things we assume when we have no particular reason to think otherwise. Later we'll see that default assumptions embody some of our most valuable kinds of commonsense knowledge: *knowing what is usual or typical.* For example, they're why we all assume that Jack has hands and feet. If such assumptions turn out to be wrong, their weak connections allow them to be easily displaced when better information comes to mind.

8.6 LEVELS

We sometimes think of memory as though it could transport us back to hear the voices of times gone by and see the sights of the past. But memory can't really take us anywhere; it can only recall our minds to prior states, to visit what *we* used to be, by putting back what was in the mind before. We introduced the level-band theory to provide a way for a memory to encompass some range or "level" of detail of descriptions, as when, in remembering that kite experience, certain aspects were recorded firmly and others weakly or not at all.

The concept of a level-band can be applied not only to descriptions of things, but also to our memories of the processes and activities we use in order to achieve our goals—that is, the mental states we re-create that once solved problems in the past. The problems we have to solve change with time, so we must adapt our old memories to our present goals. To see how level-bands can help with that, let's now return to *Play-with-Blocks*—but this time let's suppose that our child has grown to maturity and wants to build a real house. Which agents from the old building-society can still be applied to this new problem?

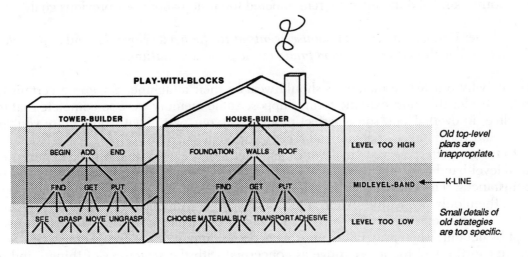

The new house-building agency can certainly use many of *Tower Builder*'s skills. It certainly will need *Add*'s lower-level skills like *Find* and *Get* and *Put*. But *House Builder* won't have so much use for *Tower Builder*'s *highest-level* agents like *Begin* and *End*—because these were specialized for making towers. Nor will it have much use for *Builder*'s *lowest-level* skills, like those in *Grasp*, because picking up such small blocks isn't the problem. But most of the skills embodied in *Builder*'s middle level-bands will still apply. These seem to embody the sort of knowledge that is most broadly and generally useful, whereas uppermost and lowest level-bands are more likely to be based on aspects of the problem that are specific to an older goal or to the particular details of the original problem. But if our memory machinery has been designed so that the contents of those distant fringes can be easily detached, the extra knowledge stored in them will rarely do much harm and can often be helpful. For example, *Tower Builder*'s fringe details could tell us what to do in case our house should grow very tall or require a high chimney.

We started out by using level-bands for *describing* things—but we ended up using them for *doing* things! In the next few sections we'll see that it is no accident that level-related ideas play many different roles in how we think.

8.7 FRINGES

It's hard to recognize a thing when you're presented with too much detail. To know that you are seeing a kite, it helps to look for paper, sticks, and string. But if you were to use a microscope, what you'd perceive would not be properties of kites at all, but merely features of particular bits of paper, sticks or string. These might allow you to identify a particular kite but not to recognize any other kite. Past a certain level of detail, the more one sees, the less one can tell what one is seeing! The same applies to memories; they should weaken their attachments at lower levels of detail.

>**Lower Band:** *Beyond a certain level of detail, increasingly complete memories of previous situations are increasingly difficult to match to new situations.*

To explain why K-lines need an upper-level fringe, let's return to that example in which our child originally learned how to build a tower—but now desires to build a house. Here, we could have another kind of difficulty if we remembered too much about our previous goals!

>**Upper Band:** *Memories that arouse agents at too high a level would tend to provide us with goals that are not appropriate to the present situation.*

To see why our K-line memories should weaken their attachments *above* a certain level of detail, consider this most extreme form. Suppose some memory were so complete that it made you relive, in every detail, some perfect moment of your past. That would erase your present "you"—and you'd forget what you had asked your memory to do!

Both fringing effects serve to make our memories more relevant to our present purposes. The central level-band helps us find general resemblances between remembered events and present circumstances. The lower fringe supplies additional details but does not force them upon us. We use them only "by default" when actual details are not supplied. Similarly, the upper fringe recalls to mind some memories of previous goals, but again, we're not forced to use them except by default, when present circumstances do not impose more compelling goals. Seen this way, we can think of the lower fringe as concerned with the *structures* of things, and we can think of the upper fringe as involved with the *functions* of things. The lower levels represent "objective" details of reality; the upper levels represent our "subjective" concerns with goals and intentions.

How could the fringes of the same K-line lie in two such different realms? Because in order to think, we need intimate connections between things and goals—between structures and their functions. What use would *thinking* be at all, unless we could relate each thing's details to our plans and intentions? Consider how often the English language employs the selfsame words for things and for their purposes. What tools would you use, when building your house, to saw and clamp and glue your wood? That's obvious: you'd use a *saw* and a *clamp* and some *glue*! Behold the wondrous force of those "meanings": no sooner do we hear the noun form of a word than our agents strain to perform the acts that correspond to it as a verb. This phenomenon of connecting *means* with *ends* is not confined to language—we'll see many other instances of it in other kinds of agencies—but language may allow such linking with the least constraint.

8.8 SOCIETIES OF MEMORIES

Yesterday, you watched Jack fly his kite. How do you remember that today? One answer would be, "*Remembering it is much like seeing it again.*" But yesterday, when you recognized that kite, you didn't really see it as something wholly new. The fact that you recognized it as a "kite" yesterday means that you already saw that kite in terms of even older memories.

This suggests two ways to make new memories of what you saw a moment ago. One scheme is shown to the left below: you simply connect a new K-line to all the agents that were recently active in your mind. The other way to make that memory is shown in the diagram to the right below: instead of attaching the new K-line to that whole multitude of separate agents, *connect it only to whichever of your older K-lines were active recently.* This will lead to a similar result, since those K-lines were involved in arousing many of the agents that were active recently. This second scheme has two advantages: it is more economical, and it leads to forming memories as organized societies.

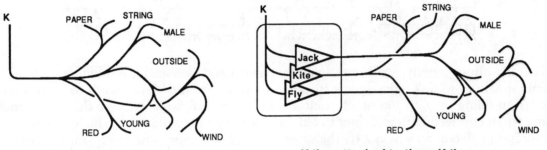

K-line attached to many agents.   **K-line attached to three K-lines.**

Consider that when you realized Jack was flying a kite, this must have involved the use of K-lines—for "Jack" and "Fly" and "Kite"—that had been formed at earlier times and were aroused by the sight of Jack flying his kite. When those three K-lines were activated, each of them in turn activated hundreds or thousands of other agents. (Your state of mind, when seeing that scene, resulted from combinations both of agents aroused directly by your senses and of agents aroused indirectly by your recognitions.) Now, our left-hand memory-scheme would need an enormous number of connections to link all those agents to the new K-line. But our right-hand scheme would obtain much the same effect by attaching the new K-line to only three old K-lines! Yet when you reactivate that K-line at some later date, it will, in turn, arouse the same K-lines for Jack, Fly, Kite, and whichever other recognitions were involved. As a result, you will reexperience many of the same recognitions as before. To that extent, you will feel and act as though you were back in the same situation again.

To be sure, these two types of memories would not produce exactly the same results. Our trick of connecting new K-lines to old ones will not recapture so many of the scene's precise, perceptual details. Instead, the kinds of mental states that this "hierarchical" type of memory produces will be based more on stereotypes and default assumptions than on actual perceptions. Specifically, *you will tend to remember only what you recognized at the time.* So something is lost—but there's a gain in exchange. These "K-line memory-trees" lose certain kinds of details, *but they retain more traces of the origins of our ideas.* These memory-trees might not serve quite so well if the original circumstances were *exactly* repeated. But that never happens, anyway—and the structured memories will be much more easily adapted to new situations.

8.9 KNOWLEDGE-TREES

If each K-line can connect to other K-lines, which, in turn, connect to others, then K-lines can form societies. But how can we make sure that this can serve our purposes, instead of becoming a great, disordered mess? What could guide them into representing useful hierarchies like these?

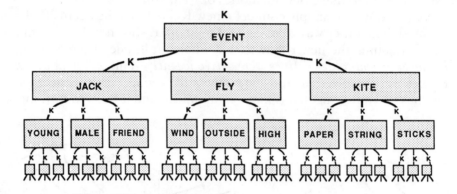

To keep things orderly, we'll now apply that level-band idea again. Remember that we first invented K-lines to link older agents together; then we invented level-bands to keep those K-lines from filling up with too much useless, unrelated stuff. Now we have the same problem again: when connecting new K-lines to old ones, we must keep them from including too much inappropriate detail. So why not try the same solution? Let's apply the level-band idea to the K-line trees themselves!

> *When making a new K-line memory, do not connect it to all the K-lines active at the time but only to those that are active within a certain level-band.*

It might be supposed that this idea would be hard to apply unless we specify what "level" means. However, something like this will happen automatically, simply because the new K-line societies will tend to inherit whatever hierarchy already existed among the original agents that become connected to those K-lines. We've actually seen two different ideas about this. In our *Kite* example, we talked about a description's "level of detail." That is, we regarded it as more elevated to talk about "*a sheet stretched across a frame*" than to discuss the paper or the sticks themselves. In our *Builder* example, we talked about goals and considered the *Tower Builder* agent itself to be a level above the agents it exploits to solve its subproblems—agents like *Begin* and *Add* and *End*.

This policy of connecting new K-lines to old ones must be used in moderation. Otherwise, *no new agents would ever be included in our memories*. Furthermore, it should not always be required to produce simple, orderly hierarchy-trees; for example, in the case of *Builder*, we found that both *Move* and *See* will often need one another's help. Eventually, all of our knowledge-structures become entangled with various sorts of exceptions, shortcuts, and cross-connections. No matter: the level-band idea will still apply in general, since most of what we know will still be mainly hierarchical because of how our knowledge grows.

8.10 LEVELS AND CLASSIFICATIONS

Isn't it interesting how often we find ourselves using the idea of *level*? We talk about a person's levels of aspiration or accomplishment. We talk about levels of abstraction, levels of management, levels of detail. Is there anything in common to all the level-things people talk about? Yes: they each appear to reflect some way to organize ideas—and each seems vaguely hierarchical. Usually, we tend to think that each of those hierarchies illustrates some kind of order that exists in the world. But frequently those orderings come from the mind and merely *appear* to belong to the world. Indeed, if our theory of K-line trees is correct, it would seem "natural" for us to classify things into levels and hierarchies—even when this does not work out perfectly. The diagram below portrays two ways to classify physical objects.

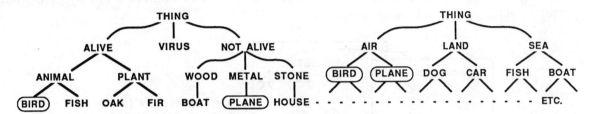

These two hierarchies split things up in different ways. The birds and airplanes are close together on one side, but far apart on the other side. Which classification is correct? Silly question! It depends on what you want to use it for. The one on the left is more useful for biologists, and the one on the right is more useful for hunters.

How would you classify a porcelain duck, a pretty decorative toy? Is it a kind of bird? Is it an animal? Or is it just a lifeless piece of clay? It makes no sense to argue about it: *"That's not a bird!" "Oh, yes, it is, and it is also pottery."* Instead, we frequently use two or more classifications at the same time. For example, a thoughtful child can play with a porcelain duck as though it were a make-believe animal, yet at the same time treat it carefully, as is appropriate when handling a delicate china object.

Whenever we develop a new skill or extend an old one, we have to emphasize the relative importance of some aspects and features over others. We can place these into neat levels only when we discover systematic ways to do so. Then our classifications can resemble level-schemes and hierarchies. But the hierarchies always end up getting tangled and disorderly because there are also exceptions and interactions to each classification scheme. When attempting a new task, we never like to start anew: we try to use what has worked previously. So we search around inside our minds for old ideas to use. Then, when part of any hierarchy seems to work, we drag the rest along with it.

8.11 LAYERS OF SOCIETIES

According to our concept of memory, the K-lines of each agency grow into a new society. So, to keep things straight, let's call the original agents *S-agents* and call their society the *S-society*. Given any S-society, we can imagine building memories for it by constructing a corresponding K-society for it. When we start making a *K-society*, we must link each K-line directly to S-agents, because there are no other K-lines we can connect them to. Later we can use the more efficient policy of linking new K-lines to old ones. But this will lead to a different problem of efficiency: the connections to the original S-agents will become increasingly remote and indirect. Then everything will begin to slow down—unless the K-society continues to make at least some new connections to the original S-society. That would be easy to arrange, if the K-society grows in the form of a "layer" close to its S-society. The diagram below suggests such an arrangement.

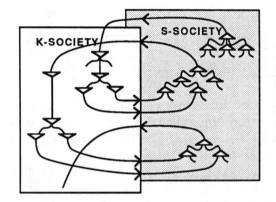

The connections in the K-society are similar to those in the S-society, except that the signals tend to flow in opposite directions.

If arranged this way, the layer pairs could form a curious sort of computer. As S-agents excite K-agents and *vice versa*, a sort of spiraling activity would ensue. Over time, the location of that activity might tend to drift upward or down and might also tend to spread out; without some control, the system might soon become chaotic. But it would be hard to control the system from within, nor would that serve the purposes of other agencies. However, we can easily imagine how yet another, third agency could confine and control the K-S system's activity—by specifying which level-band should remain active and suppressing all the rest. Indeed, that is precisely the sort of coarse control that a *B-brain* might exercise, since it could do all this without needing to understand the fine details of what is happening inside the A-brain. The third agency might simply look on and say impatiently, *"This isn't getting anywhere: move up to take a higher-level view of the situation."* Or it might say, *"That looks like progress, so move farther down and fill in more details."*

Is there any essential difference between the K- and S-societies? Not really—except that the S-society came first. Indeed, we can imagine an endless sequence of such societies, in which each new one learns to exploit the last. Later we'll propose that this is how our minds develop in infancy—as sequences of layers of societies. Each new layer begins as a set of K-lines, which starts by learning to exploit whatever skills have been acquired by the previous layer. Whenever a layer acquires some useful and substantial skill, it tends to stop learning and changing—and then yet another new layer can begin to learn to exploit the capabilities of the last. Each new layer begins as a student, learning new ways to use what older layers can already do. Then it slows its learning rate—and starts to serve both as subject and as teacher to the layers that form afterward.

CHAPTER 9

SUMMARIES

Apart from pain, whose function is obviously that of informing the higher centers of the nervous system where there is something out of order, there are many physiological mechanisms which are there for the sole reason of letting us know that something is wrong. We feel ill without knowing the reason. The very fact that we have only one term, "I feel ill," for a range of conditions based on different causes is extremely characteristic.

—KONRAD LORENZ

9.1 WANTING AND LIKING

One thing I hate is being asked questions like these:

> *Do you prefer physics to biology?*   *Did you like that play?*
> *Do you like Wagner?*   *Did you enjoy your year abroad?*

What makes us want to compress so much into such inexpressive summaries as "like," "prefer," and "enjoy"? Why try to reduce such complex things to simple values or amounts of pleasurable quality? The answer is that our measures of pleasure have many uses. They help us make comparisons, compromises, and choices. They are involved with the communication signs that we use to signify various degrees of attachment, satisfaction, and agreement. They show themselves not only in words, but also as gestures, intonations, smiles and frowns, and many other expressive signs. But we have to be careful not to accept those signs at their face value. Neither the state of the world nor that of the mind is ever so simple that it can be expressed in a single, one-dimensional judgment. No situation is ever completely satisfactory or entirely disagreeable, and our reactions of pleasure or disgust are only superficial summaries of pyramids of underlying processes. To "enjoy" an experience, some of our agents must summarize success—but other agents must be censuring their subordinates for failing to achieve *their* goals. So we ought to be suspicious when we find ourselves liking something *very* much, because that might mean some of our agencies are forcefully suppressing other possibilities.

> *The surer you are that you like what you are doing, the more completely your other ambitions are being suppressed.*

To choose between alternatives, the highest levels of the mind demand the simplest summaries. If your top-level feelings were too often "mixed," you would rarely be able to make a choice to decide which foods to eat, which paths to walk, or which thoughts to think. At the level of action, you're forced to simplify right down to expressions like "Yes" and "No." But these are not informative enough to serve the lower levels of the mind, where many processes go on at once, and every agent has to judge how well it is serving some local goals. At lower levels of the mind, there must be hosts of smaller, coexisting satisfactions and annoyances.

We often talk as though we *ought* to be controlled by what we want. Indeed we scarcely distinguish between wanting something and potentially obtaining pleasure from it; the relation between these two ideas seems so intimate that it actually feels odd to mention it. It seems so natural to want what we like and to avoid what we don't like that we sometimes feel a sense of unnatural horror when another person appears to violate that rule; then we think, *They surely wouldn't do such things unless, deep down, they really wanted to.* It is as though we feel that people *ought* to want only to do the things they like to do.

But the relation between wanting and liking is not simple at all, because our preferences are the end products of so many negotiations among our agencies. To accomplish any substantial goal, we must renounce the other possibilities and engage machinery to keep ourselves from succumbing to nostalgia or remorse. Then we use words like "liking" to express the operation of the mechanisms that hold us to our choice. *Liking's* job is shutting off alternatives; we ought to understand its role since, unconstrained, it narrows down our universe. This leads to *liking's* artificial clarity: it does not reflect what liking *is* but only shows what liking *does*.

9.2 GERRYMANDERING

We all know how accomplishment can bring satisfaction, and we tend to assume a direct connection between them. In very simple animals, where "satisfaction" means no more than meeting simple, basic needs, satisfaction and accomplishment must indeed be virtually the same. But in a complex human brain, a great many layers of agencies are interposed between the ones that deal with body needs and those that represent or recognize our intellectual accomplishments. Then what is the significance, in these more complicated systems, of those pleasant feelings of accomplishment and disagreeable sensations of defeat? They must be involved with how our higher-level agencies make summaries.

Suppose that you once had to send a present to a friend. You had to choose a gift and find a box in which to wrap it. Soon, each such job turned into several smaller ones—like finding strings and tying them. The only way to solve hard problems is by breaking them into smaller ones and then, when those are too difficult, dividing them in turn. So hard problems always lead to branching trees of subgoals and subproblems. To decide where resources should be applied, our problem-solving agents need simple summaries of how things are going. Let's suppose each agent's summary is based on other summaries it gets from the agents it supervises. Here is a pathological example of what could happen if every such summary were based on a simple majority decision:

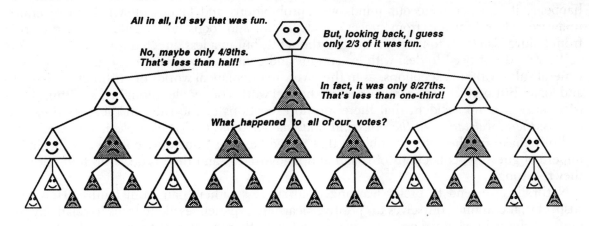

When all is done, if someone asked if you enjoyed the whole experience, you might say that it was "fun" or "terrible." But no such summary can say very much of what your agencies actually learned. Your knot-tying processes learned which actions worked and failed; your paper-folding and gift-selecting processes had other failures and accomplishments; but your overall assessment of the experience cannot reflect all those details. If the entire episode left you "unhappy," you might be less inclined to give presents in the future, but that should not have much effect on what you learned about folding paper and tying string. No single sense of "good" or "bad" can reflect much of what went on inside all your agencies; too much information must be concealed. Then why does it seem so satisfactory for us to classify our feelings into positive and negative and conclude that *on the whole* the net effect was bad or good? True, sometimes feelings are more mixed and everything seems bittersweet, but, as we'll see, there are many reasons why we have to oversimplify.

9.3 LEARNING FROM FAILURE

So far, we've talked mostly of learning from success. But consider that when you succeed, you must already have had the necessary means within your grasp. If so, then making changes in your mind might only make things worse! As people often say, *"You shouldn't argue with success."* For whenever you try to "improve" an already working procedure, you risk damaging whichever other skills depend on that same machinery.

Accordingly, it may be more important that we learn from how we fail. What should you do if some well-established method—call it **"M"**—has failed to reach a certain goal? One policy would be to alter **M**, so it won't make the same mistake again. But even that might be dangerous because it might cause **M** to fail in other ways. Besides, we might not know *how* to change **M** to remove the error. A safer way to deal with this would be to modify **M** by adding special memory devices called "censors" and "suppressors" (we'll discuss this in detail later), which remember particular circumstances in which **M** fails and later proceed to suppress **M** when similar conditions recur. Such censors would not tell you what to do, only what you shouldn't do; still, they prevent your wasting time by repeating old mistakes.

Learning has at least two sides. Some parts of our minds learn from success—by remembering when methods work. But other portions of our minds learn mainly when we make mistakes, by remembering the circumstances in which various methods failed to work. Later we'll see how this can teach not only what we shouldn't do, but also what we shouldn't *think*! When that happens, it can permeate our minds with prohibitions and taboos of which we're entirely unaware. Thus, learning from success tends to aim and focus how we think, while learning from failure also leads to more productive thoughts, but in a less directed way.

We would not need to deal with exceptions and censors if we lived in a universe of simple, general rules with no exceptions, as in the lovely mathematical worlds of arithmetic, geometry, and logic. But perfect logic rarely works in the real worlds of people, thoughts, and things. This is because it is no accident that there are no exceptions to the rules in those mathematical worlds: *there, we start with the rules and imagine only objects that obey them.* But we can't so willfully make up the rules for objects that already exist, so our only course is to begin with imperfect guesses—collections of rough and ready rules—and then proceed to find out where they're wrong.

Naturally, we tend to prefer learning from success rather than from failure. However, I suspect that confining ourselves to "positive" learning experiences alone leads to relatively small improvements in what we can already do. Probably, there is no way to avoid at least a certain degree of discomfort when we make substantial changes in how we think.

9.4 ENJOYING DISCOMFORT

*Do not become attached to the things you like, do not
maintain aversion to the things you dislike. Sorrow, fear and
bondage come from one's likes and dislikes.*
—BUDDHA

Why do children enjoy the rides in amusement parks, knowing that they will be scared, even sick? Why do explorers endure suffering and pain—knowing that their very purpose will disperse once they arrive? And what makes ordinary people work for years at jobs they hate, so that someday they will be able to—some seem to have forgotten what?

There is more to motivation than immediate reward. When we succeed at anything, a lot goes on inside the mind. For example, we may be filled with feelings of accomplishment and pride, and feel impelled to show others what we've done and how. However, it is the fate of more ambitious intellects that the sweetness of success will swiftly fade as other problems come to mind. That's good because most problems do not stand alone but are only smaller parts of larger problems. Usually, after we solve a problem, our agencies return to some other, higher-level cause for discontent, only to lose themselves again in other subproblems. Nothing would get done if we succumbed to satisfaction.

But what if a situation gets completely out of our control—and offers no conceivable escape from suffering? Then all we can do is try to construct some inner plan for tolerating it. One trick is to try to change our momentary goal—as when we say, *"It's getting there that's all the fun."* Another way is looking forward to some benefit to future Self: *"I certainly shall learn from this."* When that doesn't work, we can still resort to even more unselfish schemes: *"Perhaps others may learn from my mistake."*

These kinds of complications make it impossible to invent good definitions for ordinary words like "pleasure" and "happiness." No small set of terms could suffice to express the many sorts of goals and wants that, in our minds, compete in different agencies and on different scales of time. It is no wonder that those popular theories about reward and punishment have never actually led to explaining higher forms of human learning—however well they've served for training animals. For in the early stages of acquiring any really new skill, a person must adopt at least a partly antipleasure attitude: *"Good, this is a chance to experience awkwardness and to discover new kinds of mistakes!"* It is the same for doing mathematics, climbing freezing mountain peaks, or playing pipe organs with one's feet: some parts of the mind find it horrible, while other parts enjoy forcing those first parts to work for *them.* We seem to have no names for processes like these, though they must be among our most important ways to grow.

None of this is to say that we can discard the concepts of pleasure and liking as we use them in everyday life. But we have to understand their roles in our psychology; they represent the end effects of complex ways to simplify.

CHAPTER 10

PAPERT'S PRINCIPLE

Interviewer: Now, Adam, listen to what I say. Tell me
which is better: "a water" or "some water"?

Adam: Pop go weasel.

—ROGER BROWN AND URSULA BELLUGI,
DISCUSSING PROBLEMS WITH EXPERIMENTS
ON YOUNG CHILDREN.

10.1 PIAGET'S EXPERIMENTS

The psychologist Jean Piaget was one of the first to realize that watching children might be a way to see how mind-societies grow. In one of his classic experiments, he showed a child two matching sets of eggs and cups—and asked, *"Are there more eggs or more egg cups?"*

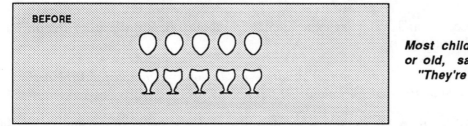

Most children, young or old, say, "They're the same."

Then he spread the eggs apart—before the child's eyes—and asked again if there were more eggs or more egg cups.

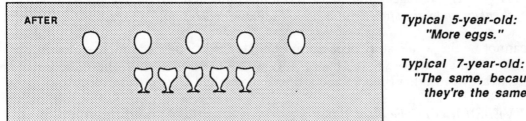

Typical 5-year-old: "More eggs."

Typical 7-year-old: "The same, because they're the same eggs."

One might try to explain this by supposing that older children are better at counting. However, this can't explain another famous experiment of Piaget's, which began by showing three jars, two filled with water. All the children agreed that the two short, wide jars contained equal amounts of liquid. Then, before their eyes, he poured all the liquid from one of the short jars into the tall, thin one and asked which jar had more liquid now.

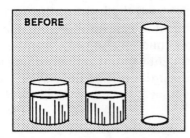 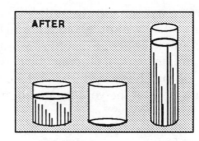

Typical 5-year-old: "More in the tall jar."

Typical 7-year-old: "The same, because it's the same water."

These experiments have been repeated in many ways and in many countries—and always with the same results: each normal child eventually acquires an adult view of quantity—apparently without adult help! The age at which this happens may vary, but the process itself seems so universal that one cannot help suspecting that it reflects some fundamental aspect of the mind's development. In the next few sections we'll examine the idea of "more" and show that it conceals the workings of a large, complex Society-of-More—which takes many years to learn.

10.2 REASONING ABOUT AMOUNTS

> **con·ser·va´tion** *n. the principle that [something] is a
> constant quantity, transformable in countless ways, but
> never increased or diminished.*
> —*Webster's Unabridged Dictionary*

What do those egg and water jar experiments say about our growth from infancy? Let's consider several explanations.

> QUANTITY: *Perhaps the younger children simply don't yet understand the basic concept of quantity: that the amount of liquid remains the same.*

In the next few sections I'll argue that we do not learn one single, underlying "concept of quantity." Instead, each person must construct a multileveled agency, which we'll call the *Society-of-More*, that finds different ways to deal with quantities.

> EXTENT: *The younger children seem unduly influenced by the larger extent of space taken up by the spread-out eggs and taller water column.*

That cannot be the whole story because most adults, too, judge that there's more water in the taller jar—if they merely see the final scene, without knowing from where the water was poured! Here are a few other theories about the younger child's judgment:

> REVERSIBILITY: *The older children pay more attention to what they think remains the same—while younger ones are more concerned with what has changed.*

> CONFINEMENT: *An older child knows that the amount of water stays the same, if none was ever added or removed or lost or spilled.*

> LOGIC: *Perhaps younger children have not yet learned to apply the kinds of reasoning that one would need to understand the concept of quantity.*

Every one of these explanations has some truth in it, but none reach the heart of the issue. It is clear that the older children know more about such matters and can do more complex kinds of reasoning. But there is ample evidence that most younger children also possess enough of the required abilities. For example, we can describe the experiment without actually doing it at all or we can perform it out of the child's sight, behind a cardboard screen. Then, when we explain what is happening, quite a few of the younger children will say, "*Of course they'll be the same.*"

Then what *is* the difficulty? Evidently, the younger children possess the ideas they need *but don't know when to apply them!* One might say that they lack adequate *knowledge about their knowledge*, or that they have not acquired the checks and balances required to select or override their hordes of agents with different perceptions and priorities. It is not enough to be able to use many kinds of reasoning; one also must know *which to use in different circumstances!* Learning is more than the mere accumulation of skills. Whatever we learn, there is always more to learn—about how to use what was already learned.

10.3 PRIORITIES

Let's try to explain the water jar experiment in terms of how a child's agencies deal with comparisons. Suppose the child begins with only three agents:

> **Tall** says, *"The taller, the more."* There's more inside a taller thing.
> **Thin** says, *"The thinner, the less."* There's less inside a thinner thing.
> **Confined** says, *"The same, because nothing was added or removed."*

How do we know children have agents like these? We can be sure that younger children have agencies like *Tall* and *Thin* because they all can make these judgments:

It is harder to know whether younger children have agents like *Confined*, but many of them can indeed explain that something remains the same when one pours a liquid back and forth. In any case, there is a conflict because the three agents give three different answers—*more*, *less*, and *same!* What could be done to settle this? The simplest theory is that the younger children have placed their agents in some "order of priority."

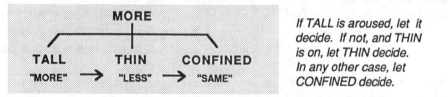

If TALL is aroused, let it decide. If not, and THIN is on, let THIN decide. In any other case, let CONFINED decide.

Such a scheme can be extremely practical, since placing all the agents in order of priority makes it easy to know which to use. For example, we often compare things by their extents— by how far they reach in space. But why put *Tall* ahead of *Wide*? People do indeed seem most sensitive to *vertical* extents. We do not know whether this is built from the start into our brains, but in any case, the bias is usually justified because *"more height"* so frequently goes along with other sorts of largenesses.

> *Who's "bigger"—you or your cousin? Stand back to back!*
> *Who's the strongest? Those adults looming way above!*
> *How to divide a liquid into equal portions? Match the levels!*

No other agent seems so good as *Tall* for making everyday comparisons. Still, no priority-scheme will always work. In the situation of the water jar experiment, *Confined* ought to come first, but the younger child's priorities lead to making the wrong judgment. One might wonder, incidentally, whether *Tall* and *Short*, or *Wide* and *Thin*, should be considered to be different agents. Logically, just one of each pair would suffice. But I doubt that in the brain it would suffice to represent *Short* by the mere inactivity of *Tall*. To adults these are "opposites," but children do not work so logically. One child I knew insisted that *knife* was the opposite of *fork*, but that *fork* was the opposite of *spoon*. Water was the opposite of *milk*. As for the opposite of *opposite*, that child considered this to be too foolish to discuss.

10.4 PAPERT'S PRINCIPLE

What should one do when different kinds of knowledge don't agree? It sometimes helps to place them in some order of priority, but as we've seen, that can still lead to mistakes. How can we make our system sensitive to different circumstances? The secret is to use the principle of noncompromise and look for help from other agencies! For help with comparing quantities, we'll need to add new "administrative agents" to our Society-of-More.

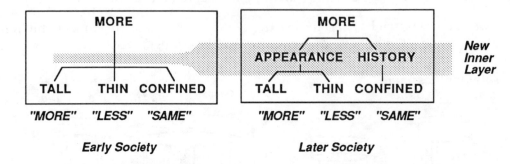

The new *Appearance* administrator is designed to say "more" when the agent *Tall* is active, to say "less" when the agent *Thin* is active, and to say nothing at all when something appears both taller and thinner. Then the other new administrator, *History*, makes the decision on the basis of what *Confined* says.

This explanation of the difference between the older and younger children was first proposed by Seymour Papert in the 1960s, when we first started to explore society of mind ideas. Most previous theories had tried to explain Piaget's experiments by suggesting that children develop different kinds of reasoning as time goes by. That certainly is true, but the importance of Papert's conception is in emphasizing not merely the ingredients of reasoning, but how they're organized: a mind cannot really grow very much merely by accumulating knowledge. It must also develop better ways to use what it already knows. That principle deserves a name.

> **Papert's Principle:** *Some of the most crucial steps in mental growth are based not simply on acquiring new skills, but on acquiring new administrative ways to use what one already knows.*

Our two new middle-level managers illustrate this idea: *Appearance* and *History* form a new, intermediate layer that groups together certain sets of lower-level skills. The choice of agents for those groups is absolutely critical. The system will work quite well if we group *Tall* and *Thin* together, so that *Confined* can take control when they conflict. But it would only make things worse if we were to group *Tall* and *Confined* together! Then what decides which groups to form? Papert's principle suggests that the processes which assemble agents into groups must somehow exploit relationships among the skills of those agents. For example, because *Tall* and *Thin* are more similar in character to one another than to *Confined*, it makes sense to group them more closely together in the administrative hierarchy.

10.5 THE SOCIETY-OF-MORE

Think how many meanings "more" must have! We seem to use a different one for every sort of thing we know.

> *More red.   More loud.   More swift.   More old.   More tall.*
> *More soft.   More cruel.   More alive.   More glad.   More wealthy.*

Each usage has a distinct sense, involving different agencies. How could all these ways to make comparisons get grouped into just one society? Here's a Society-of-More a child might use to deal with that egg cup problem.

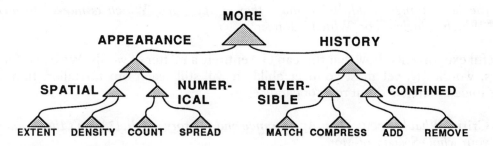

This society has two main divisions. In its *Appearance* division, a *Spatial* subdivision considers both the increased extent occupied by the spread-out eggs and also their thinned-out appearance or reduced density. In the case of those spread-out eggs, these conflict—and the Spatial agency withdraws. Then, if the child can count, *Numerical* decides; otherwise the *History* division applies some agents that use memories of recent happenings. If some of the eggs were rolled away, *Confined* would say that their amount is no longer the same; if the eggs were merely moved around, *Reversible* would claim that their amount cannot have changed.

To solve the water jar problem, the Society-of-More would need other kinds of lower-level agents:

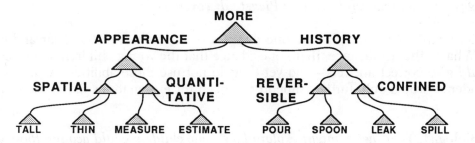

You might complain that even if we needed these hordes of lower-level agencies to make comparisons, this system has too many middle-level managers. But those mountains of bureaucracy are more than worth their cost. Each higher-level agent embodies a form of "higher-order" knowledge that helps us organize ourselves by telling us when and how to use the things we know. Without a many layered management, we couldn't use the knowledge in our low-level agencies; they'd all keep getting in one another's way.

10.6 ABOUT PIAGET'S EXPERIMENTS

Although Piaget's experiments about conservation of quantity have been confirmed as thoroughly as any in psychology, we can appreciate why many people are skeptical when they first hear of these discoveries. They contradict the traditional assumption that children are much like adults, except more ignorant. How strange it is that in all the centuries of history, these phenomena went unnoticed until Piaget—as though no one had ever watched a child carefully! But it has always been that way with science. Why did it take so long for our thinkers to discover such simple ideas as Isaac Newton's laws of motion or Darwin's idea of natural selection? Here are some frequent challenges.

> **Parent:** *Couldn't it be that younger children use words in ways that do not mean the same things to adults? Perhaps they simply take "Which is more?" to mean "Which is higher?" or "Which is longer?"*

Careful experiments show that this can't be entirely a matter of words. We can offer the same choices, wordlessly, yet most younger children will still reach out for taller, thinner jars of orange juice or stretched-out rows of candy eggs.

> **Critic:** *What happens when **Appearance** and **History** conflict? Won't that paralyze your whole Society-of-More?*

It would indeed—unless *More* has yet other levels and alternatives. And adults have other kinds of explanations—like "magic," "evaporation," or "theft." But indeed, stage magicians find that making things disappear does not entertain the youngest children; presumably they are too used to encountering the unexplainable. What happens when *More* cannot decide what to do? That depends upon the states of other agencies—including those involved in dealing with frustration, restlessness, and boredom.

> **Psychologist:** *We've heard of recent evidence that, despite what Piaget said, very young children do have concepts of quantity; many of them can even count those eggs. Doesn't that refute some of Piaget's discoveries?*

Not necessarily. Consider that no one disputes the outcomes of those jar and cup experiments. What is the significance, then, of evidence that the young children do possess methods that *could* give correct answers—and yet they do not use those abilities? As far as I can see, such evidence would only further support the need for explanations like those of Papert and Piaget.

> **Biologist:** *Your theory might explain how some children could acquire those concepts about quantities—but it doesn't explain why* all *children end up with such similar abilities! Could we be born with built-in genes that make brains do this automatically?*

This is a profound question. It is hard—but not impossible—to imagine how genes could directly influence the higher-level ideas and conceptions that we eventually learn. We'll discuss this in the appendix at the end of this book.

10.7 THE CONCEPT OF CONCEPT

In learning their Societies-of-More, children learn various skills for comparing different qualities and quantities, like number and extent. It is tempting to try to summarize all that by saying that the children are learning *something*; we could call it *the concept of quantity*. But why do we feel we have to think of what we learn as *things* or *concepts*? Why must we "thing-ify" everything?

What *is* a thing? No one doubts that a child's building-block is a thing. But is a child's love for its mother also a "thing"? We're imprisoned by our poverty of words because even though we have good ways to describe objects and actions, we lack methods for describing dispositions and processes. We can scarcely speak of what minds do except as though they were filled with things that one could see or touch; that's why we cling to terms like "concepts" and "ideas." I don't mean to say that this is always bad, for "thing-ifying" is indeed a splendid mental instrument. But for our present purpose, it is disastrous to assume that our minds contain some single "concept of quantity." At different times, a word like "more" can mean many different kinds of things. Think about each of these expressions.

More colorful. More loud. More swift. More valuable. More complicated.

We speak as though these were similar, yet each of them involves a different, hard-earned web of ways to think! The phrase "more loud" might seem at first to be merely a matter of magnitude. But consider how the sound of a distant gong seems louder than a whisper near the ear—no matter that its actual intensity is less. Your reaction to what you hear depends not only on its physical intensity, but also on what your agencies conclude about the character of its source. Thus you can usually tell whether a gong is loud but distant, rather than soft but close, by unconsciously making assumptions about the origin of that sound. And all those other kinds of "more" engage equally subtle sorts of expertise.

Instead of assuming that our children come to crystallize a single "concept of quantity," we must try to discover how our children accumulate and classify their many methods for comparing things. How do agents like *Tall*, *Thin*, *Short*, and *Wide* get formed into subagencies? To an adult, it seems natural to associate both being taller and being wider with being larger. But what prevents the child from inventing senseless "concepts" such as *"being Green and Tall and having recently been touched"*? No child has the time to generate and test all possible combinations to find which ones are sensible. Life is too short to do that many bad experiments! The secret is: *always try to combine related agents first. Tall, Thin, Short,* and *Wide* are all closely related, because they are all concerned with making comparisons between spatial qualities. In fact, they probably involve agencies that are close to one another in the brain and share so many agents in common that they'll naturally seem similar.

10.8 EDUCATION AND DEVELOPMENT

> **Parent:** *If those younger children take so long to acquire concepts like conservation of quantity, can't we help speed up their growth by teaching such things earlier?*

Such lessons just don't seem to work very well. Given enough explanation and encouragement, and enough drill and practice, we can make children *appear* to understand—yet even then they don't often apply what they've "learned" to real-life situations. Thus it seems that even when we lead them along these paths, they remain unable to use much of what we show to them until they develop inner signposts of their own.

Here's my guess about what goes wrong. Presumably the child senses that the spaced-out eggs are "more" because they stretch across a longer span. Eventually, we want that sense of greater length to be canceled out by the sense that there's more empty space between the eggs. In the more mature Papert hierarchy, this would happen automatically—but for now, the child could learn this only as a special, isolated rule. Many other problems could also be solved by making special rules for them. But to "simulate" that multilayer society, complete with middle-level agents like *Appearance* and *History*, would involve so many special rules, and so many exceptions to them, that the younger child would be unable to manage so much complexity. The result is that educational programs allegedly designed "according to Piaget" often appear to succeed from one moment to the next, but the structures that result from this are so fragile and specialized that children can apply them only to contexts almost exactly like those in which they were learned.

All this reminds me of a visit to my home from my friend Gilbert Voyat, who was then a student of Papert and Piaget and later became a distinguished child psychologist. On meeting our five-year-old twins, his eyes sparkled, and he quickly improvised some experiments in the kitchen. Gilbert engaged Julie first, planning to ask her about whether a potato would balance best on one, two, three, or four toothpicks. First, in order to assess her general development, he began by performing the water jar experiment. The conversation went like this:

> **Gilbert:** *"Is there more water in this jar or in that jar?"*
> **Julie:** *"It looks like there's more in that one. But you should ask my brother, Henry. He has conservation already."*

Gilbert paled and fled. I always wondered what Henry would have said. In any case, this anecdote illustrates how a young child may possess many of the ingredients of perception, knowledge, and ability needed for this kind of judgment—yet still not have suitably organized those components.

> **Parent:** *Why are all the agents in your societies so competitive? They're always attacking each other. Instead of making **Tall** and **Thin** cancel each other out, why can't they cooperate?*

The first part of this book has given this impression because we had to begin with relatively simple mechanisms. It is fairly easy to resolve conflicts by switching among alternatives. It is much harder to develop mechanisms that can use cooperation and compromise—because that requires more complex ways for agencies to interact. In later sections of this book we'll see how higher-level systems could make more reasonable negotiations and compromises.

10.9 LEARNING A HIERARCHY

How could a brain continue functioning while changing and adding new agents and connections? One way would be to keep each old system unchanged while building a new version in the form of a detour around or across it—but not permitting the new version to assume control until we're sure that it can also perform the older system's vital functions. Then we can cut some of the older connections.

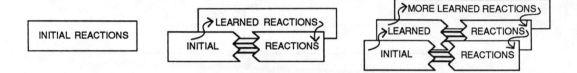

We could use this method to form our hierarchical Society-of-More:

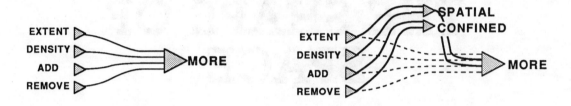

Now let's draw this in another form, as though there were no room to fit new agents in between the older ones.

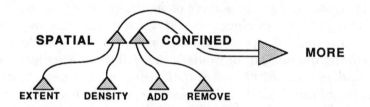

As we accumulate more low-level agents and additional intermediate layers to manage them, this grows into the very multilevel hierarchy we've seen before.

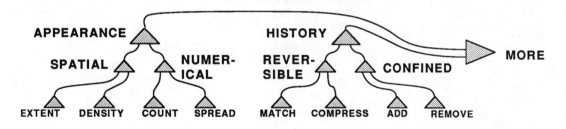

The nerve cells in an animal's brain can't always move aside to make more room for extra ones. So those new layers might indeed have to be located elsewhere, attached by bundles of connection wires. Indeed, no aspect of the brain's anatomy is more striking than its huge masses of connection bundles.

CHAPTER 11

CHAPTER 11

THE SHAPE OF SPACE

"A day of dappled seaborne clouds." The phrase and the day and the scene harmonized in a chord. Words. Was it their colours? He allowed them to glow and fade, hue after hue; sunrise gold, the russet and green of apple orchards, azures of waves, the greyfringed fleece of clouds. No, it was not their colours; it was the poise and balance of the period itself. Did he then love the rhythmic rise and fall of words better than their associations of legend and colour? Or was it that, being as weak of sight as he was shy of mind, he drew less pleasure from the reflection of the glowing sensible world through the prism of a language manycoloured and richly storied than from the contemplation of an inner world of individual emotions mirrored perfectly in a lucid supple periodic prose?

—JAMES JOYCE

11.1 SEEING RED

What possible kind of brain-event could correspond to anything like the meaning of an ordinary word? When you say "red," your vocal cords obey commands from "pronouncing agents" in your brain, which make your chest and larynx muscles move to produce that special sound. These agents must in turn receive commands from somewhere else, where other agents respond to signals from yet other places. All those "places" must comprise the parts of some society of mental agencies.

It's easy to design a machine to tell when there is something red: start with sensors that respond to different hues of light, and connect the ones most sensitive to red to a central "red-agent," making corrections for the color of the lighting of the scene. We could make this machine appear to "speak" by linking each color-agent to a device that pronounces the corresponding word. Then this machine could name the colors it "sees"—and even distinguish more hues than ordinary people can. But it would be a travesty to call this "sight," since it's nothing but a catalog that lists a lot of colored dots. It would share no human notion of what colors come to mean to us, because without some sense of texture, form, and very much more, it would have few of the qualities of our human kinds of images and thoughts.

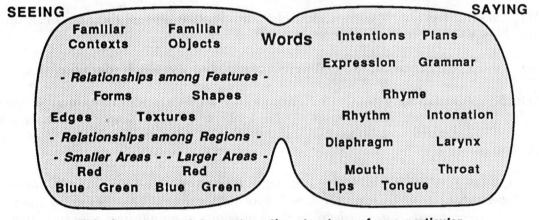

This is not meant to portray the structure of any particular society but only to suggest the variety of agencies involved.

Of course no little diagram can capture more than a fragment of any real person's thoughts about the world. But this should not be taken to mean that no machine could ever have the range of sensibilities that people have. It merely means that we aren't *simple* machines; indeed, we should understand that in learning to comprehend the qualities of vast machines, we are still in the dark ages. And in any case, a diagram can only illustrate a principle: there cannot be any compact way to represent all the details of full-grown mind-society. To talk about such complex things, we can only resort to language tricks that make our listeners' minds explore the worlds inside themselves.

11.2   THE SHAPE OF SPACE

The brain is imprisoned inside the skull, a silent, dark, and motionless place; how can it learn what it's like outside? The surface of the brain itself has not the slightest sense of touch; it has no skin with which to feel; it is only *connected* to skin. Nor can a brain see, for it has no eyes; it only is *connected* to eyes. The only paths from the world to the brain are bundles of nerves like those that come in from the eyes, ears, and skin. How do the signals that come through those nerves give rise to our sense of "being in" the outside world? The answer is that this sense is a complicated illusion. We never actually make any *direct* contact with the outside world. Instead, we work with models of the world that we build inside our brains. The next few sections try to sketch how that could come about.

The surface of the skin contains countless little touch-sensing agents, and the retina of the eye includes a million tiny light detectors. Scientists know a good deal about how these sensors send signals to the brain. But we know much less about how those signals lead to sensations of touch and of sight. Try this simple experiment:

Touch your ear.

What did that feel like? It seems impossible to answer that because there's scarcely anything to say. Now try a different experiment:

Touch your ear twice, in two different places, and also touch your nose.

Which two touches feel most similar? That question seems much easier to answer: one might say that the two ear touches feel more similar. Evidently, there is scarcely anything that one can say about a "single sensation" by itself, but we can often say much more when we can make comparisons.

> Consider the analogy to how mathematics treats a "perfect point." We shouldn't speak about its shape; it simply doesn't have a shape! But since we're used to things as having shapes, we can't help thinking of points as round, like "very tiny little dots." Similarly, we're not supposed to talk about the size of a point—since mathematical points, by definition, have no size. Still, we can scarcely help but think, in any case, "they're very small."

In fact, there's absolutely nothing to be said about a single point, except how it relates to other points. This is not because such things are too complicated to explain, but because they are too simple to explain. One cannot even speak about where a point is, by itself—since "where" has meaning only in relation to other points in space. But once we know some *pairs* of points, we can relate these to the lines that connect them, and then we can define new, different points where various pairs of lines may intersect. Repeating this can generate entire worlds of geometry. Once we understand the terrifying fact that points are nothing by themselves but exist only in relation to other points, then we can ask, as Einstein did, whether time and space are anything more than vast societies of nearnesses.

In the same way, there is little that one could say about any "single touch"—or about what any single sense-detecting agent does. However, there is much more to be said about the relations between two or more skin touches, because *the closer together two skin spots are, the more frequently they'll both be touched at the same time.*

11.3 NEARNESSES

The reason our skin can feel is because we're built with myriad nerves that run from every skin spot to the brain. In general, each pair of nearby places on the skin is wired to nearby places in the brain. This is because those nerves tend to run in bundles of parallel fibers—more or less like this:

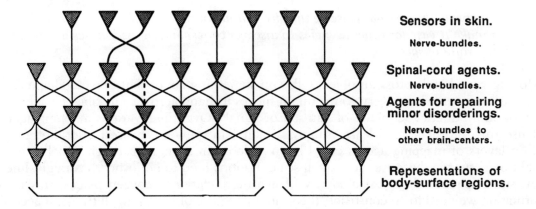

Sensors in skin.

Nerve-bundles.

Spinal-cord agents.

Nerve-bundles.

Agents for repairing minor disorderings.

Nerve-bundles to other brain-centers.

Representations of body-surface regions.

Each sensory experience involves the activity of many different sensors. In general, the greater the extent to which two stimuli arouse the same sensors, the more nearly alike will be the partial mental states those stimuli produce—and the more similar those stimuli will "seem," simply because they'll tend to lead to similar mental consequences.

> *Other things being equal, the apparent similarity of two stimuli will depend on the extent to which they lead to similar activities in other agencies.*

The fact that the nerves from skin to brain tend to run in parallel bundles means that stimulating nearby spots of skin will usually lead to rather similar activities inside the brain. In the next section we'll see how this could enable an agency inside the brain to discover the spatial layout of the skin. For example, as you move a finger along your skin, new nerve endings are stimulated—and it is safe to assume that the new arrivals represent spots of skin along the advancing edge of your finger.

Given enough such information, a suitably designed agency could assemble a sort of map to represent which spots are close together on the skin. Because there are many irregularities in the nerve-bundle pathways from skin to brain, the agencies that construct those maps must be able to "tidy things up." For example, the mapping agency must learn to correct the sort of crossing-over shown in the diagram. But that is only the beginning of the task. For a child, learning about the spatial world *beyond* the skin is a journey that stretches over many years.

11.4 INNATE GEOGRAPHY

We've seen that touching nearby spots of skin will usually give rise to similar sensations: this is because the corresponding nerves run in parallel courses and thus cause similar activities inside the brain. The reverse is also usually true: the more similar two sensations are, the closer their origins in the skin. This has an important consequence:

> *The nerve pathways that preserve the physical nearness relations of our skin-sensors can make it easy for inner agencies to discover corresponding nearnesses about the outer world of space.*

Moving your hand across an object tells you something about that object's shape. Imagine what must happen when a very young infant moves its hand across some object: each continuous motion produces a sequence of skin-sensor signals. Over time, various mapping agents can first use this information to learn, simply, which skin spots are nearest one another. Later, further layers of mapping agents could learn which skin spots lie *between* which others; this should be easy, too, because most small-scale motions tend to go in nearly straight lines. But then, since space itself is just a society of nearness relations between places, this is all the information we need to "reconstruct" the spatial structure of the skin. All this is in accord with a basic principle of mathematics:

> *Suppose you were lost in some unknown space—and could only tell which pairs of points were close to one another. That would be enough for you to figure a great deal about the space. From that alone, you could deduce if you were in a world of two dimensions or three. You could tell where there were obstacles and boundaries, holes and tunnels and bridges, and so on. You could figure out the global layout of that world from just those local bits of information about nearnesses.*

It is a wonderful fact that, in principle, one can deduce the global geography of a space from nothing more than hints about which pairs of points lie near one another! But it is another matter to actually make such maps, and no one yet knows how the brain does this. To design a machine to accomplish such tasks, one could begin with a layer of "correlation agents," one for each tiny patch of skin, each engineered to detect which other skin spots are most often aroused at nearly the same times; those will then be mapped as the nearest ones. A second layer of similar agents could then begin to make maps of larger regions, and several such layers would eventually assemble a sequence of maps on various scales, for representing several levels of detail.

If brains do something of this sort, it might illuminate a problem that has troubled some philosophers: "*Why do we all agree on what the outer world of space is like?*" Why don't different people interpret space in different, alien ways? In principle it is mathematically possible for each person to conclude, for example, that the world is three-dimensional—rather than two- or four-dimensional—just from enough experience with nearby pairs of points. However, if the wires from the skin to the brain were shuffled and scrambled around too much, we would probably never get them straightened out because the actual calculations for doing such things would be beyond our capabilities.

11.5 SENSING SIMILARITIES

*This difficulty [of making definitions] is increased by the necessity
of explaining the words in the same language, for there is often
only one word for one idea; and though it may be easy to translate
words like **bright, sweet, salt, bitter,** into another language, it is
not easy to explain them.*
— SAMUEL JOHNSON

Our ways to think depend in part on how we're raised. But at the start, much more depends upon the wiring in our brains. How do those microscopic features work to influence what happens in our mental worlds? The answer is, *our thoughts are largely shaped by which things seem most similar.* Which colors seem the most alike? Which forms and shapes, which smells and tastes, which timbres, pitches, pains and aches, which feelings and sensations seem most similar? Such judgments have a huge effect at every stage of mental growth—since *what we learn depends on how we classify.*

For example, a child who classified each fire just by the color of its light might learn to be afraid of everything of orange hue. Then we'd complain that the child had "generalized" too much. But if that child classified each flame, instead, by features that were never twice the same, that child would often be burned—and we'd complain that it hadn't generalized enough.

Our genes supply our bodies with many kinds of sensors—external event-detecting agents— each of which sends signals to the nervous system when it detects certain physical conditions. We have sensory-agents in our eyes, ears, nose, and mouth that discern light, sound, odors, and tastes; we have agents in the skin that sense pressure, touch, vibration, heat, and cold; we have internal agents that sense tensions in our muscles, tendons, and ligaments; and we have many other sensors of which we're normally unaware, such as those that detect the direction of gravity and sense the amounts of various chemicals in different parts of the body.

The agents that sense the colors of light in human eyes are much more complex than the "redness agents" of our toy machine. But this is not the reason that simple machine can't grasp what *Redness* means to us—for neither can the sense detectors in our human eyes. For just as there is nothing to say about a single point, there's nothing to be said about an isolated sensory signal. When our *Redness, Touch,* or *Toothache* agents send their signals to our brains, each by itself can only say, "I'm here." The rest of what such signals "mean" to us depends on how they're linked to all our other agencies.

In other words, the "qualities" of signals sent to brains depend only on relationships—the same as with the shapeless points of space. This is the problem Dr. Johnson faced when creating definitions for his dictionary: each separate word like "bitter," "bright," "salt," or "sweet" attempts to speak about a quality of a sensory signal. But all that a separate signal can do is announce its own activity—perhaps with some expression of intensity. Your *tooth* can't ache (it can only send signals); only *you* can ache, once your higher-level agencies interpret those signals. Beyond the raw distinctiveness of every separate stimulus, all other aspects of its character or quality—be it of touch, taste, sound, or light—depend entirely on its relationships with the other agents of your mind.

11.6 THE CENTERED SELF

How do we learn about the real, three-dimensional world? We've seen how certain agencies might map the layout of the skin. But how could we progress from that to learn about the world of space beyond the skin? One might ask why infants can't simply "look around" to see what's really going on. Unfortunately, the easy-sounding phrase "simply look" conceals too many hard problems. When you look at an object, some light from it shines into your eye and stimulates some sensors there. However, every motion of your body, head, or eye makes drastic changes to the image in your eye. How can we extract any useful information when everything changes so rapidly? Although it should be possible, in principle, to design a machine that could eventually learn to relate those motions to the resulting changes in the images, this would surely take a long time, and it appears that our brains have evolved with special mechanisms that help us compensate for motions of the body, head, and eye. This makes it easier for other agencies to learn to use visual information. Later we'll discuss some other realms of thought in which we use analogies and metaphors to change our "points of view." Perhaps those wonderful abilities evolved in similar ways, since recognizing that an object is the same when seen from different views is not so different from being able to "imagine" things that are not in view at all.

In any case, we really do not understand how the child learns to understand space. Perhaps we start by doing many small experiments that lead to our first, crude maps of the skin. Next we might start to correlate these with the motions of our eyes and limbs; two different actions that lead to similar sensations are likely to have passed through the same locations in space. A critical step would be developing some agents that "represent" a few "places" outside the skin. Once those places are established (the first ones might be near the infant's face), one could proceed to another stage: the assembly of an agency that represents a network of relationships, trajectories, and directions between those places. Once this is accomplished, the network could continue to extend to include new places and relationships.

However, this would be only the beginning. Long ago, psychologists like Freud and Piaget observed that children seem to recapitulate the history of astronomy: first they imagine the world as centered around themselves—and only later do they start to view themselves as moving within a stationary universe, in which the body is just like any other object. It takes several years to reach that stage, and even in their adolescent years, children are still improving their abilities to envision how things appear from other viewpoints.

11.7 PREDESTINED LEARNING

It would be wonderful if we could classify all behavior into two types: *"built-in"* and *"learned."* But there simply is no clear-cut boundary between heredity and environment. Later, I'll describe an agency that is sure to learn one particular thing: *to recognize human beings.* But if such an agency is destined to end up with a certain particular behavior, is it reasonable to say that it learns? Since this type of activity appears to have no common name, we'll call it "predestined learning."

Every child eventually learns to reach for food. To be sure, each different child lives through a different history of "reaching-act" experiences. Nevertheless, according to our theory of "nearness models of space," all those children will end up with generally similar results because that outcome is constrained by the nearness relations of real-world space. Why make the brain use a tedious learning process when the final outcome seems so clear? Why not build in the answer genetically? One reason could be that learning is more economical. It would require an enormous store of genetic information to force each separate nerve cell to make precisely the right connections, whereas it would require much less information to specify the construction of a learning machine designed to unscramble whatever irregularities result from a less constrained design.

This is why it isn't sensible to ask, *"Is the child's conception of space acquired or inherited?"* We acquire our conceptions of space by using agencies that learn in accord with processes determined by inheritance. These agencies proceed to learn from experience—but the outcomes of their learning processes are virtually predestined by the spatial geometry of our body parts. This kind of mixture of adaptation and predestination is quite common in biology, not only in the brain's development but in that of the rest of the body as well. How, for example, do our genes control the shapes and sizes of our bones? They may begin with some relatively precise specification of the types and location of certain early cells. But that alone would not be adequate for animals that themselves have to adapt to different conditions; therefore those early cells must themselves be programmed to adapt to the various chemical and mechanical influences that may later be imposed on them. Such systems are essential for our development, since our organs must become able to perform various tightly constrained activities, yet also be able to adapt to changing circumstances.

Perhaps the growth of the Society-of-More is another instance of predestined learning, for it seems to develop in every normal child without much outside help. It seems clear that this complex agency is not built directly by inborn genes; instead, we each discover our own ways to represent comparisons—yet we all arrive at much the same final outcome. Presumably, genetic hints must help with this by supplying new layers of agents at roughly the right times and places.

11.8 HALF-BRAINS

Let's do one more experiment: touch one ear and then touch your nose. They don't feel very similar. Now touch one ear and then the other. These touches seem more similar, although they're twice as far apart. This may be in part because they are represented in related agencies. In fact, our brains have many pairs of agencies, arranged like mirror-images, with huge bundles of nerves running between them.

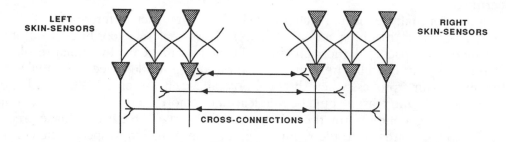

The two hemispheres of the brain look so alike that they were long assumed to be identical. Then it was found that after those cross-connections are destroyed, usually only the left brain can recognize or speak words, and only the right brain can draw pictures. More recently, when modern methods found other differences between those sides, it seems to me that some psychologists went mad—and tried to match those differences to every mentalistic two-part theory that ever was conceived. Our culture soon became entranced by this revival of an old idea in modern guise: that our minds are meeting grounds for pairs of antiprinciples. On one side stands the Logical, across from Analogical. The left-side brain is Rational; the right side is Emotional. No wonder so many seized upon this pseudoscientific scheme: it gave new life to nearly every dead idea of how to cleave the mental world into two halves as nicely as a peach.

What's wrong with this is that each brain has *many* parts, not only two. And though there are many differences, we also ought to ask about why those left-right brain halves are actually so similar. What functions might this serve? For one thing, we know that when a major brain area is damaged in a young person, the mirror region can sometimes take over its function. Probably even when there is no injury, an agency that has consumed all the space available in its neighborhood can expand into the mirror region across the way. Another theory: a pair of mirrored agencies could be useful for making comparisons and for recognizing differences, since if one side could make a copy of its state on the other side then, after doing some work, it could compare those initial and final states to see what progress had been made.

My own theory of what happens when the cross-connections between those brain halves are destroyed is that, in early life, we start with mostly similar agencies on either side. Later, as we grow more complex, a combination of genetic and circumstantial effects lead one of each pair to take control of both. Otherwise, we might become paralyzed by conflicts, because many agents would have to serve two masters. Eventually, the adult managers for many skills would tend to develop on the side of the brain most concerned with language because those agencies connect to an unusually large number of other agencies. The less dominant side of the brain will continue to develop, but with fewer administrative functions—and end up with more of our lower-level skills, but with less involvement in plans and higher-level goals that engage many agencies at once. Then if, by accident, that brain half is abandoned to itself, it will seem more childish and less mature because it lags so far behind in administrative growth.

11.9 DUMBBELL THEORIES

This fascination with left-right halves, on the part of both the lay and scientific populace, is nothing really new. It is a symptom of how we acquire various pairs of words that each divide some aspect of the world into opposing poles.

PHYSICAL   **MENTAL**

```
ACQUIRED ——————————— INNATE
CAUSE ———————————————— CHANCE
TRUE ——————————————————— FALSE
CHILD ——————————————————— ADULT
ALIVE ——————————————————— DEAD
```

Such divisions all have flaws but often give us useful ways to think. Dividing things in two is a good way to start, but one should always try to find at least a third alternative. If one cannot, one should suspect that there may not be two ideas at all, but only one, together with some form of opposite. A serious problem with these two-part forms is that so many of them are quite similar, which leads us into making false analogies. Consider how the pairs below, in which each self is split in two, lead everyone to think they share some common unity.

```
THOUGHT ——————————— FEELING
LOGIC ——————————————— INTUITION
LITERAL——————————— ANALOGICAL
REASON ————————————— EMOTION
DELIBERATE ———————— SPONTANEOUS
SCIENTIFIC ——————————— ARTISTIC
QUANTITY ——————————— QUALITY
```

The items on the left are seen as neutrally objective and mechanical and only found within the head. We think of Thought and its associates as accurate, but rigid and insensitive. The items on the right are seen as matters of the heart—as vital, warm, and individual; we like to believe that Feeling is the better judge of the things that ought to matter most. Cool Reason, by itself, seems too impersonal, too far from flesh; Emotion lies much closer to the heart, but it, too, can be treacherous when it grows so intense that reason gets completely overwhelmed.

How marvelous this metaphor! How could it work so well, unless it had some basic truth? But wait: whenever any simple idea appears to explain so many things, we must suspect a trick. Before we're drawn into dumbbell schemes, we owe it to ourselves to try to understand their strange attractiveness, in order that we not be deceived, as Wordsworth said, by

> . . . *some false secondary power, by which,*
> *In weakness, we create distinctions, then*
> *Deem that our puny boundaries are things*
> *Which we perceive, and not which we have made.*

CHAPTER 12

LEARNING MEANING

How many times in the course of my life had I been disappointed by reality because, at the time I was observing it, my imagination, the only organ with which I could enjoy beauty, was not able to function, by virtue of the inexorable law which decrees that only that which is absent can be imagined.

—MARCEL PROUST

12.1 A BLOCK-ARCH SCENARIO

Our child, playing with some blocks and a toy car, happens to build this structure. Let's call it a *Block-Arch*.

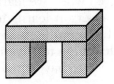

Two standing blocks and a lying block.

Block-Arch seems to cause a strange new phenomenon: when you push the car through it, your arm gets trapped! Then, in order to complete that action, you must release the car—and reach around to the other side of the arch, perhaps by changing hands. The child becomes interested in this "*Hand-Change*" phenomenon and wonders how *Block-Arch* causes it. Soon the child finds another structure that seems similar—except that *Hand-Change* disappears because you can't even push the car through it. Yet both structures fit the same description!

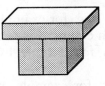

Two standing blocks and a lying block.

But if *Block-Arch* causes *Hand-Change*, then this can't be a block-arch. So the child must find some way to change the mental description of *Block-Arch* so it won't apply to this. What is the difference between them? Perhaps this is because those standing blocks now touch one another, when they didn't touch before. We could adapt to this by changing our description of *Block-Arch*: "*There must be two standing blocks and a lying block. The standing blocks must not touch.*" But even this does not suffice, because the child soon finds yet another structure that matches this description. Here, too, the *Hand-Change* phenomenon has disappeared; now you *can* push the car through it without letting go!

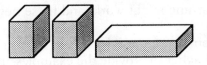

Two standing blocks and a lying block.
The standing blocks must not touch.
They must support the lying block.

Again we must change our description to keep this from being considered a *Block-Arch*. Finally the child discovers another variation that *does* produce *Hand-Change*:

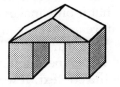

Two standing blocks and something else.
The standing blocks must not touch.
They must support the other thing.
The other thing may be A WEDGE OR A BLOCK.

Our child has constructed for itself a useful conception of an arch, based entirely upon its own experience.

12.2 LEARNING MEANING

What is learning, anyway? That word is certainly hard to define. The child in our *Block-Arch* scenario has found one way to learn one sense of what some adults mean by "arch." But we can't assume that the same kinds of processes are involved when we learn to recite a poem, to use a spoon, and to tie a shoe. What happens when a person learns to read, learns to add numbers, learns a new language, learns to anticipate the dispositions of a friend, or learns to build a tower that will stand? If we tried to find a single definition for "learning" to span so many kinds of processes, we'd end up with some phrase too broad to have much use—like this:

"Learning" is making useful changes in the workings of our minds.

The problem is that we use the single word "learning" to cover too diverse a society of ideas. Such a word can be useful in the title of a book, or in the name of an institution. But when it comes to studying the subject itself, we need more distinctive terms for important, different ways to learn. Even that one *Block-Arch* scene reveals at least four different ways to learn. We'll give them these new names:

Uniframing combining several descriptions into one, for example, by observing that all the arches have certain common parts.

Accumulating collecting incompatible descriptions, for example, by forming the phrase "block or wedge."

Reformulating modifying a description's character, for example, by describing the separate blocks rather than the overall shape.

Trans-framing bridging between structures and functions or actions, for example, by relating the concept of arch to the act of changing hands.

These words will be explained in the sections that follow. It seems to me that the older words used in psychology—such as *generalizing, practicing, conditioning, memorizing,* or *associating*—are either too vague to be useful or have become connected to theories that simply aren't sound. In the meantime, the revolutions of computer science and Artificial Intelligence have led to new ideas about how various kinds of learning might work, and these new ideas deserve new names.

Our *Block-Arch* scenario is based on a computer program developed by Patrick Winston in 1970. Winston's program required an external teacher to provide the examples and to say which of them were arches and which were not. In my unprogrammed version of this, the teacher has been replaced by the concern of some agency inside the child to account for the emergence of that mysterious *Hand-Change* phenomenon: why do certain structures force you to let go of the toy car, while other structures don't? We thus assume that the child is led to learn for itself in order to account for strange events. One might complain that it only makes learning harder to explain, to make it depend upon the child's curiosity. But if we are ever really to understand how our minds grow, we must first face reality: people just don't learn so well unless they're interested or concerned. The older theories of learning and remembering never got very far because in trying to oversimplify, they lost essential aspects of the context. It wouldn't be much use to have a theory in which knowledge is somehow stored away—without a corresponding theory of how later to put that knowledge back to work.

12.3 UNIFRAMES

The child in our *Block-Arch* scene examined several different arrangements of blocks—yet ended up describing them as all the same! The great accomplishment was in discovering how to describe all the different instances of arch with the selfsame phrase, *"a top supported by two standing blocks that do not touch."* I'll use the new word "uniframe" for this—a description constructed to apply to several different things at once. How does a person make a uniframe?

Our child's *Block-Arch* uniframe was constructed in several steps—and each step used a different learning-scheme! The first step dissects the scene into blocks with specific properties and relationships; some were "lying down" or "standing up," and some were touching or supporting other ones. Next, we required our uniframe to insist that the arch top *must* be supported by the standing blocks: let's call this *enforcement*. Then, we required our uniframe to reject structures in which the two standing blocks touch one another; we could call this *prevention*: a way to keep from accepting an undesired situation. Finally, we required our uniframe to be neutral about the arch top's shape in order to keep from making distinctions we don't consider relevant. Let's call that *tolerance*.

How does a person know how to choose which features and relations to *enforce*, *prevent*, or *tolerate*? When we compared the two structures below, we enforced the relation that A is supported by B and C. But think of all the other differences we could have emphasized instead.

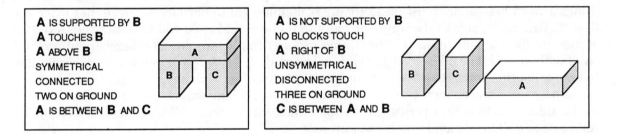

Was it wasteful to use only one of these facts when we could have used them all? Should we learn to exploit all the information we can get? No! There are good reasons not to notice too much, for every seemingly essential fact can generate a universe of useless, accidental, and even misleading facts.

Most differences are redundant. Most of the rest are accidents.

For example, suppose that we already know A is supported by B. There is then no need to remember that A touches B or that A is above B—because these are things that we can figure out. For a different kind of example, suppose we knew that A was *not* supported by B. It then seems unnecessary to remember that "A was to the right of B." Common sense can tell us that if A is not on B, it must lie somewhere else. However, at least in the present context, it does not matter whether that "somewhere else" is to the right; another time it might just as likely lie to the left. If we stored such details too recklessly, our minds would get cluttered up with useless facts.

But how can we judge which facts are useful? *On what basis can we decide which features are essential and which are merely accidents?* Such questions can't be answered as they stand. They make no sense apart from how we want to use their answers. There is no single secret, magic trick to learning; we simply have to learn a large society of different ways to learn!

12.4 STRUCTURE AND FUNCTION

*When the eye or the imagination is struck with any uncommon
work, the next transition of an active mind is to the means by
which it was performed.*
— SAMUEL JOHNSON

Suppose an adult watched our child and said, "I see you've built an arch." What might the child think this means? To learn new words or new ideas, one must make connections to other structures in the mind. "I see you've built an arch" should make the child connect the word "arch" to agencies embodying descriptions of both the *Block-Arch* and the *Hand-Change* phenomena—since those are what is on the child's mind.

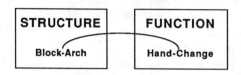

But one can't learn what something means merely by tying things to names. Each word-idea must also be invested with some causes, actions, purposes, and explanations. Consider all the things a word like "arch" must mean to any real child who understands how arches work and how they're made, and all the ways one can use them! A real child will have noticed, too, that arches are like variants of many other things experienced before, like "bridge without a road," "wall with door," "tablelike," or "shaped like an upside-down U." We can use such similarities to help find other things to serve our purposes: to think of an arch as a passage, hole, or tunnel could help someone concerned with a transportation problem; describing an arch as "top held up by sides" could help a person get to something out of reach. Which kind of description serves us best? That depends upon our purposes.

Among our most powerful ways of thinking are those that let us bring together things we've learned in different contexts. But how can one think in two different ways at once? By building, somewhere inside the mind, some arches of a different kind:

```
┌─────────────────┐ ┌─────────────────┐
│ STRUCTURE       │ │ FUNCTION        │
│    ARCH           BRIDGE            │
│  APERTURES        PASSAGES          │
│  LEGS                               │
│    COLUMNS         SUPPORTS         │
│                       DOOR          │
│  HOLE            WINDOW             │
└─────────────────┘ └─────────────────┘
                              ETC.
```

Is that a foolish metaphor—to talk of building bridges between places in the mind? I'm sure it's not an accident that we so often frame our thoughts in terms of familiar spatial forms. Much of how we think in later life is based on what we learn in early life about the world of space.

12.5 THE FUNCTIONS OF STRUCTURES

Many things that we regard as physical are actually psychological. To see why this is so, let's try to say what we mean by "chair." At first it seems enough to say:

"A chair is a thing with legs and a back and seat."

But when we look more carefully at what we recognize as chairs, we find that many of them do not fit this description because they don't divide into those separate parts. When all is done, there's little we can find in common to all chairs—except for their intended use.

"A chair is something you can sit upon."

But that, too, seems inadequate: it makes it seem as though a chair were as insubstantial as a wish. The solution is that we need to combine at least two different kinds of descriptions. On one side, we need structural descriptions for recognizing chairs when we see them. On the other side we need functional descriptions in order to know what we can *do* with chairs. We can capture more of what we mean by interweaving both ideas. But it's not enough merely to propose a vague association, because in order for it to have some use, we need more intimate details about *how* those chair parts actually help a person to sit. To catch the proper meaning, we need connections between parts of the chair structure and the requirements of the human body that those parts are supposed to serve. Our network needs details like these:

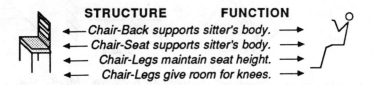

Without such knowledge, we might just crawl under the chair or try to wear it on our head. But with that knowledge we can do amazing things, like applying the concept of a chair to see how we could sit on a box, even though it has no legs or back!

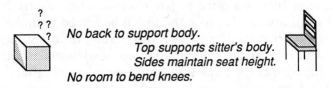

Uniframes that include structures like this can be powerful. For example, such knowledge about relations between structure, comfort, and posture could be used to understand when a box could serve as a chair: that is, only when it is of suitable height for a person who does not require a backrest or room to bend the knees. To be sure, such clever reasoning requires special mental skills with which to redescribe or "reformulate" the descriptions of both box and chair so that they "match" despite their differences. Until we learn to make old descriptions fit new circumstances, our old knowledge can be applied only to the circumstances in which it was learned. And that would scarcely ever work, since circumstances never repeat themselves perfectly.

12.6 ACCUMULATION

Uniframing doesn't always work. We often try to make an everyday idea precise—but just can't find much unity. Then, we can only accumulate collections of examples.

It certainly is hard to find any properties that all these share. Coins are hard and round and flat. Bills are thin and flexible. Bullion has unusual weight, and credits aren't even physical. We recognize them all as media of trade—but that won't help us recognize the things themselves. The situation is the same for furniture. It's not so hard to say what furniture is *for*—"things that equip a room for living in." But when it comes to the objects themselves, it's even hard to find a uniframe for "chair." Again, its function-role seems clear—"a thing one can sit upon." The problem is that one can sit on almost anything—a bench, a floor, a tabletop, a horse, a stack of bricks, a rock. Even defining *Arch* has problems, since many things we recognize as arches just don't match our *Block-Arch* uniframe:

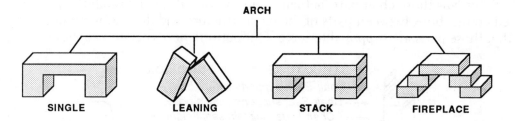

All these shapes could be described as *"shape with hole"* or *"blocks that bridge across a gap."* But those descriptions would also admit many things we don't want to regard as arches. The simplest way to learn, when one can't find a uniframe, is to accumulate descriptions of experiences.

At first it may seem simpler to accumulate examples than to find more uniform ways to represent them. But later there's a price to pay for this: when we try to *reason about things*, accumulations can be nuisances—because then we'll be forced to find a different argument or explanation to justify each separate example. Most likely, different parts of our brains have evolved to use both kinds of strategies. Accumulations need not take longer to manipulate if all the examples can be handled at the same time, by separate agents that don't interfere with one another. But once those processes begin to need each other's help, the whole society's efficiency will decline rapidly. Perhaps that slowing-down itself might be the stimulus that makes us start to try to unify—at least for processes we use frequently.

A simpler theory of when we start new uniframes would be that in the brain, there is an architectural constraint on how many K-lines are directly accessible to various types of agents. For example, the agents in a certain agency might be able to accumulate no more than about seven branches for each classification in a certain hierarchy. When more than that accumulate, the agency would be forced either to merge some examples into uniframes or to turn for help from somewhere else.

12.7 ACCUMULATION STRATEGIES

Let's make a dumbbell theory of some people's personalities.

Uniframers *disregard discrepancies in favor of imagined regularities. They tend to be perfectionists but also tend to think in terms of stereotypes. This sometimes leads to recklessness because they have to reject some evidence in order to make their uniframes.*

Accumulators *are less extreme. They keep collecting evidence and hence are much less prone to make mistakes. But then they're also slower to make discoveries.*

Of course these imaginary personalities are only caricatures, and everyone blends both extremes. Most people find some reasonable compromise, though a few of us lean more in one direction than the other. I'm sure we all use mixtures of different learning strategies—accumulations of descriptions, K-lines, uniframes, or whatever. On the surface, it might seem easier to make accumulations than to make uniframes—*but choosing what to accumulate may require deeper insight*. In any case, whenever an accumulation becomes too large and clumsy, we try to replace some groups of its members with a uniframe. But even when we succeed in finding a suitably compact uniframe, we can expect it, too, to accumulate exceptions eventually, since first descriptions rarely work for all our later purposes.

For example, when a child first encounters dogs, an attempt might be made to create a uniframe that catalogs those animals' parts—eyes, ears, teeth, head, body, tail, legs, and so on. But the child will eventually have to learn that even here there are exceptions.

Furthermore, that uniframe won't help answer the child's most urgent questions about any one dog in particular: Is it friendly? Does it have a loud bark? Is it the kind that tends to bite? Each such concern could require building a different kind of hierarchy-tree.

This leads to an inescapable difficulty. Our various motives and concerns are likely to require incompatible ways to classify things. You can't predict a dog's bite from its bark. Each of the classifications we build must embody different kinds of knowledge, and we can rarely use more than a few of them at once. When we have a goal that is simple and clear, we may be able to select one particular kind of description that makes the problem easy to solve. But when goals of several types conflict, it is harder to know just what to do.

12.8 PROBLEMS OF DISUNITY

When should you accumulate, and when should you make uniframes? The choice depends upon your purposes. Sometimes it is useful to regard things as similar because they have similar forms, but sometimes it makes more sense to group together things with similar *uses*. At one moment you may wish to emphasize a similarity; at the next moment, you may want to emphasize a distinction. Often, we have to use both uniframes and accumulations in combination. In *Block-Arch*, for example, we found that there could be two different kinds of arch tops—the block and the wedge. Accordingly, when we used the phrase "block or wedge," we actually inserted a "subaccumulation" into our uniframe.

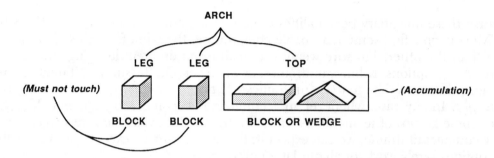

Accumulations rarely seem quite satisfactory because we feel ideas should have more unity. We wouldn't have a word for chair or arch or currency if they meant nothing more than lists of unrelated things. If each did not involve some unifying thought, we'd never think to make those lists in the first place! Why is it so hard to describe their essences? In the next few sections we'll discover a number of reasons for this. Here is one of them:

> *Many good ideas are really two ideas in one—which form a bridge between two realms of thought or different points of view.*

Whenever we build a bridge between structure and function, one end of that bridge may represent a goal or use, while the other end describes what we might use to gain those ends. But it is rare for those structures to correspond neatly to those functions. The problem is that we usually find many different ways to achieve any goal. This means that we'll find an accumulation on the structural side of the bridge. For example, if you want to reach something high up, you can stand on a chair, reach with a stick, or ask someone taller to get it for you. Similarly, an accumulation of functions or goals can be found for any structure. My colleague Oliver Selfridge once wrote an entire book entitled *Things to Do with a Stick*.

> *Our different worlds of ends and means don't usually match up very well. So when we find a useful, compact uniframe in one such world, it often corresponds to an accumulation in our other worlds.*

We encountered this problem earlier. When we classified birds as *animals* while classifying airplanes as *machines*, we thereby forced disunity upon the class of things that fly. Later, when we come to theories about metaphors, we'll see that such problems are almost inevitable because we know only a very few—and, therefore, very precious—schemes whose unifying powers cross many realms.

12.9 THE EXCEPTION PRINCIPLE

What should one do with a law or rule that doesn't always work? We saw one way when we developed our uniframe for the *Block-Arch*. We simply kept changing it to fit each new example. But what if, after all that work, there still remain exceptions that don't fit?

> **The Exception Principle:** *It rarely pays to tamper with a rule that nearly always works. It's better just to complement it with an accumulation of specific exceptions.*

All children learn that birds can fly and that animals that swim are fish. So what should they do when told that penguins and ostriches are birds that cannot fly, or that whales and porpoises are animals that swim but aren't fish? What should the children do with uniframes that no longer work so well? The exception principle says: Do not change them too hastily. We should never expect rules to be perfect but only to say what is typical. And if we try to modify each rule, to take each exception into account, our descriptions will become too cumbersome to use. It's not so bad to start with "Birds can fly" and later change it into *"Birds can fly, unless they are penguins or ostriches."* But if you continue to seek perfection, your rules will turn into monstrosities:

> *Birds can fly, unless they are penguins and ostriches, or if they happen to be dead, or have broken wings, or are confined to cages, or have their feet stuck in cement, or have undergone experiences so dreadful as to render them psychologically incapable of flight.*

Unless we treat exceptions separately, they'll wreck all the generalizations we may try to make. Consider why the commonsense idea of *fish* is so useful. It is an accumulation of general information about a class of things that have much in common: animals that live in the water, have a certain sort of streamlined shape, and move by wriggling their bodies and fanning the water with various finlike appendages. However, a biologist's idea of *fish* is very different, being more involved with the origins and internal mechanisms of those animals. This leads to emphasizing aspects less evident to the eye: if whales have lungs where trout have gills, they must be uniframed in different ways. Children are disturbed to hear that whales are not fish because they are usually more concerned with uses and appearances than with origins and mechanisms. They're more likely to want classifications like these:

> *What do those animals do? Where do they live? Are they easy to catch? Are they dangerous? Are they useful? What do they eat? How do they taste?*

The power of ordinary words like "fish" comes from how we make them span so many meaning-worlds at once. However, in order to do this, we have to be able to tolerate many exceptions. We almost never find rules that have no exceptions—except in certain special, artificial worlds that we ourselves create by making up their rules and regulations to begin with. Artificial realms like mathematics and theology are built from the start to be devoid of interesting inconsistency. But we must be careful not to mistake our own inventions for natural phenomena we have discovered. To insist on perfect laws in real life is to risk not finding any laws at all. Only in the sciences, where every exception must be explained, does it make sense to pay that price.

Why can one build a tower or arch of stone or brick, but not of water, air, or sand? To answer that amounts to asking, *"How do towers work?"* But asking that seems quite perverse, because the answer seems so obvious: *"Each block holds the next one up, and that's all there is to it!"* As we've said before:

>*An idea will seem self-evident—once you've forgotten learning it!*

We often use words like "insight" or "intuition" to talk about understandings that seem especially immediate. But it is bad psychology to assume that what seems "obvious" is therefore simple or self-evident. Many such things are done for us by huge, silent systems in our mind, built over long forgotten years of childhood. We rarely think about the giant engines we've developed for understanding space, which work so quietly that they leave no traces in our consciousness. How towers work is something everyone has known for so long that it seems odd to mention it:

>*A tower's height depends only upon the heights of its parts! None of the other properties of the blocks matter—neither what they cost, nor where they've been, nor what you think of them. Only lifting counts—so we can build a tower by thinking only about actions that increase its height.*

This makes tower building easy, by letting us separate the basic building plan from all the small details. *To build a tower of a certain height, just find enough "pieces of height" and stack them up by lifting actions.* But towers have to stay up, too. So the next problem is to find actions we can take to make our tower stable. Here we can use a second, wonderful principle:

>*A tower is stable if each block is properly centered on the last. Because of this, we can build a tower by first lifting each block vertically and then adjusting it horizontally.*

Notice that this second kind of action—adjusting for stability—requires only horizontal movements, which do not affect the tower's height at all. This explains why towers are so easy to build. To increase a tower's height, you need only vertical lifting actions. The second-rank goal, stability, requires only horizontal sliding motions, which don't interact with height at all —provided the blocks have horizontal surfaces. This lets us achieve our tower-building goal simply by doing "first things first."

To adults it is no mystery that height and width are independent of each other. But this is not so evident in infancy: to understand the world of space and time, each child must make many such discoveries. Still, the division into Lifting and Sliding has a special importance; there are an infinity of ways to move around inside the world: how could a person ever learn them all? The answer: We don't need to learn them all, because we can learn to deal with each dimension separately. Lifting has a special significance because it isolates the vertical dimension from the others and relates it to ideas about balancing. The complementary operations of Sliding can then be divided into two remaining dimensions: either to push and pull or to move from side to side. One way to Lift and two ways to Slide makes three—and that is just enough to move around in a three-dimensional world!

12.11 HOW CAUSES WORK

It's wonderful when we can find out something's cause. A tower is high because each of its separate blocks contribute height; it stands because those blocks are adequately firm and wide. A baby cries because it wants food. A stone falls down because it's pulled by gravity. Why can we explain so many things in terms of causes and effects? Is it because there is a cause for everything—or do we merely learn to ask only about the kinds of happenings that have causes? Is it that causes don't exist at all but are inventions of our minds? The answer is all of the above. Causes are indeed made up by minds—but only work in certain parts of certain worlds.

What are causes, anyway? The very concept of a cause involves a certain element of style: a causal explanation must be brief. Unless an explanation is compact, we cannot use it as a prediction. We might agree that X causes Y, if we see that Y depends more on X than on most other things. But we wouldn't call X a cause of Y if describing X involved an endless discourse that mentioned virtually everything else in the world.

> There can't be any "causes" in a world in which everything that happens depends more or less equally upon everything else that happens.

Indeed, it wouldn't make any sense to talk about a "thing" in such a world. Our very notion of a "thing" assumes some constellation of properties that stays the same (or changes in ways we can predict) when other things around it change. When *Builder* moves a block, that block's location will change—but not its color, weight, material, size, or shape. How convenient that our world lets us change a thing's location while leaving so many other properties unchanged! This lets us predict the effect of motions so well that we can chain them into combinations never tried before—yet still predict their principal effects. Furthermore, because our universe has three "dimensions," we can easily predict the effect of combining several actions from knowing only their effects in each of those three dimensions.

Why *does* a block retain its size and shape when it is moved? It is because we're fortunate enough to live within a universe in which effects are localized. A solid object with a stable shape can exist only because its atoms "stick together" so tightly that when you move some of them, the others are pulled along. But this can happen only in a universe whose force laws work in close accord with the "nearnesses" of time and space—in other words, a universe in which entities that are far apart have much less effect on each other than ones that are close together. In worlds without constraints like that, there could be no "things" or "causes" for us to know.

> To know the cause of a phenomenon is to know, at least in principle, how to change or control some aspects of some entities without affecting all the rest.

The most useful kinds of causes our minds can discern are predictable relationships between the actions we can take and the changes we can sense. This is why animals tend to evolve sensors that detect stimuli that can be affected by those animals' own actions.

12.12 MEANING AND DEFINITION

mean·ing *n. 1. that which exists in the mind, view, or contemplation
as a settled aim or purpose; that which is meant or intended to be done;
intent; purpose; aim; object. 2. that which is intended to be, or in
fact is, conveyed, denoted, signified, or understood by acts or language;
the sense, signification, or import of words; significance; force.*
—Webster's Unabridged Dictionary

What is a meaning? Sometimes we're told a definition of a word, and suddenly, we know a way to use that word. But definitions do not often work so well. Suppose you had to explain what "game" means. You could start like this:

> **GAME:** *An activity in which two teams compete to make a ball do something that results in a winning score.*

This fits a certain range of games—but what of games that just use words, or keep no scores, or lack the element of competition? We can capture the nature of more kinds of games by using other definitions, but nothing seems to catch them all. We simply cannot find much in common to everything we call a game. Yet one still feels there is a certain unity that underlies the idea of a game. For example, we feel that we could recognize new games, and that "game" is more than an arbitrary accumulation.

But now let's turn our attention away from the physical aspects of games and focus on the *psychological purposes* that games can serve. Then it is much easier to find some qualities that are common to most adult games:

> **GAME:** *An activity that is engaging and diverting, deliberately detached from real life.*

This second kind of definition treats a game, not as a kind of object-thing, but as a process in the mind. At first this might seem somewhat strange, but really it is nothing new—even our first definition already contained psychological elements, concealed in the words "competing" and "winning." When seen this way, different kinds of games seem much more similar. This is because they all serve common purposes—despite the great diversity of their physical appearances. After all, there is virtually no limit to the variety of physical objects or structures that could be used to accomplish the same psychological purpose—in this case, to make an activity diverting (whatever that might mean). Naturally, then, it would be hard to specify the range of all the possible physical forms of games.

Of course, it is no great surprise to find that "game" has a more psychological character than does "brick," which we can define in physical terms without referring to our goals. But most ideas lie in between. We saw this in the case of "chair," which we cannot describe without referring both to a physical structure and to a psychological function.

12.13 BRIDGE-DEFINITIONS

At last we're coming close to capturing the meanings of things like chairs and games. We found that structural descriptions are useful, but they always seem too specific. Most chairs have legs, and most games have scores—but there are always exceptions. We also found purposeful descriptions to be useful, but they never seemed specific enough. *"Thing you can sit upon"* is too general to specify a chair, since you can sit on almost anything. *"Diverting activity"* is too broad for game—since there are many other ways to turn our minds from serious things. In general, a single definition rarely works.

>**Purposeful definitions are usually too loose.**
> *They include many things we do not intend.*
>**Structural definitions are usually too tight.**
> *They reject many things we want to include.*

But we can often capture an idea by squeezing in from several sides at once, to get exactly what we need by using two or more different kinds of descriptions at the same time.

>*Our best ideas are often those that bridge between two different worlds!*

I don't insist that every definition combine just these particular ingredients of structure and purpose. But that specific mixture does have a peculiar virtue: it helps us bridge between the "ends" we seek and the "means" we have. That is, it helps us connect things we can recognize (or make, find, do, or think) to problems we want to solve. It would be of little use to know that X's "exist" without some way to find and use them.

When we discussed accumulation, we saw that the concepts of "furniture" and "money" have reasonably compact functional definitions but accumulate many structural descriptions. Conversely, the concepts of "square" or "circle" have compact structural definitions but accumulate endless collections of possible uses.

To learn to use a new or unfamiliar word, you start by taking it to be a sign that there exists, inside some other person's mind, a structure you could use. But no matter how carefully it is explained, you must still rebuild that thought yourself, from materials already in your own mind. It helps to be given a good definition, but still you must mold and shape each new idea to suit your own existing skills—hoping to make it work for you the way it seems to work for those from whom you learn.

What people call "meanings" do not usually correspond to particular and definite structures, but to connections among and across fragments of the great interlocking networks of connections and constraints among our agencies. Because these networks are constantly growing and changing, meanings are rarely sharp, and we cannot always expect to be able to "define" them in terms of compact sequences of words. Verbal explanations serve only as partial hints; we also have to learn from watching, working, playing—and thinking.

SEEING AND BELIEVING

Cezanne said, "Though the world appears
Complex, it's made of cubes and spheres,
Along with cylinders and cones:
Four fundamentals that, like bones
In flesh, uphold whatever drapes
Variety upon their shapes."

"They're doubly basic," Freud said. "These
Are more than just geometries:
Your simple solids symbolize
The organs that attract our eyes;
The only subject of the arts
Is men's and women's private parts."

The body can as well express
Our sadness and our happiness
And even sex's mindless dance
Portrays the spirit's circumstance
Of oscillation to and fro
Between the cosmic Yes and No.

"The world," van Gogh said, "is a face
In which I see my soul's grimace."
But has reality become
Merely emotion's medium?
O universe of forms, I ask
Are you a mirror, or a mask?

—THEODORE MELNECHUK

13.1 REFORMULATION

Imagine all the kinds of arches one can build.

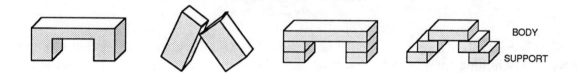

How could we capture what's common to so many things with just one single uniframe? Impossible—if we were forced to think of them in terms of blocks and how they're placed. Not one of the expressions we used before applies to all of them: neither "three blocks," nor "two blocks standing up," nor "the supports must not touch." How could we make our minds perceive all these arches as the same? One way would be to draw this imaginary line:

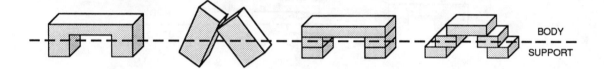

Now, suddenly, all those different arches fit one single frame—of a single Body with two Supports. There are two different ideas here. The first is the idea of dividing an object's description into an "essential" portion, namely the "body," and some auxiliary portions, which correspond to the support. Later we'll see that this is a powerful idea in its own right. The second idea is even more powerful and general: after failing to find a unified description of all those arches, we abandoned the method we were using—and adopted, instead, a quite different style of description. In a word, we *reformulated* the problem in new terms. We started by using a language that was based on expressing the precise shapes of individual blocks. We replaced this by another language in which we can speak of shapes and outlines that are not confined to those of the blocks themselves.

Reformulation is clearly very powerful—but how does one do it? How do people find new styles of description that make their problems seem easier? Does this depend upon some mysterious kind of insight or upon some magically creative gift—or do we simply come upon them by accident? As I said when discussing creativity, these seem to me mere matters of degree, since people are always making reformulations of various sorts. Even when we contemplate those rarest and most revolutionary new ideas that come like revelations, suddenly to shed new light on entire fields of thought—like evolution, gravity, or relativity—we usually see by hindsight that these were variants of things that people knew before that time. Then we have to ask, instead, for reasons why those reformulations were so long postponed.

13.2 BOUNDARIES

In the sky there is no distinction of east and west;
people create distinctions out of their own minds and then
believe them to be true.
—BUDDHA

What is creativity? How do people get new ideas? Most thinkers would agree that some of the secret lies in finding "new ways to look at things." We've just seen how to use the Body-Support concept to reformulate descriptions of some spatial forms, and soon we'll see some other ways to reformulate in terms of strength, containment, cause, and chain. But first let's look more carefully at how we made those four different arches seem the same, by making each of them seem to match *"a thing supported by two legs."* In the case of *Single-Arch*, we did this by imagining some boundaries that weren't really there: this served to break a single object into three.

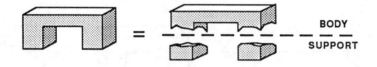

However, we dealt with *Tower-Arch* by doing quite the opposite: we treated some real boundaries as though they did not exist:

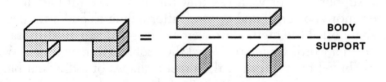

How cavalier a way to treat the world, to see three different things as one and to represent one thing as three! We're *always* changing boundaries! Where does an elbow start or end? When does a youth become an adult? Where does an ocean change into a sea? Why must our minds keep drawing lines to structure our reality? The answer is that unless we made those mind-constructed boundaries, we'd never see any "thing" at all! This is because we rarely see anything twice as exactly the same. Each time we're almost certain to be looking from a somewhat different view, perhaps from nearer or farther, higher or lower, in a different color or shade of light, or against a different background. For example, consider these two appearances of the same table.

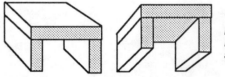

These are quite different when described in terms of the actual lines and surfaces. But when described in terms of body and support, both pictures are the same!

Unless the mind could thus discard the aspects of each scene that are not essential to its present purposes, we could never learn anything. Otherwise, our recollections would rarely match appearances. Then nothing could make any sense—since nothing would seem permanent.

13.3 SEEING AND BELIEVING

A child was asked to draw a person.

Where is the body? Why are the arms and legs connected to the head?

When questioned, many young children actually prefer these to the drawings most adults like.

We normally assume that children see the same as we do and only lack our tricky muscle skills. But that doesn't explain why so many children produce this particular kind of drawing, nor why they seem so satisfied with them. In any case, this phenomenon makes it seem very unlikely that a child has a realistic, picturelike "image" in mind.

Now let's consider a different idea. We'll suppose that the child does not have anything like a picture in mind, but only some network of relationships that various "features" must satisfy. For example, a child's "person-drawing" feature-network might consist of the following features and relations:

HEAD	*Large closed figure.*
EYES	*Two circles, high in head.*
MOUTH	*Object centered below eyes.*
BODY	*Large closed figure.*
ARMS	*Two lines, attached high on body.*
LEGS	*Two lines, attached low on body.*

To convert this description into an actual drawing, the child must employ some sort of "drawing procedure." Here's one in which the process simply works its way down the feature list, like a little computer program:

1. *Consider the next feature on the list.*
2. *IF such a feature is already drawn, go to step 3. Otherwise draw it.*
3. *IF list is finished, stop. Otherwise, go back to step 1.*

When the child starts to draw, the first item on the list is "large closed figure." Since there isn't any such thing yet, the child draws one: that's the head. Next the eyes and mouth get drawn. But then, when it comes to drawing the body feature, step 2 of the procedure finds that a "large closed figure" has already been drawn. Accordingly, nothing new is required, and the procedure simply advances to step 3. As a result, the child goes on to attach the arms and legs to the feature that has been assigned to both the body and the head.

An adult would never make such a "mistake," since once some feature has been assigned to represent a head, that feature is thereafter regarded as "used up" or "occupied" and cannot represent anything else. But the child has less capacity or inclination for "keeping track." Accordingly, since that "large closed figure" satisfies the description's requirements for both the head and the body—albeit at different moments of time—there is no cause for discontent. The little artist has satisfied all the conditions required by its description!

13.4 CHILDREN'S DRAWING-FRAMES

That body-head drawing seems very wrong to most adults, yet it seems to please many children. *Does it really look like a person to those children?* That seems like a simple question, but it is not—for we must remember that a child is not a single agent and that various other agencies inside a child's mind may not be satisfied at all. At the moment, those other agencies are not in control and have little effect. Yet if some creature came on the scene that really looked like that, most children would be terrified. It does not make much sense to speak of what a person "really" sees, because we have so many different agencies.

Older children tend to draw more distinct body parts and other features such as fingers and toes, hair, eyebrows, and clothes.

What happened in the intervening years to make the older children draw the body separately? This could come about without even making any change in the list of features and relations we used in the previous section. It would need only a small change in step 2 of our drawing procedure:

1. *Consider the next feature on the list.*
2. **Draw such a feature, even if a similar one has already been drawn.**
3. *IF list is finished, stop. Otherwise, go back to step 1.*

This ensures that every feature mentioned in the list will be represented only once in the drawing, even if two such features look alike. Of course, this requires some ability to count each feature only once, and never twice. How interesting that in order to make mature, realistic drawings, the child could exploit the same kind of ability it must acquire in order to count things properly!

To be sure, we could explain the child's progress in other ways. That new and "more realistic" picture could come from adding a neck to the feature list, for that *would* demand a separate body and head. It might suffice simply to impose an additional constraint or relationship: *that the head be above the body.* One might argue that the younger child never had a clear concept of a separate and distinct body feature in the first place; after all, there are many things that you can do with your arms and legs or with your head—but your body only gets in the way.

In any case, after mastering the art of making these body-head drawings, many children seem to progress rather slowly in the art of making personal portraits, and these types of "childish" drawings often persist for some years. I suspect that after children learn to make recognizable figures, they usually move on to face the problems of representing much more complicated scenes. As they do this, we should continue to appreciate how well children deal with the problems they set for themselves. They may not meet our own grown-up expectations, but they often solve their own versions of the problems we pose.

13.5 LEARNING A SCRIPT

An expert is one who does not have to think. He knows.
—FRANK LLOYD WRIGHT

What will our portrait-drawing child try next? Some children keep working to improve their person pictures. But most of them go on to put their newfound skills to work at drawing more ambitious scenes in which two or more picture-people interact. This involves wonderful problems about how to depict social interactions and relationships—and these more ambitious projects lead the child away from being concerned with making the pictures of the individual more elaborate and realistic. When this happens, the parent may feel disappointed at what seems to be a lack of progress. But we should try to appreciate the changing character of our children's ambitions and recognize that their new problems may be even more challenging.

This doesn't mean that drawing learning stops. Even as those children cease to make their person pictures more elaborate, the speed at which they draw them keeps increasing, and with seemingly less effort. How and why does this happen? In everyday life, we take it for granted that "practice makes perfect," and that repeating and rehearsing a skill will, somehow, automatically cause it to become faster and more dependable. But when you come to think of it, this really is quite curious. You might expect, instead, that the more you learned, the *slower* you would get—from having more knowledge from which to choose! *How does practice speed things up?*

Perhaps, when we practice skills we can already perform, we engage a special kind of learning, in the course of which the original performance process is replaced or "bridged-across" by new and simpler processes. The "program" to the left below shows the many steps our novice portrait drawer had to take in order to draw each childish body-face. The "script" to the right shows only those steps that actually produce the lines of the drawing; this script has only half as many steps.

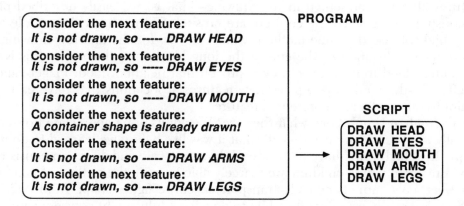

The people we call "experts" seem to exercise their special skills with scarcely any thought at all—as though they were simply reading preassembled scripts. Perhaps when we "practice" to improve our skills, we're mainly building simpler scripts that don't engage so many agencies. This lets us do old things with much less "thought" and gives us more time to think of other things. The less the child has to think of where to put each arm and leg, the more time remains to represent what that picture-person is actually doing.

13.6   THE FRONTIER EFFECT

We can get more insight about children from another experiment of Piaget's. A child is shown a short block resting on a longer one and is asked to draw the scene. Next the child is asked to draw a sketch of what might happen if we pushed the upper block a little to the right. At first, the result is more or less what we'd expect.

Original   Moved to right

But when we ask the child to do the same repeatedly, we see a strange result. The top block suddenly grows shorter as it meets the edge of the long block!

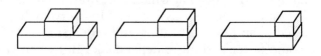

To understand what happened, just put yourself in the child's place. You've started to draw the upper edge of the short box, but how do you decide where to stop?

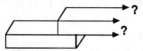

Younger children don't yet possess much ability to draw lines in good proportion. Instead, they tend to use procedures that locate each new feature in some recognizable relationship to other features already represented in the drawing—that is, to "easily described places" that have previously been depicted. Since there are no such features near the middle of the long block, the child will use the same method, whatever it is, for the first few drawings. But it is easy to describe the location of the end of the long block, and that's why this is where the younger children tend to stop, once they approach that neighborhood. Piaget called this the "frontier effect"—the tendency to place new features at locations that have easily described relationships to other, already represented features.

Why can't children simply copy what they see? We adults don't appreciate how complicated copying really is—because we can't recall what it was like before we learned to do it. To make a good copy, the child would have to draw each line to a scale and direction consistent with all the others. But these young children are scarcely able to trace the outline of an object with a finger; they certainly cannot mentally transport an entire figure shape from one location to another. So it is actually easier for the child to do what adults might consider more "abstract": first to construct a mental description of the relations involved in the scene, and then to design a drawing-scheme to represent those relationships. *It can require more skill to produce what we regard as a simple copy or imitation than to produce what we consider to be an "abstract" representation!*

13.7 DUPLICATIONS

Sometimes when we watch a scene, the wholes we see "add up" exactly to the sum of their separate parts. But other times, we do not mind if certain things get counted twice. In the first figure below, we divide an arch into body and support with no concern that both parts have exactly the same boundaries. In the second figure, we seem to see two complete arches, despite the fact that there aren't enough legs to make two *separate* arches.

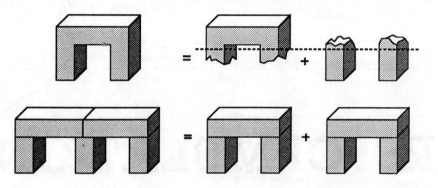

Sometimes it is vital to keep track, to count each thing exactly once. But in other situations, no harm will come from counting twice. It is efficient to use the same block twice when making viaducts for roads for cars. But if we tried to use the same five blocks to build two separate bridges, we'd end up short. Different kinds of goals require different styles of description. When we discussed the concept of *More*, we saw how each child must learn when to describe things in terms of appearance and when to think in terms of past experience. The double-arch problem also offers a choice of description styles. If you plan to build several separate things, you'd better keep count carefully or take the risk of running out of parts. But if you do that *all* the time, you'll miss your chance to make one object serve two purposes at once.

We could also formulate this as a choice between structural and functional descriptions. Suppose we tried to make a match between the structural elements of the double-arch and those of two separate block-arches. In one way to do that, we would first assign three blocks to each arch—and then verify that each arch consists of a top supported by two blocks that do not touch. Then, of course, we'll only find a single three-block arch. There isn't any second arch, since there are only two blocks left.

On the other hand, we could base our description of the double-arch scene on the more functional body-support style of description. According to that approach, we must focus first on the most "essential" parts. The most essential part of an arch is its top block—and we do indeed find two blocks that could serve as top blocks. Then we need only verify that each of them is supported by two blocks that do not touch—and this is indeed the case. In a function-oriented approach, it seems natural to count carefully the most essential elements, but merely to verify that the functions of the auxiliary elements will somehow be served. The functional type of description is easier to adapt to the purposes of higher-level agencies. This doesn't mean that functional descriptions are necessarily better. They can make it hard to keep track of real constraints; hence they have a certain tendency to lead toward overoptimistic, wishful thought.

CHAPTER 14

REFORMULATION

*Until that scholastic paradigm [the medieval 'impetus' theory]
was invented, there were no pendulums, but only swinging
stones, for scientists to see. . . . Do we, however, really need to
describe what separates Galileo from Aristotle, or Lavoisier from
Priestly, as a transformation of vision? Did these men really see
different things when looking at the same sorts of objects? . . .
I am acutely aware of the difficulties created by saying that
when Aristotle and Galileo looked at swinging stones, the first
saw constrained fall, the second a pendulum. Nevertheless,
I am convinced that we must learn to make sense of sentences
that at least resemble these.*

—THOMAS KUHN

14.1 USING REFORMULATIONS

What can we do when we can't solve a problem? We can try to find a new way to look at it, to describe it in different terms. Reformulation is the most powerful way to attempt to escape from what seems to be a hopeless situation. Thus, when we couldn't find anything common to all those different kinds of arches, we changed our way of looking at them. We moved from the world of rigid, geometric block descriptions to a less constrained domain of body-support descriptions—and there we found a way to make a uniframe for all of them: a span supported by a pair of legs. But think of all the other ways a person might describe an arch.

Aesthetic: *A pleasing, shapely form.*
Dynamical: *The top will fall if either leg is removed.*
Topological: *The arch surrounds a hole in space.*
Geometrical: *The three blocks form an "inverted U" shape.*
Architectural: *The arch's top could be the base of something else.*
Constructional: *Place two blocks, then place another across their tops.*
Circumventional: *Can be used as a detour, to go around an obstacle.*
Transportational: *Can be used as a bridge, to go from one place to another.*

Each of these involves a different "realm" of thought with its own style for describing things. And every different realm of thought can bring new kinds of skills to bear on a problem. We each learn various ways to reason about paths and obstacles; we each learn ways to deal with vertical support; with doors and windows; with boxes and bridges and tunnels; with stacks and rows and stairs and ramps.

To an outsider, it may seem that a creative inventor (or designer or thinker) must possess an endless source of novel ways to deal with things. Yet inside the inventor's mind, that all might stem from variations on far fewer themes. Indeed, in that inventor's view, those styles of thought may seem so clear (and those inventions all so similar) that the question turns the other way: *"Why can't outsiders understand how to think about these simple kinds of problems?"* In the long run, the most productive kinds of thought are not the methods with which we solve particular problems, but those that lead us to formulating useful new kinds of descriptions.

How do we reformulate? Each new technique presumably begins by exploiting methods already learned in other, older agencies. So new ideas often have roots in older ones, adapted for new purposes. In the next section, we'll see how that body-support idea has counterparts in virtually every realm of thought. Toward the end of this book, we'll speculate about how those various realms themselves evolve inside the mind.

14.2 THE BODY- SUPPORT CONCEPT

We were able to uniframe many kinds of arches by dividing each into a Body and a Support. See how well that technique works on many other sorts of things.

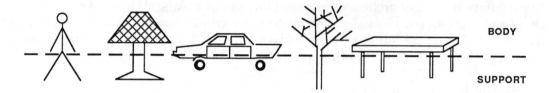

What makes such simple cuts seem meaningful? It is because we can imagine purposes for each. In everyday life, there is a special significance to dividing a table into "top" and "legs." This is because the tabletop serves our principal use for a table, as *thing to put things on.*" The table's legs serve only secondary purposes: without those legs, the top would fall—but without its top, the table has no use at all. And it would make no sense to imagine dividing that table in half, vertically, to see it as *two stuck-together, L-shaped parts.*

This must be one reason why the body-support idea seems so universal. It is not merely a matter of physical support: the more profound idea is that of building a mental bridge between a thing and a purpose. This is why bridge-definitions are so useful: they help us connect structural descriptions to psychological goals. But the point is that it is not enough just to link together descriptions from two different worlds—"*top supported by legs*" and "*thing to put things on.*" It is not enough simply to know that tables keep things off the floor. To use that knowledge, we must also know *how* it is done: that things have to be put *on* the table, rather than, for example, between the table's legs.

This is where the body-support representation helps us to classify our knowledge. The "body" represents those parts of a structure that serve as the direct instrument for reaching the goals and the "support" represents all the other features that merely serve that instrument. Once we can classify the tabletop as the "body" of the table, we will tend to think only of using the tabletop for keeping things off the floor. Of course, we would gain even more power by understanding *how* those supports assist the body's goal; that is, by understanding that the table's legs are for keeping the tabletop itself away from the floor. A good way to understand *that* is to have a representation of what might happen if one of the table's legs failed to perform its function.

Difference in Support Structure	Functional Effect on Body
Remove left leg.	Left side of top falls.
Remove right leg.	Right side of top falls.

To understand how something works, it helps to know how it can fail.

14.3 MEANS AND ENDS

How do we connect the things we have with the goals we want to achieve? The answer: We have many ways! Each use or purpose may suggest some corresponding way to split things up —and in each such view there will seem to be some "most essential" parts. These are the ones that, in such a view, appear to serve the goal directly; the rest will seem like secondary parts that only support the role of the main parts. We do this not only in the physical realm, but in many other realms as well.

```
FUNCTION        CONCLUSION              BODY
    |        ENDS     |       EFFECT     |
STRUCTURE      |   PREMISE      |    SUPPORT
            MEANS              CAUSE
```

Each of these dumbbell distinctions has its own style for distinguishing essential parts from supportive parts. And even in the world of physical things, we can apply these different mental views in different ways. For example, there are many ways to describe the act of standing on a table in order to be able to reach higher.

Support *Tables hold things away from the floor.*
Function *Tables are for supporting things.*
Conclusion *If you put something on a table, its height increases.*
Cause-Effect *I can reach higher because I start higher.*
Means-Ends *If I want to reach higher, I can stand on a table.*

Even when we simply put something on a table, we're likely to employ several such descriptions at the same time—perhaps in different sections of the mind. The quality of our understanding depends upon how well we move between those different realms. In order to translate easily from one of them to another, we must discover systematic cross-realm correspondences. However, finding these is rare. Usually, the situation is like that we found for chairs and games: each description-element in one world corresponds to a hard-to-describe accumulation of structures in the other world. What is remarkable about the body-support concept is how often it leads to systematic cross-realm correspondences. For example, we can use it to translate "*supported by,*" in the architectural realm, into "*horizontal surface underneath*" in the geometrical realm. To be sure, this correspondence fails to represent the possibility of supporting an object by suspending it from above. But some exceptions are inevitable.

Our systematic cross-realm translations are the roots of fruitful metaphors; they enable us to understand things we've never seen before. When something seems entirely new in one of our description-worlds, it may turn out that when translated to some other world it resembles something we already know.

Now, before you turn to the following page, try to solve this puzzle.

Puzzle: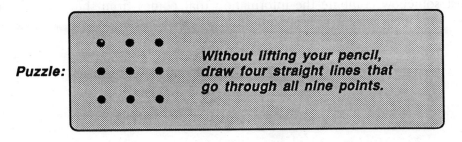

14.4   SEEING SQUARES

Most people find the nine-dot problem hard to solve because they assume that the dots form a square that bounds the working space. Indeed, the problem is insoluble unless the drawing can extend outside that area. Thus the problem is easier if one does not perceive those dots as forming a square. We often self-impose assumptions that make our problems more difficult, and we can escape from this only by reformulating those problems in ways that give us more room.

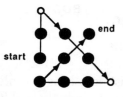

The difficulty of this problem seems more ethical than conceptual. It feels like "cheating" to go outside the square.

Really, there was never any square in the first place—that is, in the literal sense of "a rectangle with equal sides." What makes us see so many different sorts of things as though they were squares?

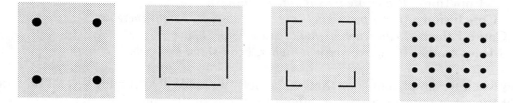

Some of these squares have no corners, others have no edges, and some of them have neither corners nor edges! What makes us see them all as squares? Psychologists have long wondered how we recognize such similarities but often forgot to ask how we recognize the very simplest forms of squares in the first place. Which comes first in recognition, specific features or global shapes? It must depend upon one's state of mind. The way we perceive the world, from one moment to another, depends only in part on what comes from our eyes: the rest of what we think we see comes from inside our brain; we respond not only to visual features, but also to our remembrances of things we've seen before and to our expectations of what we ought to see.

It is tempting to assume that our visual processes work only in one direction, bringing information from the world into the mind:

World ⟶ Sensation ⟶ Perception ⟶ Recognition ⟶ Cognition

But this does not explain how what we see is influenced by what we expect to see. Human vision must somehow combine the information that comes from the outer world with the structures in our memories. The situation must be more like this:

Sensation ⟶ Description ⟵ Expectation

14.5 BRAINSTORMING

Once you set that nine-dot problem into a larger frame, you could solve it in a routine way, with only a little thought. What lets you reformulate such complex scenes so easily—once you've thought of doing it? It must be that your mind is constantly preparing ways to do such things by building up connections between different kinds of descriptions. Then, when you finally change your view to find another way to look at things, you can apply a lifetime of experience as easily as turning on a switch.

This takes us back to the question of when to try to be a Reductionist or a Novelist. How do you decide when to quit after investing a great deal of effort in doing something a certain way? It would be bad to discard all that work just before you might find the answer—but there's no way to be sure when that will happen. Should people always try to break their well established, self-made mental bonds and try to think in less constricted ways? Of course not. It will usually do more harm than good.

However, when you're really stuck, you may as well try wilder ways to find some new ideas. You might even consider using one of the systematic, therapylike disciplines that go under names like *brainstorming, lateral thinking, meditation,* and so forth. These can help, when people get badly stuck, by encouraging the search for new formulations. However, when you switch to unfamiliar views of things you may get new ideas, but you also put yourself in the situation of a novice; you become less able to judge which new ideas are likely to be compatible with any of your older skills.

In any case, you must not be too quick to think, *How stupid it was not to see that right away!* Remember the principle of exceptions: it may be rash to change yourself too much just to accommodate a single strange experience. To take every exception seriously is to risk the loss of general rules that previous experience has shown to work most frequently. You must also be particularly wary of methods you can *always* use:

> *Quit what you're doing.*
> *Find an easier problem.*
> *Take a rest. You'll feel better afterward.*
> *Simply wait. Eventually the situation will change.*
> *Start over again. Things may work out better the next time.*

Such methods are *too* general; they're things that one can always do, but they do not apply especially well to any particular problem. Sometimes they can help us get unstuck, but they must be barred from usual thought—or at least be given low priority. It isn't any accident that the things that we can "always" do are just the ones we should rarely do.

14.6 THE INVESTMENT PRINCIPLE

To him that has, more shall be given; but from him that has not,
the little that he has shall be taken away.
—ST. MATTHEW

Some ideas acquire undue influence. The prominence of the body-support idea is well-deserved; no other scheme compares to its ability to help us link things into causelike chains. But there are other, not so honorable ways for ideas to gather influence.

> **The Investment Principle:** *Our oldest ideas have unfair advantages over those that come later. The earlier we learn a skill, the more methods we can acquire for using it. Each new idea must then compete against the larger mass of skills the old ideas have accumulated.*

This is why it's so much easier to do new things in older ways. Each new idea, however good in principle, seems awkward until we master it. So old ideas keep gaining strength, while new ones can rarely catch up. Furthermore, our oldest and best-developed skills will be the first to spread to other realms of thought where again they'll start out far enough ahead to keep any new ideas from taking root.

In the short run, you will usually do better by using an old idea than by starting out anew. If you can already play the piano well, it is easy to start playing the organ in the same way. The many superficial similarities will make it hard for you to tell which aspects of your old skills are unsuitable, and the easiest course is to keep applying your old technique, trying to patch each flaw until none show. In the long run, you'd probably do better by starting fresh with a new technique—and then borrowing what you can from your older skills. The trouble is that *we're almost always immersed in the "short run."* So the principles both of investment and of exception make us reluctant to tamper with our well-established skills and uniframes lest we endanger all that we have built upon those old foundations. I don't mean to say there's anything wrong, in principle, with using what you are comfortable with and already know. But it is dangerous to support your old ideas merely by accumulating ways to sidestep their deficiencies. That only increases the power of your old ideas to overcome new ones and could lead your style of thought to base itself yet all the more, as time goes by, upon less and less.

Evolution illustrates how processes can become enslaved by the investment principle. Why do so many animals contain their brains inside their heads—as with fish, amphibians, reptiles, birds, and bats? This arrangement was inherited long before our earliest aquatic ancestor first crawled upon the land three hundred million years ago. For many of those animals—wood-peckers, for example—another arrangement might serve at least as well. But once the pattern of centralizing so many functions in the head was established, it carried with it great networks of dependencies involving many aspects of anatomy. Because of this, any mutation that changed any part of that arrangement would disrupt many other parts and lead to dreadful handicaps, at least in the short run of evolution. And because evolution is so inherently short-sighted, it would not help if, over longer spans of time, such changes could lead to advantages. Perhaps the best example of this can be seen in the fact that virtually every detail of every plant and animal on earth is written in terms of a genetic code that has scarcely changed a single bit in a billion years. It does not seem to be a particularly efficient or reliable code, yet so many structures have been based on it that all living things are stuck with it! To change a single detail of that code would cause so many proteins to get tangled up that not a single cell could live.

14.7 PARTS AND HOLES

As an example of reformulation, we'll represent the concept of a box in the form of a machine that has a goal. We can use this to understand the *Hand-Change* phenomenon. What makes a *Block-Arch* trap a person's arm so that there's no way to escape except to withdraw? One way to explain this is to imagine the arch as made of four potential obstacles—that is, if we include the floor.

An obstacle is an object that interferes with the goal of moving in a certain direction. To be trapped is to be unable to move in any acceptable direction. Why does the block-arch form a trap? The simplest explanation is that each of its four sides is a separate obstacle that keeps us from escaping in a certain direction. (For our present purposes, we'll regard moving the hand forward or backward as unacceptable.) Therefore we're trapped, since there are only four acceptable directions—up, down, left, or right—and each of them is separately blocked. Psychologically, however, there's something missing in that explanation: it doesn't quite describe our sense of being trapped. When you're caught inside a box, you feel as though something is trying to keep you there. The box seems more than just its separate sides; you don't feel trapped by any particular side. It seems more like a conspiracy in which each obstacle is made more effective because of how all the other obstacles work together to keep you from going around it. In the next section we'll assemble an agency that represents this active sense of frustration by showing how those obstacles cooperate to keep you in.

In order to represent this concept of trap or enclosure, we'll first need a way to represent the idea of a container. To simplify matters, instead of trying to deal with a genuine, six-sided, three-dimensional boxlike container, we'll consider only a two-dimensional, four-sided rectangle. This will let us continue to use our *Block-Arch* uniframe, together with that extra side to represent the floor.

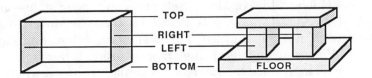

Why focus so sharply on the concept of a container? Because without that concept, we could scarcely understand the structure of the spatial world. Indeed, every normal child spends a great deal of time learning about space-surrounding shapes—as physical implements for containing, protecting, or imprisoning objects. But the same idea is also important not only physically, but psychologically, as a mental implement for envisioning and understanding other, more complicated structures. This is because the idea of a set of "all possible directions" is one of the great, coherent, cross-realm correspondences that can be used in many different realms of thought.

14.8 THE POWER OF NEGATIVE THINKING

When life walls us in, our intelligence cuts an opening, for, though
there be no remedy for an unrequited love, one can win release from
suffering, even if only by drawing from the lessons it has to teach.
The intelligence does not recognize in life any closed situations
without an outlet.

—MARCEL PROUST

How do boxes keep things in? Geometry is a fine tool for understanding shapes, but alone, it can't explain the *Hand-Change* mystery. For that, you also have to know how *moving* works! Suppose you pushed a car through that *Block-Arch*. Your arm would be imprisoned until you pulled it out. How can you comprehend the cause of this? The diagram below depicts an agency that represents the several ways an arm can move inside a rectangle. The top-level agent *Move* has four subagents: *Move-Left*, *Move-Right*, *Move-Up*, and *Move-Down*. (As before, we'll ignore the possibility of moving in and out, in three dimensions.) If we connect each of these sub-agents to the corresponding side of our four-sided box frame, each agent will be able to test whether the arm can move in the corresponding direction (by seeing whether there is an obstacle there). Then, if every direction is blocked, the arm can't move at all—and that's what we mean by being "trapped."

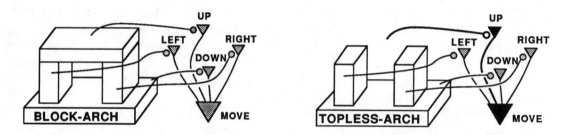

The "---o" symbol indicates that each box-frame agent is connected to *inhibit* the corresponding subagent of *Move*. An obstacle to the left puts *Move-Left* into a can't-move state. If all four obstacles are present, then all four box-frame agents will be activated; this will inhibit all of *Move*'s agents—which will leave *Move* itself in a can't-move state—and we'll know that the trap is complete. However, if we saw a *Topless-Arch*, then the *Move-Up* agent would not be inhibited, and *Move* would not be paralyzed! This suggests an interesting way to find an escape from a *Topless-Arch*. First you *imagine* being trapped inside a box-frame—from which you know there's no escape. Then, since the top block is actually missing, when your vision system looks for actual obstacles, there will be no signal to inhibit the *Move-Up* agent. Accordingly, *Move* can activate that agent, and your arm will move upward automatically to escape!

This method has a paradoxical quality. It begins by assuming that escaping is impossible. Then this pessimistic mental act—imagining that one's arm is trapped—leads directly to finding a way out. We usually expect to solve our problems in more positive, goal-directed ways, by comparing what we have with what we wish—and then removing the differences. But here we've done the opposite. We compared our plight, not with what we want, but with a situation even worse—the least desirable antigoal. Yet even that can actually help, by showing how the present situation fails to match that hopeless state. Which strategy is best to use? Both depend on recognizing differences and on knowing which actions affect those differences. The optimistic strategy makes sense when one sees several ways to go—and merely has to choose the best. The pessimistic strategy should be reserved for when one sees no way at all, when things seem really desperate.

14.9 THE INTERACTION-SQUARE

What's so special about moving left or right or up or down? At first one might suppose that these ideas work only for motions in a two-dimensional space. But we can also use this square-like frame for many other realms of thought, *to represent how pairs of causes interact*. What is an interaction, anyway? We say that causes interact if, when combined, they lead to effects that neither can cause separately. For example, by combining horizontal and vertical motions, we can move to places that can't be reached with either kind of motion by itself. We can represent the effects of such combinations by using a diagram with labels like those on a compass.

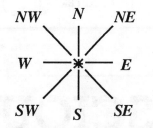

Many of our body joints can move in two independent directions at once—not the knee, but certainly the wrist, shoulder, hip, ankle, thumb, and eye. How do we learn to control such complicated joints? My hypothesis is that we do this by training little interaction-square agencies, which begin by learning something about each of the nine possible motion combinations. I suspect that we also base many of our nonphysical skills on interaction-square arrays *because that is the simplest way to represent what happens when two causes interact*. (There is even some evidence that many sections of the brain are composed of square arrays of smaller agencies.)

Consider that the *Spatial* agency in our Society-of-More is not really involved with space at all, but with interactions between agents like *Tall* and *Thin*. If you were told that one object **A** is both taller and wider than another object **B**, you could be sure that there is "more" of **A**. But if you were told that **A** is taller and thinner than **B**, you couldn't be sure which one is "more." An interaction-square array provides a convenient way to represent all the possible combinations:

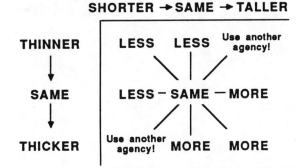

If square-arrays can represent how *pairs* of causes interact, could similar schemes be used with three or more causes? That might need too many "directions" to be practical. We'd need twenty-seven directions to represent three interacting causes this way, and eighty-one to represent four. Only rarely, it seems, do people deal with more than two causes at a time; instead, we either find ways to reformulate such situations or we accumulate disorderly societies of partially filled interaction-squares that cover only the most commonly encountered combinations.

CHAPTER 15

CONSCIOUSNESS AND MEMORY

But if this is true, are we not led into what philosophers call an infinite regress, the explanation of one thing in terms of an earlier which again needs the same type of explanation?
If Constable saw the English landscape in terms of Gainsborough's paintings, what about Gainsborough himself? We can answer this. Gainsborough saw the lowland scenery of East Anglia in terms of Dutch paintings which he arduously studied and copied. . . . And where did the Dutch get their vocabulary? The answer to this type of question is precisely what is known as the "history of art."

—E. H. GOMBRICH

15.1 MOMENTARY MENTAL STATE

We normally assume that consciousness is knowing what happens in our minds right at the present time. In the next few sections, I'll argue that consciousness does not concern the present, but the past: it has to do with how we think about the records of our recent thoughts. But how can thinking about thoughts be possible at all?

> There's something queer about describing consciousness: whatever people mean to say, they just can't seem to make it clear. It's not like feeling confused or ignorant. Instead, we feel we know what's going on but can't describe it properly. How could anything seem so close, yet always keep beyond our reach?

There is a simple sense in which thinking about a thought is not so different from thinking about an ordinary thing. We know that certain agencies must learn to recognize—and even name—the feel of touching a hand or an ear. Similarly, there must be other agencies that learn to recognize events *inside* the brain—for example, the activities of the agencies that manage memories. And those, I claim, are the bases of the awarenesses we recognize as consciousness.

There is nothing peculiar about the idea of sensing events inside the brain. Agents are agents—and it is as easy for an agent to be wired to detect a *brain-caused brain-event*, as to detect a *world-caused brain-event*. Indeed, only a small minority of our agents are connected directly to sensors in the outer world, like those that send signals from the eye or skin; most of the agents in the brain detect events inside the brain. But here we're especially concerned with agents that are engaged in using and changing our most recent memories. These lie at the roots of consciousness.

Why, for example, do we become less conscious of some things when we become more conscious of others? Surely this is because some resource is approaching some limitation—and I'll argue that it is our limited capacity to keep good records of our recent thoughts. Why, for example, do thoughts so often seem to flow in serial streams? It is because whenever we run out of room, the records of our recent thoughts must then displace those of our older ones. And why are we so unaware of how we get our new ideas? Because whenever we solve hard problems, our short-term memories become so involved with doing *that* that they have neither time nor space for keeping detailed records of what they themselves have done.

What happens when we try to think about our most recent thoughts? We examine our recent memories. But these were already involved in what we were "thinking" about—*and any self-inspecting probe is prone to change just what it's looking at.* Then the system is likely to break down. It is hard enough to describe something with a stable shape; it is even harder to describe something that changes its shape before our eyes; and it is virtually impossible to speak of the shapes of things that change into something else each time we try to think of them. And that's what happens when we try to think about our present thoughts—since each such thought must change our mental state! Would any process not become confused that alters what it's looking at? In such a fix, how could one ever hope to be articulate?

15.2 SELF-EXAMINATION

What do we mean by words like "sentience," "consciousness," or "self-awareness"? They all seem to refer to the sense of feeling one's mind at work—but beyond that, it is hard to say whether there are any differences in what they mean. For instance, suppose that you had just smiled, and someone asked if you had been conscious of this. It would scarcely matter how that question was posed:

> *"Did you just smile?"*
> *"Do you realize that you just smiled?"*
> *"Do you remember smiling?"*
> *"Were you conscious of doing so?"*
> *"Were you aware of it?"*

Each of these questions really asks what you can say about your recent mental past. In order for you to reply truthfully, *"Yes, I know I smiled,"* your speaking-agencies must use some records about the recent activity of certain agents. But what about all the other activities involved in everything you say and do? If you were truly self-aware, wouldn't you know all those other things as well? There is a common myth that what we view as consciousness is measurelessly deep and powerful—yet actually, we scarcely know a thing about what happens in the great computers of our brains. How can we think, not knowing what it is to think? How can we get such good ideas, yet not be able to say what ideas are or how they're made?

Why is it so hard to talk about our present state of mind? We've already seen several reasons for this. One is that the time-delays between the different parts of a mind mean that the concept of a "present state" is not psychologically sound. Another reason is that each attempt to reflect upon our mental state will change that state, and this means that trying to know our state is like photographing something that is moving too fast: such pictures will always be blurred. In any case, we aren't much concerned in the first place with learning how to describe our mental states; instead, we're more engaged with practical things, like making plans and carrying them out.

How much genuine self-insight is possible for us? I'm sure our memory-machinery provides some useful clues, if only we could learn to interpret them. But it is unlikely that any part of the mind can ever obtain complete descriptions of what happens in the other parts, because, it seems, our memory-control systems have too little temporary memory even to represent their *own* activities in much detail.

15.3 MEMORY

In order for a mind to think, it has to juggle fragments of its mental states. Suppose you want to rearrange the furniture inside a room you know. Your attention keeps shifting, first to one corner, then to another, next to the center of the room, and then, perhaps, to how the light falls on some object on a shelf. Different ideas and images interrupt each other. At one moment it seems as though your entire mind were focused on one small detail; at another moment you might dwell on why you are thinking about that room in the first place; then you might find yourself comparing or contrasting two different rearrangements of that scene: "*If that couch were over here, there would be room for guests to chat—but no, that would block the path so that they wouldn't be able to enter.*"

How do our various agencies keep track of imaginary changes in scenes? Where do the different versions go when out of mind, and how do we get them back again? They must be stored as memories. But what do we mean by *that*? Some readers may be surprised to learn that biologists still have no well-established theory of what happens in our brains when memories are formed. Psychologists, however, do agree that there must be at least two different mechanisms. We appear to have "long-term memories," which can persist for days or years or all one's life. We also have "short-term memories," which last only for seconds or minutes. In the next few sections we'll talk mostly about the uses of these transient traces of our recent thoughts. For example, whenever we get stuck in the course of solving a problem, we need to be able to backtrack, modify our strategy, and try again. To do this we need those short-term memories, if only not to repeat the same mistake.

How much do we remember? Sometimes we surprise ourselves by remembering things we didn't know we knew. Could this mean that we remember *everything*? Some older theories in psychology have supposed this to be true, and there are many legends of persons having fabulous abilities. For example, we often hear about people with "photographic memories" that enable them to quickly memorize all the fine details of a complicated picture or a page of text in a few seconds. So far as I can tell, all of these tales are unfounded myths, and only professional magicians or charlatans can produce such demonstrations.

In any case, I suspect we never really remember very much about a particular experience. Instead, our various agencies selectively decide, unconsciously, to transfer only certain states into their long-term memories—perhaps because they have been classified as useful, dangerous, unusual, or significant in other respects. It would be of little use for us simply to maintain vast stores of unclassified memories if, every time we needed one, we had to search through all of them. Nor would it be of any use for them all to flood at once into our agencies. Instead, each of us must develop fruitful and effective ways to organize our memories—but how that's done is inaccessible to consciousness. What barriers keep us from knowing such things? The next few sections sketch out some theories, both about how our memory-systems work and why we can't find this out directly by examining our own thoughts.

15.4 MEMORIES OF MEMORIES

Ask anyone for memories from childhood, and everyone will readily produce a handful of stories like this:

> *My neighbor's father died when I was four. I remember sitting with my friend in front of their house, watching people come and go. It was strange. No one said anything.*

It's hard to distinguish memories from memories of memories. Indeed, there's little evidence that *any* of our adult memories really go way back to infancy; what seem like early memories may be nothing more than reconstructions of our older thoughts. For one thing, recollections from our first five years seem strangely isolated; if we ask what happened earlier that day, the answer almost always is, *"I can't remember **that**."* Furthermore, many of those early memories involve incidents so significant that they probably occupied the child's mind repeatedly over a period of years. Most suspicious of all is the fact that such recollections are frequently described as seen through other, older eyes—with the narrator portrayed *inside* the scene, right near the center of the stage. Since we never actually see ourselves, these *must* be reconstructed memories, rehearsed and reformulated since infancy.

I suspect that this "amnesia of infancy" is no mere effect of decay over time but an inevitable result of growing out of infancy. A memory is not a separate entity, apart from how it works upon the mind. To remember an early experience, you must be able not only to "retrieve" some old records, but to reconstruct how your earlier mind reacted to them—and to do that, you would have to become an infant again. To outgrow infancy, you have to sacrifice your memories because they're written in an ancient script that your later selves can no longer read.

We reconstruct our recent memories as well, since they portray less what we saw than what we recognized. From every moment to the next, your mental state is shaped not only by signals from the outer world, but by agents activated by the memories these evoke. For example, when you see a chair, what makes it appear to you to be a chair—rather than an assortment of sticks and boards? It must evoke some memories. Only a part of your impression comes from agents activated directly by your vision; most of what your higher-level agencies experience comes from the memories those vision-agents activate. Usually, we have no conscious sense of this happening, and we never use words like "memory" or "remembering" when the process works quickly and quietly; instead, we speak of "seeing" or "recognizing" or "knowing." This is because such processes leave too few traces for the rest of the mind to contemplate; accordingly, such processes are unconscious, because consciousness requires short-term memory. It is only when a recognition involves substantial time and effort that we speak of "remembering."

Then what do we mean by "memory"? Our brains use many different ways to store the traces of our pasts. No single word can describe so much, unless it is used only in a general, informal sense.

> *Memories are processes that make some of our agents act in much the same ways they did at various times in the past.*

15.5 THE IMMANENCE ILLUSION

*Everyone will readily allow that there is a considerable difference
between the perceptions of the mind, when a man feels the pain of
excessive heat, or the pleasure of moderate warmth, and when he
afterwards recalls to his memory this sensation, or anticipates it by
his imagination. These faculties may mimic or copy the
perceptions of the senses; but they never can entirely reach the force
and vivacity of the original sentiment. . . . The most lively
thought is still inferior to the dullest sensation.*
—DAVID HUME

We like to think of memories as though they could restore to us things we've known in the past. But memories can't really bring things back; they only reproduce some fragments of our former states of mind, when various sights, sounds, touches, smells, and tastes affected us. Then what makes some recollections seem so real? The secret is that real-time experience is just as indirect! The closest we can come to apprehending the world, in any case, is through the descriptions our agents make. In fact, if we inquired instead about why real things seem real, we would see that this depends, as well, on memories of things we've already known.

For instance, when you see a telephone, you have a sense not only of the aspects you can see—its color, texture, size, and shape—but also of how it feels to hold the instrument to your ear. You also seem to know at once what telephones are *for*: that you speak into *here* and listen *there*; that when it rings you answer it; that when you want to call, you dial it. You have a sense of what it weighs, of whether it is soft or hard, of what its other side is like—although you haven't even touched it yet. These apprehensions come from memories.

> **The Immanence Illusion:** *Whenever you can answer a question without a noticeable delay, it seems as though that answer were already active in your mind.*

This is part of why we feel that what we see is "present" in the here and now. But it isn't really true that whenever a real object appears before our eyes, its full description is instantly available. Our sense of momentary mental time is flawed; our vision-agencies begin arousing memories *before* their own work is fully done. For example, when you see a horse, a preliminary recognition of its general shape may lead some vision-agents to start evoking memories about horses before the other vision-agents have discerned its head or tail. Perceptions can evoke our memories so quickly that we can't distinguish what we've seen from what we've been led to recollect.

This explains some of the subjective difference between seeing and remembering. If you first *imagined* a black telephone, you probably would not find it hard to reimagine it as red. But when you *see* a black telephone and then attempt to think of it as red, your vision-systems swiftly change it back! So the experience of seeing things has a relatively rigid character, in contrast to the experience of imagining things. Every change that the rest of your mind tries to impose upon your vision-agencies is resisted and usually reversed. Perhaps it is this descriptive rigidity that we identify with "vividness" or "objectivity." I do not mean to suggest that this is usually an illusion, because it often truly reflects the persistency and permanence of actual physical objects. Sometimes, though, our sense of objectivity can get reversed—as when an attitude or memory becomes more stable and persistent than what it represents. For example, our attitudes toward things we love or loathe are often much less changeable than those things themselves—particularly in the case of other people's personalities. In instances like these, our private memories can be more rigid than reality.

15.6   MANY KINDS OF MEMORY

We often talk of "memory" as though it were a single definite thing. But everyone has many kinds of memories. Some things we know seem totally detached from time, like such facts as that twelve inches make a foot or that a bull has dangerous horns. Other things we know seem linked to definite spans of time or space, like memories of places where we've lived. Still other recollections seem like souvenirs of episodes we can reexperience: *"Once, when visiting my grandparents, I climbed an old apple tree."*

> *A brain has no single, common memory system. Instead, each part of the brain has several types of memory-agencies that work in somewhat different ways, to suit particular purposes.*

Why do we have so many kinds of memory? If memories are records of our mental states of earlier times, how are those records stored and kept? A popular image of memories is that they are like objects we store away in various "places" in the brain. But then what are those places like? How do memories get into them and come out again? And what takes place inside of the vaults in which they're stored? Are memory banks like freezer chests where time stands still, or do their contents interact? How long can our old memories remain; do some of them grow old and die, do they get weak and fade away or just get lost and never found?

We have the impression that even our long-term memories become harder to recall as time goes on, and that might lead us to suppose that they have some inherent tendency to fade away. But even that is uncertain; it could simply be because so many other memories begin to interfere with them. Most likely, some types of memory mechanisms retain the records of sensations only for seconds; we use others to adopt habits, goals, and styles that we hold only for days or weeks; and we make personal attachments that endure through many months or years. Yet suddenly, from time to time, we'll modify some memories that seemed, till then, quite permanent.

More evidence that there are many kinds of memory has come from accidental injuries to brains. One injury may cause the loss of abilities to deal with names; another injury can make you lose some capacity to recognize faces or to remember musical tunes; still other kinds of injuries can leave unchanged what you have learned in earlier times but keep you from learning anything more in some particular domain. There is evidence that long-term memories can never form at all unless their short-term antecedents are permitted to persist for certain intervals; this process can also be blocked by various drugs and injuries, and this is why some people can never recollect what happened in the minutes before a brain concussion.

Finally, it appears that there are strong limitations on how rapidly we can construct our long-term memories. Despite all the legends about prodigies, there seems to be no evidence from any well-designed experiments that any human being can continue to construct long-term memories, over any substantial interval of time, more than two or three times faster than the average person.

15.7 MEMORY REARRANGEMENTS

Let's return to moving mental furniture. What would we need to imagine moving things around a room? First we'd need some way to represent how objects are arranged in space. In our *Block-Arch* scenario, the scene was represented in terms of the shapes of the objects and the relations between them. In the case of a room scene, you might also relate each object to the walls and corners of the room; you might notice that the couch is about midway between a table and a chair, and that all three are lined up near a certain wall.

Once we have a method for representing rooms, we also need techniques for manipulating these representations. How could we envision the result of exchanging that couch and chair? Let's oversimplify, and suppose that this can be done simply by exchanging the states of two agencies—an agency **A**, which represents the couch, and another agency **B**, which represents the chair. To exchange these states, let's assume that both agencies have access to two "short-term memory-units," called M-1 and M-2, which can record the states of agencies. Then we can exchange the states of **A** and **B**, first by storing away the states **A** and **B**, and then by restoring them in reverse order. In other words, we could use the following simple four-step "script":

1. *Store the state of **A** in M-1.*
2. *Store the state of **B** in M-2.*
3. *Use M-2 to determine the state of **A**.*
4. *Use M-1 to determine the state of **B**.*

A "memory-control script" like this can work only if we have memory-units that are small enough to pick out couch-sized portions of the larger scene. M-1 and M-2 would not do the job if they could store only descriptions of entire rooms. In other words, we have to be able to connect our short-term memories only to appropriate aspects of our current problems. Learning such abilities is not simple, and perhaps it is a skill some people never really master. What if we wanted to rearrange three or more objects? As a matter of fact, it is possible to produce *any* rearrangement whatsoever, using only operations that exchange two objects at a time! When you approach an unfamiliar kind of problem, it's best to start by making only one or two changes at a time. Then, in the course of becoming an expert, you discover schemes that make several useful changes in memory at once.

Our pair-exchanging script needs more machinery. Because each memory-unit must wait until the previous step is finished, the timing of each script step may have to depend on various "condition sensors." Shortly we'll see that even this is not enough to solve hard problems: our memory-control processes also need ways to *interrupt* themselves while they call on other agencies or memories for help. Indeed, the problems we must solve when managing our memories are surprisingly like those we face when dealing with things in the outside world.

15.8 ANATOMY OF MEMORY

What controls the working of the mind from one moment to the next? How do we keep our place when doing complicated jobs, so that when interrupted from outside—or by another thought from inside—we can "get back" to where we were, instead of having to start all over again? How do we keep in mind which things we've tried and what we've learned along the way, so that we don't go round and round in loops?

No one yet knows how memories control themselves inside our brain; perhaps each major agency has somewhat different processes, each suited to the special kinds of jobs it does. The diagram below suggests some of the sorts of memory-machinery we'd expect to find inside a typical large agency.

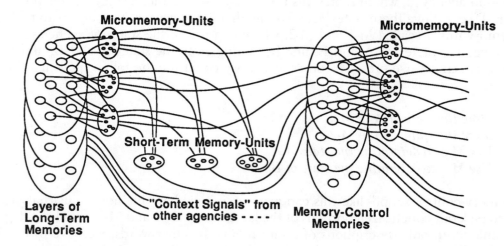

We'll assume that every substantial agency has several "micromemory-units," each of which is a sort of temporary K-line that can quickly store or restore the state of many of the agents in that agency. Each agency also has several "short-term memory-units," which can, in turn, store or restore the states of the micromemories themselves. When any of these temporary memory-units are reused, the information that was stored in them is erased—unless it has somehow been "transferred" into more "permanent" or "long-term" memory-systems. There is good evidence that, in human brains, the processes that transfer information into long-term memory are very slow, requiring time intervals that range from minutes to hours. Accordingly, most temporary memories are permanently lost.

A growing child acquires many ways to control all these mechanisms. Accordingly, our diagram includes the flow of information among the other agencies. Since this memory-controlling agency must also learn and remember, our diagram includes a memory-system for it as well.

Imagine that you plan to take a trip. You start to think of how you'll pack your traveling case and start some spatial problem-solving agency—call it *Packer*—to see how to fit the larger items in. Then you interrupt yourself to think about your smaller things, perhaps of how to pack your jewelry in a smaller box. Now *Packer* has to reapply itself to a new and different box-packing problem. The problem of keeping track of what is happening is hard enough when one agency calls on another one for help. Until the other's job is done, the first agency has to keep some temporary record of what it was doing. In *Packer*'s case the problem is even worse because it interrupts *itself* to pack the smaller box. And here is the important point: when that second packing job is done and we return to the first, we mustn't go all the way back to the very beginning, or else we would be caught in a circular loop. *Instead, we must return to the point where we were when we interrupted ourselves*—which means the system needs some memory to keep track of what it was doing before. This is exactly the same problem we mentioned long ago when *Find* and *See* had several different jobs to do at the same time.

Why do we so often get confused when we're interrupted? Because then we have to keep our place in several processes at once. To keep things straight, our memory-control machinery needs intricate skills. Yet psychologically, we're unaware that ordinary thinking is so complicated. If someone asked, *"What was your mind just doing?"* you might say something like this:

> *"I was thinking about packing that suitcase, and started to wonder if the umbrella might fit. I remembered that on an earlier trip, I managed to fit my camera tripod in the same case, and I tried to compare the umbrella and the tripod in my mind, to see which one was longer."*

Now, this might be a true account of some of the things you were thinking about. But it says little of how the mental work was actually done. To understand how thinking works, we'd really need descriptions of the processes themselves:

> *"A few moments ago, I activated two micromemory-units inside **Packer,** one of my space-arranging agencies, while also activating one of **Packer**'s memory-control scripts. This script proceeded to use the information inside those two micro-memory-units as cues to fetch certain partial states from the long-term memory-system attached to **Packer.** Next, the script controlling **Packer**'s memory-system requested a certain higher-level planning-agency to record most of **Packer**'s present state. Then it interchanged the contents of the two active micromemory units, and then used other cues to fetch another, second script from long-term memory—and thus erased the present copy of itself. The last step of that second script caused yet another micromemory-unit to restore **Packer** to its previous state, so that the original script could continue on its interrupted course. Then . . ."*

But no one ever says such things. The processes are too many levels away from those we use to work the short-term memories involved with language and consciousness. We couldn't think so if we wanted to—without knowing more about the anatomy of our memory-machinery. Even if we had ways to represent those processes at higher levels, our memory-controls would probably be overloaded by attempting, both at the same time, to solve a difficult problem and to remember everything that was done while solving it.

15.10 LOSING TRACK

Whenever we solve complicated problems, we get into situations in which our agencies must keep account of many processes at once. In computer programs, the many "subjobs" often seem to pile up like the blocks of a tower. Indeed, computer programmers often use the word "stack" to describe such situations. But I doubt that untrained human minds use anything so methodical; in fact, we simply aren't very good at dealing with the kinds of situations that need such memory-stacks. This could be why we get confused when hearing sentences like this:

This is the malt that the rat that the cat that the dog worried killed ate.

The very same words can be rearranged to make an equivalent sentence anyone can understand:

This is the dog that worried the cat that killed the rat that ate the malt.

The first sentence is hard to understand because so many verb processes interrupt one another that when the end of the sentence comes, three similar processes are still active—but they have lost track of what roles should be assigned to all the remaining nouns, namely, the rat, cat, and malt. Why do visual processes so rarely encounter similar difficulties? One reason is that our visual-systems can support more simultaneously operating processes than our language-systems can, and this reduces the need for any process to interrupt another one. A second reason is that the vision-agencies can choose for themselves the sequence in which they attend to details, whereas language-agencies are controlled by the person who is speaking.

It takes each person many years to learn to use those memory-systems well. Younger children certainly cannot keep track as well as adults. It's generally of little use to ask a pair of two-year-olds to play together or to take turns at using a toy. We consider them to be too self-centered and impatient for that. Surely much of their undisciplined impulsiveness comes from desires that are less regulated than our own. But that impatience could also stem from insecurity about memory: *the child may fear that what it wants will slip from mind while other thoughts are entertained.* In other words, the child who is asked to "take turns" might fear that by the time its turn arrives, it may not want the object anymore.

When people ask, "Could a machine ever be conscious?" I'm often tempted to ask back, "Could a person ever be conscious?" I mean this as a serious reply, because we seem so ill equipped to understand ourselves. Long before we became concerned with understanding how we work, our evolution had already constrained the architecture of our brains. However, we can design our new machines as we wish, and provide them with better ways to keep and examine records of their own activities—and this means that machines are potentially capable of far more consciousness than we are. To be sure, simply providing machines with such information would not automatically enable them to use it to promote their own development, and until we can design more sensible machines, such knowledge might only help them find more ways to fail: the easier to change themselves, the easier to wreck themselves—until they learn to train themselves. Fortunately, we can leave this problem to the designers of the future, who surely would not build such things unless they found good reasons to.

15.11 THE RECURSION PRINCIPLE

Let's consider one last time how a mind could juggle nonexistent furniture inside an imaginary room. To compare different arrangements, we must somehow maintain at least two different descriptions in the mind at once. Can we store them in different agencies, both active at the same time? That would mean splitting our space-arranging-agency into two different smaller portions, each working on one of those descriptions. On the surface, there's nothing clearly wrong with that. However, if those smaller agencies became involved with similar jobs, then they, in turn, would also have to split in two. And then we'd have to do each of those jobs with but one-quarter of a mind. If we had to divide agencies into smaller and smaller fragments, each job would end up with no mind at all!

At first this might seem to be an unusual situation. But it really is very common. The best way to solve a hard problem is to break it into several simpler ones, and break those into even simpler ones. Then we face the same issue of mental fragmentation. Happily, there is another way. We can work on the various parts of a problem in serial order, one after another, using the same agency over and over again. Of course, that takes more time. But it has one absolutely fundamental advantage: each agency can apply its full power to every subproblem!

> **The Recursion Principle:** *When a problem splits into smaller parts, then unless one can apply the mind's full power to each subjob, one's intellect will get dispersed and leave less cleverness for each new task.*

Indeed, our minds don't usually shatter into helpless fragments when problems split into parts. We *can* imagine how to pack a jewelry box without forgetting where it will fit into a suitcase. This suggests that we can apply our full space-arranging resources to each problem in turn. But how, then, do we get back to the first problem, after we've thought about the other ones, without having to start all over again? To common sense the answer seems clear: we simply *"remember where we were."* But this means that we must have some way to store, and later re-create, the states of interrupted agencies. Behind the scenes, we need machinery to keep track of all the incomplete accomplishments, to remember what was learned along the way, to compare different results, and to measure progress in order to decide what to do next. All this must go on in accord with larger, sometimes changing plans.

The need to recall our recent states is why our "short-term memories" *are* short-term memories! In order to do their complex jobs so quickly and effectively, each micromemory-device must be a substantial system of machinery, with many intricate and specialized connections. If so, our brains cannot afford to make too many duplicate copies of that machinery, so we must reuse what we have for different jobs. Each time we reuse a micromemory-device, the information stored inside must be erased—or moved to another, less costly place. But that would also take some time and interrupt the flow of thought. Our short-term memories must work too fast to have any time for consciousness.

CHAPTER 16

EMOTION

Each emotion has its own world view.
Love includes and merges and nurtures
Joy is light and dances even with the eyes
Grief heavy, hopeless, and the lungs empty, as is the heart
and Hate wants to destroy, to kill,
that is its very nature
    and almost involuntarily, matter of factly,
    another part of me says
    "es ist das hier ja nicht unbekannt."

—MANFRED CLYNES

16.1 EMOTION

Why do so many people think emotion is harder to explain than intellect? They're always saying things like this:

> *"I understand, in principle, how a computer might solve problems by reasoning. But I can't imagine how a computer could have emotions, or comprehend them. That doesn't seem at all the sort of thing machines could ever do."*

We often think of anger as nonrational. But in our Challenger scenario, the way that *Work* employs *Anger* to subdue *Sleep* seems no less rational than using a stick to reach for something beyond one's grasp. *Anger* is merely an implement that *Work* can use to solve one of its problems. The only complication is that *Work* cannot arouse *Anger* directly; however, it discovers a way to do this indirectly, by turning on the fantasy of Professor Challenger. No matter that this leads to states of mind that people call emotional. To *Work* it's merely one more way to do what it's assigned to do. *We're always using images and fantasies in ordinary thought.* We use "imagination" to solve a geometry problem, plan a walk to some familiar place, or choose what to eat for dinner: in each, we must envision things that aren't actually there. The use of fantasies, emotional or not, is indispensable for *every* complicated problem-solving process. We always have to deal with nonexistent scenes, because only when a mind can change the ways things *appear to be* can it really start to think of how to change the ways things *are*.

In any case, our culture wrongly teaches us that thoughts and feelings lie in almost separate worlds. In fact, they're always intertwined. In the next few sections we'll propose to regard emotions not as separate from thoughts in general, but as varieties or types of thoughts, each based on a different brain-machine that specializes in some particular domain of thought. In infancy, these "protospecialists" have little to do with one another, but later they grow together as they learn to exploit one another, albeit without understanding one another, the way *Work* exploits *Anger* to stop *Sleep*.

Another reason we consider emotion to be more mysterious and powerful than reason is that we wrongly credit it with many things that reason does. We're all so insensitive to the complexity of ordinary thinking that we take the marvels of our common sense for granted. Then, whenever anyone does something outstanding, instead of trying to understand the process of thought that did the real work, we attribute that virtue to whichever superficial emotional signs we can easily discern, like motivation, passion, inspiration, or sensibility.

In any case, no matter how neutral and rational a goal may seem, it will eventually conflict with other goals if it persists for long enough. No long-term project can be carried out without some defense against competing interests, and this is likely to produce what we call emotional reactions to the conflicts that come about among our most insistent goals. The question is not whether intelligent machines can have any emotions, but whether machines can be intelligent without any emotions. I suspect that once we give machines the ability to alter their own abilities we'll have to provide them with all sorts of complex checks and balances. It is probably no accident that the term "machinelike" has come to have two opposite connotations. One means completely unconcerned, unfeeling, and emotionless, devoid of any interest. The other means being implacably committed to some single cause. Thus each suggests not only inhumanity, but also some stupidity. Too much commitment leads to doing only one single thing; too little concern produces aimless wandering.

16.2 MENTAL GROWTH

In ancient times it was believed that the newborn mind started out just like a full-grown mind, except for not yet being filled with ideas. Thus children were seen as ignorant adults, conceived with all their future aptitudes. Today, there are many different views. Some modern theories see a baby's mind as starting with a single Self whose problem is to learn to distinguish itself from the rest of the world. Others see the infant's mind as a place containing a horde of mind-fragments, mixed together in a disconnected and incoherent confusion in which each must learn to interact and cooperate with the others so that they can grow together to form a more coherent whole. Yet another image sees the child's mind as growing through a series of layerlike construction stages in which new levels of machinery are based and built upon the older ones.

How *do* our minds form? Is every person born containing a hidden, built-in intellect just waiting to reveal itself? Or must minds grow in little steps from emptiness? The theories of the next few sections will combine ingredients from both these conceptions. We'll start by envisioning a simple brain composed of separate "proto-specialists," each concerned with some important requirement, goal, or instinct, like food, drink, shelter, comfort, or defense. But there are reasons why those systems must be merged. On one side, we need administrative agencies to resolve conflicts between the separate specialists. On the other side, each specialist must be able to exploit whatever knowledge the others gain.

For a relatively simple animal, a loose-knit league of nearly separate agencies with built-in goals might suffice for surviving in a suitable environment. But human minds don't merely learn new ways to reach old goals; we can also learn new kinds of goals. This enables us to live within a broader range of possible environments, but that versatility comes with its own dangers. If we could learn new goals without constraint, we'd soon fall prey to accidents—both in the world and inside our own minds. At the simplest levels, we have to be protected against such accidents as learning not to breathe. On higher levels, we need protection against acquiring lethal goals like learning to suppress our other goals entirely—the way that certain saints and mystics do. What sorts of built-in self-constraints could guide a mind toward goals that will not cause it to destroy itself?

No possible inheritance of built-in genes can tell us what is good for us—because, unlike all other animals, we humans make for ourselves most of the problems we face. Accordingly, each human individual must learn new goals from what we call the traditions and heritages of our peers and predecessors. Consequently our genes must build some sort of "general-purpose" machinery through which individuals can acquire and transmit goals and values from one generation to another. How could brain-machines transfer things like values and goals? The next few sections suggest that this is done by exploiting the kinds of personal relationships we call emotional, such as fear and affection, attachment and dependency, or hate and love.

16.3 MENTAL PROTO-SPECIALISTS

Suppose you had to build an artificial animal. First you'd make a list of everything you want your animal to do. Then you'd ask your engineers to find a way to meet each need.

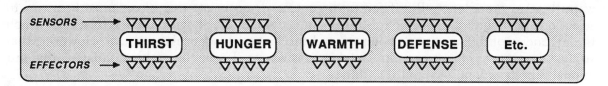

This diagram depicts a separate agency for each of several "basic needs." Let's call them "proto-specialists." Each has a separate mini-mind to do its job and is equipped with special sensors and effectors designed to suit its specific needs. For example, the proto-specialist for *Thirst* might have a set of parts like these:

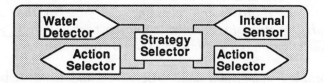

It would not usually be practical to make an animal that way. With all those separate specialists, we'd end up with a dozen different sets of heads and hands and feet. Not only would it cost too much to carry and feed all those organs; they'd also get in one another's way! Despite that inconvenience, there actually do exist some animals that work this way and thus can do many things at once. Genetically, the swarms of social ants and bees are really multibodied individuals whose different organs move around freely. But most animals economize by having all their proto-specialists share common sets of organs for their interactions with the outer world.

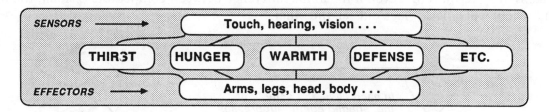

Another kind of economy comes from allowing the proto-specialists to share what they learn. Whether you seek warmth, safety, nutrition, or companionship—eventually you'll have to be able to recognize and act in order to acquire the objects you need. So even though their initial goals are entirely different, all those different proto-specialists will end up needing to solve the same sorts of "subproblems"—such as finding ways around obstacles and deciding how to conserve limited resources. Whenever we try to solve problems of increasing complexity, whatever particular techniques we already know become correspondingly less adequate, and it becomes more important to be able to acquire *new* kinds of knowledge and skills. In the end, most of the mechanisms we need for any highly ambitious goal can be shared with most of our other goals.

When a dog runs, it moves its legs.
When a sea urchin runs, it is moved by its legs.
—Jakob von Uexküll

16.4 CROSS-EXCLUSION

An ordinary single-bodied animal can only move in one direction at a time, and this tends to constrain it to work toward only one goal at a time. For example, when such an animal needs water urgently, its specialist for "thirst" takes control; however, if cold is paramount, finding warmth takes precedence. But if several urgent needs occur at once, there must be a way to select one of them. One scheme for this might use some sort of central marketplace, in which the urgencies of different goals compete and the highest bidder takes control. However, that strategy is prone to fall into a funny, fatal indecisiveness. To see the problem, imagine that our animal is both very hungry and very thirsty.

Suppose that our animal's hunger is, at first, just slightly more urgent than its thirst. So it sets out on a trek toward the North Plain, where food is usually found. When it arrives and takes a bite of food, its thirst instantly takes precedence over its need for food!

Now that thirst has top priority, our animal sets out on the long journey toward South Lake. But once it arrives and takes one satisfying sip, the balance instantly tips back to food! Our animal is doomed to journey back and forth, getting hungrier and thirstier. Each action only equalizes ever-growing urgencies.

This would be no problem at a dinner table, where food and drink are both within easy reach. But under natural conditions, no animal could survive the waste of energy, when every minor fluctuation caused a major change in strategy. One way to manage this would be to use that "marketplace" infrequently—but that would make our animal less capable of dealing with emergencies. Another way is to use an arrangement called *cross-exclusion*, which appears in many portions of the brain. In such a system, each member of a group of agents is wired to send "inhibitory" signals to all the other agents of that group. This makes them competitors. When any agent of such a group is aroused, its signals tend to inhibit the others. This leads to an avalanche effect—as each competitor grows weaker, its ability to inhibit its challengers also weakens. The result is that even if the initial difference between competitors is small, the most active agent will quickly "lock out" all the others.

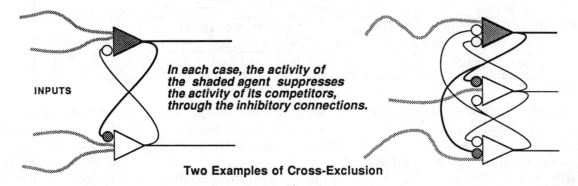

INPUTS

In each case, the activity of the shaded agent suppresses the activity of its competitors, through the inhibitory connections.

Two Examples of Cross-Exclusion

Cross-exclusion arrangements could provide a basis for the principle of "noncompromise" in regions of the brain where competitive mental agents lie close together. Cross-exclusion groups can also be used to construct short-term memory-units. Whenever we force one agent of such a group into activity, even for a moment, it will remain active (and the others will remain suppressed) until the situation is changed by some other strong external influence. Weaker external signals will have scarcely any effect at all because of resistance from within. Why call this a *short-term* memory if it can persist indefinitely? Because when it *does* get changed, no trace will remain of its previous state.

EMOTION

16.5 AVALANCHE EFFECTS

Few of the schemes we've discussed would actually work if they were built exactly as they were described. Most of them would soon break down because virtually all their agents would become engaged into unconstrained activity. Suppose each typical agent tends to arouse several others. Then each of those would turn on several more—and each of those would turn on yet others; the activity would spread faster than a forest fire. But all that activity would accomplish nothing, since all those agents would interfere with each other and none of them could gain control of the resources they need. Indeed, this is more or less what happens in an attack of epilepsy.

Similar problems arise in every biological system. Each simple principle or mechanism must be controlled to operate within some limited range. Even little groups of genes embody schemes to regulate the quantities of proteins they cause to be manufactured inside every cell. We find the same pattern repeated on every scale. Every biological tissue, organ, and system is regulated by several kinds of control mechanisms, and wherever these fail we find disease. What normally protects our brains from such avalanches of activity? The cross-exclusion scheme is probably the most usual way to regulate the levels of activities within our agencies. But there are also several other frequently encountered schemes to prevent explosions.

> **Conservation:** *Force all activities to depend upon some substance or other kind of quantity of which only a certain amount is available. For example, we controlled our* Snarc *machine by setting a limit on the total electric current available to all the agents; this permitted only a few of them to be active at any particular moment.*

> **Negative Feedback:** *Supply a "summary" device that estimates the total activity in the agency and then broadcasts to that agency an "inhibitory" signal whose strength is in proportion to that total. This will tend to damp down incipient avalanches.*

> **Censors and Suppressors:** *The "conservation" and "feedback" schemes tend to be indiscriminate. Later we'll discuss methods that are more sensitive and versatile in learning to recognize—and then to avoid—specific patterns of activity that have led to trouble in the past.*

These methods are simple enough to be applied inside small societies, but they are not versatile enough to solve all the management difficulties that can arise in the more complex societies we need for learning to solve harder problems. Fortunately, systems built upon larger scales may be able to apply their own enhanced abilities to managing themselves—by formulating and solving their own self-regulation problems. In the next few sections we'll see how such capacities could grow in the course of several stages of development. Not all of this need happen independently inside each separate child's mind, because that child's family and cultural community can develop self-regulation schemes of great complexity. All human communities seem to work out policies for how their members ought to think, in forms that are thought of as common sense or as law, religion, or philosophy.

Imagine that a thirsty child has learned to reach for a nearby cup. What keeps that child, afterward, from reaching for a cup in every other circumstance—say, when it is lonely or when it is cold? How do we keep separate what we learn for satisfying different goals? One way is to maintain a separate memory bank for every distinct goal.

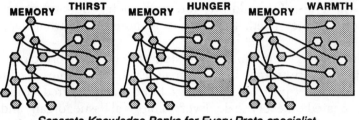

MEMORY THIRST MEMORY HUNGER MEMORY WARMTH

Separate Knowledge Banks for Every Proto-specialist

To make this work, we must restrict each specialist to learn only when its own goal is active. We can accomplish that by building them into a cross-exclusion system so that, for example, *Hunger's* memories can be formed only when *Hunger* is active. Such a system will never get confused about which memories to use. When it is hungry it will do only what it learned to do at previous times when it was hungry; it won't eat when it is thirsty or drink when it is hungry. But it would be too extravagant to have to keep completely different memories for every goal— since, as we said, most real-world goals engage the same kinds of knowledge about the world. Wouldn't it be better if all those specialists could share a common, general-purpose memory?

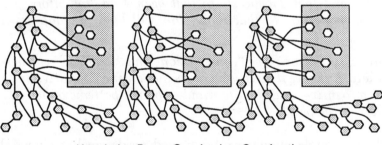

Knowledge-Bases Growing into One Another

This would lead to problems, too. Whenever any specialist tried to rearrange some memories to its own advantage, it might damage structures upon which the others have come to depend. There would be too many unpredictable interactions. How could specialists cooperate and share what they have learned? If they were like people, they could communicate, negotiate, and organize. But because each separate specialist is much too small and specialized to understand how the others work, the best each can do is learn to exploit what the others can do, without understanding how they do it.

16.7 EXPLOITATION

How could any specialist cooperate when it doesn't understand how the others work? We manage to do our worldly work despite that same predicament; we deal with people and machines without knowing how *their* insides work. It's just the same inside the head; each part of the mind exploits the rest, not knowing how the other parts work but only what they seem to do.

Suppose *Thirst* knows that water can be found in cups—but does not know how to find or reach for a cup; these are things only *Find* and *Get* can do. Then *Thirst* must have some way to exploit the abilities of those other agents. *Builder*, too, has a similar problem because most of its subagents cannot communicate directly with one another. It would be easy for *Thirst* or *Builder* simply to turn on other agents like *Find* and *Get*. But how will those subordinates know *what* to find or get? Must *Thirst* transmit to *Find* a picture of a cup? Must *Builder* send a picture of a brick? The trouble is that neither *Builder* nor *Thirst* is the sort of agent to contain the kind of knowledge required by *Find*—namely, the visual appearances of things. That kind of knowledge lies inside the memory-machinery of *See*. However, *Thirst* can achieve its drinking goal by activating *two* connections: one to cause *See* to "hallucinate" a cup and another connection to activate *Find*. *Find* itself can activate *Get* later. This should suffice for *Thirst* to locate and obtain a cup—if there is one in sight.

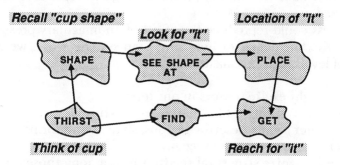

This scheme could be unreliable. If *See* became concerned with another object at that moment, *Get* would acquire the wrong object. Infants often disappoint themselves this way. Still, this scheme has the kind of simplicity one needs when starting to build any larger skill: one needs a process that *sometimes* works before one can proceed to improve it.

This is merely a sketch of how to build an automatic "getting machine." We'll return to this idea much later, when we discuss language, because what *Thirst* and *Builder* have to do resembles what people do when using words. When you say to another person, *"Please pass the cup,"* you don't emit a picture of a cup but merely send a signal that exploits the other person's memory.

Achieving a goal by exploiting the abilities of other agencies might seem a shabby substitute for knowing how to do the work oneself. Yet this is the very source of the power of societies. No higher-level agency could ever achieve a complex goal if it had to be concerned with every small detail of what each nerve and muscle does. Unless most of its work were done by other agencies, no part of a society could do anything significant.

16.8 STIMULUS VS. SIMULUS

We've just seen how one agency could exploit another one by focusing its attention on some object in the outer world. Thus *Thirst* can make *Get* reach for a cup—provided there's a cup in view. But what about that fantasy about Professor Challenger, in which there was no real villain on the scene, but just a memory? Apparently one agency can activate another merely by imagining a stimulus! One way to do this would be for *Anger* to somehow construct an artificial picture that other agencies like *See* could "see." If this were done with enough detail, the other agents couldn't tell that the image wasn't genuine. However, to construct the sorts of images we see on television screens, we'd have to activate a million different sensory nerves, which would require a huge amount of machinery. Besides, we could do more with less:

> A *fantasy need not reproduce the fine details of an actual scene. It need only reproduce that scene's effect on other agencies.*

Fantasies usually "depict" occurrences we've never seen. They need no detailed, realistic images—since the higher levels of the mind don't really "see" things anyway! Instead, they deal with summaries of signals that come from sensory experience and are condensed at several levels along the way. In the fantasy of Professor Challenger, there was no need to see any of the actual features of Challenger himself; it was enough to reproduce some sense of how his presence once affected us.

What kind of process could reproduce the effect of an imaginary presence? Although scientists don't yet know the fine details of how our vision-systems work, we can assume that they involve a number of levels, perhaps like this:

-- First, rays of light excite sensors in our retinas.
--- Then, other agents detect boundaries and textures.
---- Then, yet other agents describe regions, shapes, and forms.
----- Then, some memory-frames recognize familiar objects.
------ Next, we recognize structural relationships among those objects.
------- Finally, we relate these structures to functions and goals.

Accordingly, it would be possible to produce illusions by operating at *any* of these levels. Most difficult of all would be to construct a "picture-image" by arousing, from inside the brain, the million lowest-level sensor-agents involved in real-world vision. Perhaps the simplest way of all would be to force just the highest-level vision-agents into whichever states would result from seeing a certain scene: this would only require some suitable K-lines. Let's call this a *simulus*— a reproduction of only the higher-level effects of a stimulus. A simulus at the very highest levels could lead a person to recollect virtually no details about a remembered object or event, yet be able to apprehend and contemplate its most significant structures and relationships while experiencing a sense of its presence. A simulus may have many advantages over a picture-image. Not only can it work more swiftly while using less machinery, but we can combine the parts of several simuli to imagine things we have never seen before—and even to imagine things that couldn't possibly exist.

16.9 INFANT EMOTIONS

A child forsaken, waking suddenly,
Whose gaze afeard on all things round doth rove,
And seeth only that it cannot see
The meeting eyes of love.
                                        —GEORGE ELIOT

Some readers may be horrified at picturing a baby's mind as made up of nearly separate agencies. But we'll never understand how human natures grow without some theories for how they start. One evidence for separateness is how suddenly infants switch from smiles of contentment to shrieks of hunger-rage. In contrast to the complex mixtures of expressions that adults show, young children seem usually to be in one or another well-defined state of activity —contentment, hunger, sleepiness, play, affection, or whatever. Older children show less sudden mood changes, and their expressions suggest that more different things are happening at once. Our minds may thus originate as sets of relatively simple, separate need machines. But soon enough each becomes enmeshed in all the rest of our growing machinery.

How should we interpret an infant's apparent single-mindedness? One explanation of those striking shifts in attitude is that one agency attains control and forcibly suppresses the rest. Another view is that many processes continue at once—but only one at a time can be expressed. It would be more efficient to keep the whole array of proto-specialists at work. Then each would be more ready to assume control in case of an emergency.

What would be the advantage in a mechanism that makes a baby conceal that mixture of emotions, expressing only one of them at a time? Perhaps that artificial sharpening promotes the child's welfare by making it easier for the parent to respond to whichever problem has the greatest urgency. It's hard enough to know what infants want, yet think how much harder it would be if they confronted us with complicated expressions of mixed feelings! Those infants' very lives—and, in turn, our own lives—depend upon their expressing themselves clearly. To achieve that clarity, their agencies must be equipped with powerful cross-exclusion devices to magnify small differences that make it clear which needs come first. This leads to simple "summaries"—which manifest themselves as drastic changes in appearance, voice, and mood that others can interpret easily. And this is why, under circumstances in which adults merely frown, babies tend to shriek.

Given that those signs are clear, what forces us to respond to them? To help their offspring grow, most animals evolve two matching schemes: communication is a two-way street. On one side, babies are equipped with cries that can arouse parents far away, out of sight, or sound asleep—for along with sharpening those signs, cross-exclusion also amplifies their intensity. And on the other side, adults are made to find those signals *irresistible:* there must be special systems in our brains that give such messages a high priority. To what might those baby-watching agents be connected? My guess is that they're wired to the remnants of the same proto-specialists that, when aroused, caused us as infants to cry in the first place. This leads adults to respond to babies' cries by attributing to them the same degrees of urgency that we ourselves would have to feel to make us shriek with similar intensity. This drives the babies' caretakers to respond to their needs with urgent sympathy.

16.10 ADULT EMOTIONS

Since emotions are few and reasons are many (said the
robot, Giskard), the behavior of a crowd can be more
easily predicted than the behavior of one person can.
—ISAAC ASIMOV

What are emotions, anyway? Our culture sees this question as a deep and ancient mystery. How could the idea of society of mind contribute to what our ancestors have said? Common-sense psychology has not even reached a consensus on *which* emotions exist.

Restlessness	Fear	Gladness	Jealousy	Sorrow
Curiosity	Hate	Enthusiasm	Ambition	Thirst
Infatuation	Anger	Admiration	Laziness	Disgust
Impatience	Love	Boredom	Contempt	Hunger
Excitement	Greed	Reverence	Anxiety	Lust

If there exists anger, what constitutes rage? How does fear relate to fright, terror, dread, dismay, and all such other awful things? How does love relate to reverence or to attachment or infatuation? Are these just various degrees of intensity and direction, or are they genuinely different entities that happen to be neighbors in an uncharted universe of affections? Are hate and love quite separate things, or, similarly, courage and cowardice—or are these merely pairs of extremes, each just the absence of its peer? What are emotions, anyway, and what are all the other things we label moods, feelings, passions, needs, or sensibilities? We find it hard to agree on the meanings of words like these, presumably because few of them actually correspond to clearly distinct mental processes. Instead, when we learn such words, we each attach to them variously different and personal accumulations of conceptions in our minds.

Infants' early emotion signs clearly signify their needs. We later learn to use such signals in more exploitative ways. Thus you can learn to use affection or anger as a social coin in trade for various accommodations; for example, one can pretend to be angry or pleased, or even offer —that is, threaten or promise—*to become* angry or affectionate in certain circumstances. Our culture is ambivalent about such matters; on one side we're taught that emotions should be natural and spontaneous; on the other side we're told that we must learn to regulate them. We recognize in deeds (though not in words) that feeling may be easier to understand and modify than other parts of intellect. We censure those who fail to learn to control their emotions but merely pity those whose problem-solving capabilities are poor; we blame for "lack of self-control," but not for "weakness of intelligence."

Our earliest emotions are built-in processes in which inborn proto-specialists control what happens in our brains. Soon we learn to overrule those schemes, as our surroundings teach us what we *ought* to feel. Parents, teachers, friends, and finally our self-ideals impose upon us new rules for how to use the remnants of those early states: they teach us how and when to feel and show each kind of emotion sign. By the time we're adults, these systems have become too complicated to understand. By the time we've passed through all those stages of development, our grown-up minds have been rebuilt too many times to remember or understand much of how it felt to be an infant.

CHAPTER 17

CHAPTER 17

DEVELOPMENT

To the child, Nature gives various means of rectifying any mistakes he may commit respecting the salutary or hurtful qualities of the objects which surround him. On every occasion his judgments are corrected by experience; want and pain are the necessary consequences arising from false judgment; gratification and pleasure are produced by judging aright. Under such masters, we cannot fail but to become well informed; and we soon learn to reason justly, when want and pain are the necessary consequences of a contrary conduct.

In the study and practice of the sciences it is quite different; the false judgments we form neither affect our existence nor our welfare; and we are not forced by any physical necessity to correct them. Imagination, on the contrary, which is ever wandering beyond the bounds of truth, joined to self-love and that self-confidence we are so apt to indulge, prompt us to draw conclusions that are not immediately derived from facts. . . .

—A. LAVOISIER

17.1 SEQUENCES OF TEACHING-SELVES

Up to this point we've portrayed the mind as made of scattered fragments of machinery. But we adults rarely see ourselves that way; we have more sense of unity. In the next few sections we'll speculate that this coherency is acquired over many "stages of development." Each new stage first works under the guidance of previous stages, to acquire some knowledge, values, and goals. Then it proceeds to change its role and becomes a teacher to subsequent stages.

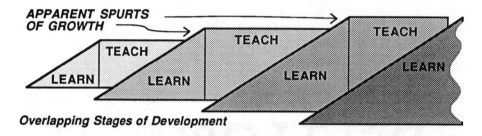

Overlapping Stages of Development

How could an early stage teach anything to a later one when it knows less than its student does? As every teacher knows, this is not as hard as it might seem. For one thing, it is usually easier to *recognize* a solution to a problem than to discover a solution; this is what we called the "puzzle principle." A teacher need not know how to solve a problem to be able to reward a student for doing so or to help the student search for solutions by imparting ways to sense when progress has been made. Even better is for the teacher to impart new goals to the student.

How could an early stage of development affect the goals of a later stage? One simple way would be to give each later stage some access to the goals of earlier stages; however, those early goals would then remain infantile. How could later stages develop more advanced goals? Shortly we'll see an astonishing answer: it is not necessary to formulate more advanced goals at "higher levels" of organization because they are likely to develop spontaneously, as subgoals of relatively simple goals.

In any case, it wouldn't be safe to send the student into the world equipped with systems that have not yet been tried and tested. A safer strategy would be to keep each new stage suppressed —that is, *incapable of controlling the child's actual behavior*—until it passes tests to verify that it is at least as capable as its predecessor. This could explain some of those apparently sudden "spurts" in our children's development—for example, in episodes of rapid growth of language skills. That apparent speed could be illusory if it were merely the end result of longer, hidden projects carried out silently inside the mind.

Returning to our sense of Self, how could so many steps and stages lead to any sense of unity? Why wouldn't they lead us, instead, to feel increasingly fragmentary and dispersed? I suspect the secret is that after each old stage's work is done, its structure still remains available for further use. These remnants of our prior selves provide us with a powerful resource: whenever one's present mind gets confused, it can exploit what once was used by earlier minds. Even though we weren't as smart then as we are now, we can be sure that every stage once had, in its turn, some workable ways to manage things.

> *One's present personality cannot share many of the thoughts of all one's older personalities—and yet it has some sense that they exist. This is one reason why we feel that we possess an inner Self—a sort of ever-present person-friend, inside the mind, whom we can always ask for help.*

17.2 ATTACHMENT-LEARNING

Suppose a child were playing in a certain way, and a stranger appeared and began to scold and criticize. The child would become frightened and disturbed and try to escape. But if, in the same situation, the child's parent arrived and proceeded to scold and criticize, the result would be different. Instead of being frightened, the child would feel guilty and ashamed, and instead of trying to escape, the child would try to change what it was doing, in attempts to seek reassurance and approval.

I suspect that these two scenarios engage different learning mechanisms. In the encounter with the forbidding visitor, the child might learn *"I should not try to achieve my present goal in this kind of situation."* But when scolded by someone to whom the child is "attached," the child might learn *"I ought not to want to achieve that goal at all!"* In the first case, it is a matter of learning *which* goal to pursue in which circumstance; in the second instance, it is more a question of what goals one should have. If my theory is right, the presence of the attachment-person actually switches the effect of learning over to different sets of agents. To see the difference, let's make a small reformulation of the concept of a difference-engine to represent three different kinds of learning that an infant might use.

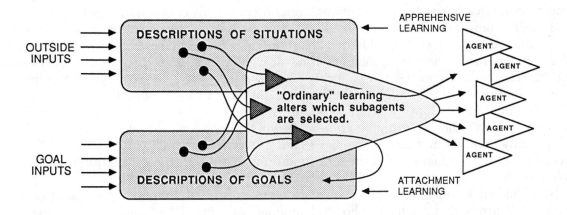

Key fig. 17-2, from p. 159

In the case of ordinary forms of failure or success signals, the learner modifies *the methods used to reach the goal.*

In the case of fear-provoking disturbances, the learner may modify *the description of the situation itself.*

In the case of attachment-related failure or reward signals, the learner modifies *which goals are considered worthy of pursuit.*

So far as I know, this is a new theory about attachment. It asserts that there are particular types of learning that can proceed only in the presence of the particular individuals to whom one has become attached.

17.3 ATTACHMENT SIMPLIFIES

No form of behavior is accompanied by stronger feeling than is attachment behavior. The figures towards whom it is directed are loved and their advent is treated with joy. So long as a child is in the unchallenged presence of a principal attachment-figure, or within easy reach, he feels secure. A threat of loss creates anxiety, and actual loss, sorrow; both, moreover, are likely to arouse anger.
— JOHN BOWLBY

Most higher animals have evolved instinctive "bonding" mechanisms that keep the youngsters close to the parents. Human infants, too, are born with tendencies to form special attachments; all parents know their powerful effects. Early in life, most children become attached to one or a few family members or caretakers, sometimes so firmly that for several years such children may never stray more than a few meters from the attachment-figure. During those years, a prolonged separation of the child from those particular persons may be followed by an enduring depression or disturbance, during which the child's personality does not develop normally.

What is the function of childhood attachment? The simplest explanation is that it evolved to keep children within a safe sphere of nurture and protection. But according to our theory, our human bond machinery has the additional function of forcing children to acquire values, goals, and ideals from particular older individuals. Why is this so important? Because even though there are many ways a child could learn about ordinary causes and effects, *there is no way for a child to construct a coherent system of values—except by basing it upon some already existing model.* The task of constructing a "civilized personality" must be far beyond the inventive power of any single individual. Furthermore, if too wide a variety of adult models were available, it would be too hard to build a coherent personality of one's own, because one would have to pick and choose fragments from all those different personalities—and this might lead to so many conflicts and inconsistencies that many of them would cancel each other out. It would simplify the child's task if the attachment mechanism restricted attention to only a few role models.

How did our attachment-bonds evolve? In many species of animals, attachment occurs so swiftly and firmly that scientists who study animal behavior call it "imprinting." Presumably, the machinery that makes us learn our parents' goals is descended from the mechanisms of our animal ancestors. Presumably our infantile attachment-bonds form as soon as various inborn systems learn to distinguish the parents' individual peculiarities—first by senses of touch, taste, and smell; then by sound of voice and, finally, by sight of face.

Once those attachment-bonds are formed, a child won't react in the same way to the faces and voices of strangers and parents, for these have different effects on how we learn. The effect of an attachment-person's affection or rejection is not like that of ordinary "success-failure" goal-rewards—which merely teach us what to do in order to achieve our goals. Attachment-related signals seem to work directly on those goals themselves—and thus can modify our personalities. Attachments teach us ends, not means—and thus impose on us our parents' dreams.

17.4 FUNCTIONAL AUTONOMY

We've talked about some ways to learn goals from other people. But how do we come to make goals for ourselves? It seems simple enough always to move from goal to subgoal—but how could one go the other way, moving outward to find new kinds of goals? Our answer may seem strange at first: there is a sense in which we never really need to invent new "high-level" goals at all. This is because, in principle, at least, it is enough to keep inventing lower-level subgoals for problems that we have to solve! Here is why this need not limit our ambitions:

> **Functional Autonomy.** *In the course of pursuing any sufficiently complicated problem, the subgoals that engage our attentions can become both increasingly more ambitious and increasingly detached from the original problem.*

Suppose a baby's initial goal was to reach a certain cup. This could lead to the subgoal of learning how to move the arm and hand efficiently, which, in turn, could lead to sub-subgoals of learning to move around obstacles. And this could keep growing into increasingly general and abstract goals of learning how to understand and manage the physical world of space and time. Thus one can begin with a lowly goal, yet end up with some sub-subgoals that lead our minds into the most ambitious enterprises we can conceive.

This can also happen in the social realm. The same baby can form, instead, the subgoal of engaging another person's help in bringing it that drinking cup. This can lead to trying to find more effective ways to influence that other person—and thus the child could become concerned with representing and predicting the motives and dispositions of other people. Again, a relatively modest drinking goal can lead to a larger competence—this time in the realm of comprehending social interactions. An initially simple concern with personal comfort becomes transformed into a more ambitious, less self-centered enterprise.

> *Virtually any problem will be easier to solve the more one learns about the context world in which that problem occurs. No matter what one's problem is, provided that it's hard enough, one always gains from learning better ways to learn.*

Many of us like to believe that our intellectual enterprises lie on higher planes than our everyday activities. But now we can turn that academic value-scheme upon its head. When we get right down to it, our most abstract investigations can be seen as having origins in finding means to ordinary ends. These turn into what we regard as noble qualities when they gain enough functional autonomy to put their roots aside. In the end, our initial goals matter scarcely at all, because no matter what our original objectives, we can gain more by becoming better able to predict and control our world. It may not even matter whether an infant was initially inclined to emulate or to oppose a parent, or was first moved primarily by fear or by affection. The implements of accomplishment are much the same in either case. Knowledge is power. Whatever one's goals, they will be easier to achieve if one can become wise, wealthy, and powerful. And these in turn can best be gained by understanding how things work.

17.5 DEVELOPMENTAL STAGES

On the surface, the theories of Jean Piaget and Sigmund Freud might seem to lie in different scientific universes. Piaget seems to be concerned almost wholly with intellectual matters, while Freud studies emotional mechanisms. Yet the differences are not really clear. It is widely understood that emotional behavior depends on unconscious machinery, but we do not so often recognize that ordinary "intellectual" thinking also depends on mechanisms that are equally hidden from introspection.

In any case, despite their differences, both these great psychologists asserted that every child proceeds through "stages" of mental development. And surely every parent notices how children sometimes seem to stay the same but at other times appear to change more rapidly. Rather than review particular theories of how children progress through stages, let's look at the concept of "stage" itself.

Why can't we grow by steady, smooth development?

I'll argue that nothing so complex as a human mind can grow, except in separate steps. One reason is that it is always dangerous to change a system that already works. Suppose you discover a new idea or way to think that seems useful enough to justify building more skills that depend on it. What happens if, later, it should turn out that this idea has a serious flaw? How could you restore your previous abilities? One way might be to maintain such complete records that you could "undo" all the changes that were made—but that wouldn't work if those changes had already made your quality of thought so poor that you couldn't recognize how poor it had become. A safer way would be to keep some older versions of your previous mind intact as you constructed each new version. Then you could "regress" to a previous stage in case the new one failed, and you could also use it to evaluate the performance of the new stage.

Another conservative strategy is never to let a new stage take control of actual behavior until there is evidence that it can outperform its predecessor. What would an outside observer see if a child employed this strategy? One would observe only "plateaus," during which there were few apparent changes in behavior, followed by "spurts of growth" in which new capacities emerge suddenly. Yet that appearance would be illusory, since the actual times of development would occur within those silent periods. This scheme has the great advantage of permitting the child to continue to function during mental growth and, thus, maintain "business during renovations." Each working version can hold still while new ones safely move ahead.

This applies to every large organization, not only to those involved in a child's development. Given a community that is already functioning, it is always dangerous to make more than a few changes at once. Each change is prone to have some harmful side effects on other systems that depend on it. Some of those side effects may not become apparent until so many of them have accumulated that the system has deteriorated past any point of turning back. Accordingly, it is better to stop from time to time to make inspections and repairs. The same is true for learning any complex skill; unless your goal is held unchanged for long enough, you won't have time enough to learn the skills required to accomplish it. It simply isn't practical to make minds grow by steady, smooth development.

17.6 PREREQUISITES FOR GROWTH

What controls the pace of mental growth? Although some aspects of development depend on external circumstances and others seem to happen only by chance, certain aspects of our growth seem almost to proceed relentlessly from stage to stage, as though those stages were predestined. This brings us back to asking why development proceeds in stages at all.

One reason a skill may grow in steps is that it needs "prerequisites." You cannot start to build a house by placing its roof on top; first you have to build some walls. That's not an arbitrary rule; it's inherent in the enterprise. It is the same for mental skills; some processes cannot be learned until certain other processes become available. Many of Piaget's theories were based on his suspicion that certain concepts had prerequisites. For example, he argued that a child must possess ideas about which operations are *reversible* before that child can grow good concepts about how quantities are conserved. Hypotheses like these led Piaget to do his great experiments. But consider how easily those experiments could have been done a thousand years before; the only equipment they required were children, water, and various jars. Were Piaget's ideas prerequisites for conceiving those experiments?

To build a good Society-of-More, it simply would not be practical for a child to introduce those middle-level agents *Appearance* and *History* until some lower-level agents such as *Tall*, *Thin*, *No Loss*, and *Reversible* had become available. Before that stage, there would be nothing for those managers to do! To be sure, that isn't strictly true, just as one *could* start to construct a house with a roof, by using temporary scaffolding and later building the house's sides. We can never be absolutely *sure* of what a skill's prerequisites must be—and this will always complicate psychology.

The reason we know so little about how children's minds grow is that we can't observe the processes that are responsible. It could take several years to refine a new agency, and during that time, the child's behavior will be dominated by other processes in other agencies, which are themselves growing through their own, overlapping stages of development. A serious problem for the psychologist is that *certain types of mental growth can never be directly observed at all*. This applies, in particular, to those all-important "B-brain" processes with which we learn new ways to learn. Only the indirect products of this ever appear in the child's actual behavior, and even these may not become overt until long after that higher-level growth has occurred. Perhaps most difficult of all is detecting the development of suppressors and censors. See 27.2. It is hard enough to analyze what people do, but it is almost impossible to recognize the things they *never* do.

To make matters worse, many of the "stages of development" that we actually observe do not really exist. From time to time, each parent has the illusion that a child has suddenly changed, when this is only the result of not observing several smaller, real changes in the past. In such a case, if there exists a "stage of growth," it is inside the parent's mind, and not in the child at all!

17.7 GENETIC TIMETABLES

When we first introduced Papert's principle—that is, the idea of growing by inserting new levels of management into old agencies—we did not ask *when* new layers should be built. If managers are inserted too soon, when their workers are still immature, little will get done. And if those managers come too late, that, too, would delay the mental growth. What could ensure that managers are not engaged too late or too soon? We all know children who seem to have matured too quickly or too slowly, in various respects, to match other areas of their growth. In an ideal system, each developing agency would be controlled by another agency equipped to introduce new agents just when they're needed—that is, when enough has been learned to justify the start of another stage. In any case, it would surely be disastrous if all our potential capacity to learn became available too soon. If every agent could learn from birth, they'd all be overwhelmed by infantile ideas.

One way to regulate such things would be to actuate new agencies at genetically predetermined times. At various stages of biological "maturity," certain classes of agents would be enabled to establish new connections, while others would be forced to slow their growth by making permanent connections that, till then, had been reversible. Could any clockwork scheme like this be guaranteed to work? Consider the fact that most of our children acquire agents like *Reversible* and *Confined* before they are five years old. For *those* children, at least, it would suffice to activate new intermediate-level agents at that age, so those children could proceed to build agents like *Appearance* and *History*. However, children who weren't ready yet would then be slightly handicapped by being forced to build some less-than-usually effective Societies-of-More. Nor would that rigid maturation-scheme serve well those children who had already moved "ahead of schedule." It would be better to have systems in which the timing of each stage depends on what has actually happened earlier.

One way a stagelike episode might start could stem from what we called the *investment principle*: once a certain skill surpasses all its close competitors, it becomes increasingly likely to be employed—and thereby increases its opportunities to develop even further. This self-enhancing effect can cause a spurt of rapid progress in which a particular skill quickly comes to dominate the scene. One way a stagelike episode might end could stem from what we called the *exception principle*. To see how this could happen, suppose that a certain agency develops so useful a way to do some job that many other agencies soon learn to exploit that capability. The more those other agencies become dependent on that skill, the more disruption will result from every further "improvement" in it—since it now has more customers to please! Even increasing the speed of one process could damage other agencies that depend upon how long it takes to work. Thus, once a scheme persists for long enough, it gets to be extremely hard to change—*not because of limitations inherent in itself or in the agency that developed it, but because of how the rest of the society depends upon its present form.*

Once it becomes too hard to change an old agency, it is time to build another one; further progress may require revolution rather than evolution. This is another reason why a complex system must be grown in a sequence of separate steps.

17.8 ATTACHMENT-IMAGES

Guilt is the gift that keeps on giving.
—JEWISH PROVERB

All people talk of goals and dreams, of personal priorities, of goods and bads, rights and wrongs, virtues and depravities. What makes our ethics and ideals develop in our children's minds?

In one of the theories of Sigmund Freud, an infant becomes enamored of one or both parents, and somehow this leads the baby into absorbing or, as Freud put it, "introjecting" the goals and values of those love-objects. Thenceforth, throughout later life, those parent-images persist inside the grown-up child's mind, to influence whatever thoughts and goals are considered worthy of pursuit. We are not compelled to agree with all of Freud's account, but we have to explain why children develop models of their parents' values at all. So far as the child's safety is concerned, it would suffice for attachment to keep the child in the parents' physical vicinity. What could be the biological and psychological functions of developing complicated self-ideals?

The answer seems quite clear to me. Consider that our models of ourselves are so complex that even adults can't explain them. How could a fragmentary infant mind know enough to build such a complicated thing—without some model upon which to base it? We aren't born with built-in Selves—but most of us are fortunate enough to be born with human caretakers. Then, our attachment mechanisms force us to focus on our parents' ways, and this leads us to build crude images of what those parents themselves are like. That way, the values and goals of a culture pass from one generation to the next. They are not learned the way skills are learned. We learn our earliest *values* under the influence of attachment-related signals that represent, not our own success or failure, but our parents' love or rejection. When we maintain our standards, we feel virtuous rather than merely successful. When we violate those standards, we feel shame and guilt rather than mere disappointment. This is not just a matter of words: those things are not the same; it is like the difference between ends and means.

How could coherence be imposed upon a multitude of mindless agencies? Freud may have been the first to see that this could emerge from the effects of infant attachment. It was several more decades before psychologists recognized that separating children from their attachments can have devastating effects on the growth of their personalities. Freud also observed that children frequently reject one parent in favor of the other, in a process that suggests the cross-exclusiveness of sexual jealousy; he called this the Oedipus complex. It seems plausible that something of this sort ought to happen regardless of any connection between attachment and sexuality. If a developing identity is based upon that of another person, it must become confusing to be attached to two dissimilar adult "models." This might lead a child to try to simplify the situation by rejecting or removing one of them from the scene.

Many people dislike the thought of being dominated from within by the image of a parent's wish. Yet, in exchange, that slavery is just what makes us relatively free (as compared with other animals) from being forced to obey so many other kinds of unlearned, built-in instinct-goals.

17.9 DIFFERENT SPANS OF MEMORIES

Everyone can master a grief but he who has it.
—WILLIAM SHAKESPEARE

Consider the plight of a mother with a new infant. Her baby will demand her time for many years. Sometimes she must wonder, *"How does this baby justify such sacrifice?"* Various answers come to mind: *"Because I love it." "Because someday it will care for me." "Because it's here to carry on our line."* But reasoning rarely brings answers to such questions. Usually, those questions simply fade away as parents continue to nurture their children as though they were parts of their own bodies. Sometimes, though, strains may overwhelm the mechanisms that protect each child from harm, and this results in tragedies.

These complex parent-to-child and child-to-parent bonds must be based on certain types of memory. Some memories are less changeable than others, and I suspect that attachment-bonds involve memory-records of a type that can be rapidly formed but then become peculiarly slow to change. On the child's side, perhaps these bonds are descended from the forms of learning called "imprinting," with which many kinds of infant animals quickly learn to recognize their parents. On the parents' side, the adult animals of many species will reject infants not involved in bonding shortly after birth; then foster-parenting becomes impossible. Why should bonding memories be so hard to change? In animals, there usually are evolutionary disadvantages to raising the offspring of unrelated individuals. Human infants must develop under the additional constraint of requiring constant adult models as a basis for their personalities. Similar goal-affecting bonds could explain the often irresistible force of "peer pressure" in later life. Perhaps all such attachment-bonds exploit the same machinery.

Many animals form other kinds of social bonds as well, like those in which an individual selects a mate and then remains attached to it for life. Many people do this, too, and a number of the ones who don't have been observed to select, instead, from among alternatives of seemingly similar appearance or character—as though those persons were attached, if not to individuals, to certain constant prototypes. Other people frequently find themselves enslaved by infatuations that some parts of their minds find unwelcome but cannot prevent or overcome; once formed, those memory-bonds will only slowly fade away. The time spans of our different sorts of memories evolved to suit, not our own needs, but those of our ancestors.

We all know the seemingly inexorable time span of mourning, in which it often takes so long to accept the loss of those we love. Perhaps this, too, reflects the slowness of attachment-change, though it is only one factor. This could also be partially responsible for the prolonged psychological disability that can follow the experience of a physical, emotional, or sexual assault upon a person. One might ask, since there are so many other devastating aspects of such an experience, why it should involve any connection with attachment memory. I suspect that any form of intimacy, however unwelcome, has effects upon machinery shared by both attachment and sexuality, and is liable to disturb or disrupt the machinery with which we make relationships in ordinary life. No matter how brief that violent episode, it may lead to long derangements in our usual relationships, in part because those agencies are slow to change. It doesn't help very much for the victim to try to view the situation neutrally, because the rest of the mind cannot control those agencies; only time can reconstruct their normal functioning. It is an injury more terrible than loss of sight or limb, to lose the normal use of the agencies with which one builds one's own identity.

17.10 INTELLECTUAL TRAUMA

One of Freud's conceptions was that the growth of many individuals is shaped by unsuspected fears that lurk in our unconscious minds. These powerful anxieties include the dread of punishment or injury or helplessness or, worst of all, the loss of the esteem of those to whom we are attached. Whether this is true or not, most psychologists who hold this view apply it only to the social realm, assuming that the world of intellect is too straightforward and impersonal to be involved with such feelings. But intellectual development can depend equally upon attachments to other persons and can be similarly involved with buried fears and dreads.

Later, when we discuss the nature of humor and jokes, we'll see that many of the consequences of both social and intellectual failures are rather similar. A major difference is that in the social world, only other persons can inform us about our violations of taboos—whereas within the realm of intellect, we can often detect our own deficiencies. A tower-building child needs no teacher to complain when a misplaced block spoils all the work. Nor does a thinking child's mind need anyone to tell it when some paradox engulfs and whirls it into a frightening cyclone. By itself, the failure to achieve a goal can cause anxiety. For example, surely every child must once have thought along this line:

> *Hmmm. Ten is nearly eleven. And eleven is nearly twelve. So ten is nearly twelve. And so on. If I keep on reasoning this way, then ten must be nearly a hundred!*

To an adult, this is just a stupid joke. But earlier in life, such an incident could have produced a crisis of self-confidence and helplessness. To put it in more grown-up terms, the child might think, *I can't see anything wrong with my reasoning—and yet it led to bad results. I merely used the obvious fact that if A is near B, and B is near C, then A must be near C. I see no way **that** could be wrong—so there must be something wrong with my mind.* Whether or not we can recollect it, we must once have felt some distress at being made to sketch the nonexistent boundaries between the oceans and the seas. What was it like to first consider *"Which came first, the chicken or the egg?"* What came before the start of time; what lies beyond the edge of space? And what of sentences like *"This statement is false,"* which can throw the mind into a spin? I don't know anyone who recalls such incidents as frightening. But then, as Freud might say, this very fact could be a hint that the area is subject to censorship.

If people bear the scars of scary thoughts, why don't these lead, as our emotion-traumas are supposed to do, to phobias, compulsions, and the like? I suspect the answer is that they do—but disguised in forms we don't perceive as pathological. Every teacher knows and loathes how certain children turn away from learning things they believe they cannot learn: *"I simply can't. I'm just no good at that."* Sometimes this might represent only a learned way to avoid the shame and stress that came from social censure of failures in the past. But it might equally represent a reaction to the *nonsocial* stress that came from having been unable to deal with certain ideas themselves. Today, we generally regard emotional incompetence as an illness to be remedied. However, we generally accept incompetence of intellect as a normal, if unfortunate, deficiency in "talents," "aptitudes," and "gifts." Accordingly, we say things like *"That child isn't very bright,"* as though that person's poverty of thought were part of some predestined fate—and, therefore, isn't anyone's fault.

17.11 INTELLECTUAL IDEALS

If the mind were an ego-personality, it could do this and that as it would determine, but the mind often flies from what it knows is right and chases after evil reluctantly. Still, nothing seems to happen exactly as its ego desires. It is simply the mind clouded over by impure desires, and impervious to wisdom, which stubbornly persists in thinking of "me" and "mine."
—BUDDHA

How do we deal with thoughts that lead to frightening results? What should one think about the "nearly" paradox that threatens to imply that all things, large and small, might be the same size? One strategy would be to constrain that kind of reasoning, by learning never to chain together more than two or three such *nearness* links. Then, perhaps, one might proceed to generalize that strategy, in fear that it's unsafe to chain together too many instances of *any* form of inference.

But what could the phrase *"too many"* mean? There is no universal answer. Just as in the case of *More*, we have to learn this separately in each important realm of thought: *what are the limitations of each type and style of reasoning?* Human thought is not based on any single and uniform kind of "logic," but upon myriad processes, scripts, stereotypes, critics and censors, analogies and metaphors. Some are acquired through the operation of our genes, others are learned from our environments, and yet others we construct for ourselves. But even inside the mind, no one really learns alone, since every step employs many things we've learned before, from language, family, and friends—as well as from our former Selves. Without each stage to teach the next, no one could construct anything as complex as a mind.

There is another way our intellectual growth is not so different from our emotional development: we can make *intellectual* attachments, too, and want to *think* the way certain other persons do. These intellectual ideals may stem from parents, teachers, and friends; from persons one has never met, such as writers; even from legendary heroes who did not exist. I suspect we depend as much on images of how we ought to think as we do on images of how we ought to feel. Some of our most persistent memories are about certain teachers, but not about what was taught. (At the moment I'm writing this, I feel as though my hero Warren McCulloch were watching disapprovingly; he would not have liked these neo-Freudian ideas.) No matter how emotionally neutral an enterprise may seem, there's no such thing as being "purely rational." One must always approach each situation with *some* personal style and disposition. Even scientists have to make stylistic choices:

> *Is there enough evidence yet, or should I seek more?*
> *Is it time to make a uniframe—or should I accumulate more examples?*
> *Can I rely on older theories here, or should I trust my latest guess?*
> *Should I try to be Reductionist or Novelist?*

At every step, the choices we make depend on what we have become. Our sciences, arts, and moral skills do not originate from detached ideals of truth, beauty, or virtue but stem partly from our endeavors to placate or please the images established in earlier years. Our adult dispositions thus evolve from impulses so infantile that we would surely censure them, if they were not by now transformed, disguised, or—as Freud said—"sublimated."

REASONING

Machines—with their irrefutable logic, their cold preciseness of figures, their tireless, utterly exact observations, their absolute knowledge of mathematics—they could elaborate any idea, however simple its beginning, and reach the conclusion. Machines had imagination of the ideal sort—the ability to construct a necessary future from a present fact. But Man had imagination of a different kind; the illogical, brilliant imagination that sees the future result vaguely, without knowing the why, nor the how; an imagination that outstrips the machine in its preciseness. Man might reach the conclusion more swiftly, but the machine always reached it eventually, and always the right conclusion. By leaps and bounds man advanced. By steady, irresistible steps the machine marched forward.

—JOHN W. CAMPBELL, JR.

18.1 MUST MACHINES BE LOGICAL?

What's wrong with the old arguments that lead us to believe that if machines could ever think at all, they'd have to think with perfect logic? We're told that by their nature, all machines must work according to rules. We're also told that they can only do exactly what they're told to do. Besides that, we also hear that machines can only handle quantities and therefore cannot deal with qualities or anything like analogies.

Most such arguments are based upon a mistake that is like confusing an agent with an agency. When we design and build a machine, we know a good deal about how it works. When our design is based on neat, logical principles, we are likely to make the mistake of expecting the machine to *behave* in a similarly neat and logical fashion. But that confuses what the machine does inside itself—that is, how it "works"—with our expectations of how it will appear to behave in the outer world. Being able to explain in logical terms how a machine's parts work does not automatically enable us to explain its subsequent *activities* in simple, logical terms. Edgar Allan Poe once argued that a certain chess-playing "machine" had to be fraudulent because it did not always win. If it were really a machine, he argued, it would be perfectly logical—and therefore could never make any mistakes! What is the fallacy in this? *Simply that there is nothing to prevent us from using logical language to describe illogical reasoning.* To a certain extent it's true that machines can do only what they are designed to do. But this does not preclude us, when once we know how thinking works, from designing machines that think.

When do we actually use logic in real life? We use it to simplify and summarize our thoughts. We use it to explain arguments to other people and to persuade them that those arguments are right. We use it to reformulate our own ideas. But I doubt that we often use logic actually to solve problems or to "get" new ideas. Instead, we formulate our arguments and conclusions in logical terms *after* we have constructed or discovered them in other ways; only then do we use verbal and other kinds of formal reasoning to "clean things up," to separate the essential parts from the spaghettilike tangles of thoughts and ideas in which they first occurred.

To see why logic must come afterward, recall the idea of solving problems by using the generate and test method. In any such process, logic can be only a fraction of the reasoning; it can serve as a test to keep us from coming to invalid conclusions, but it cannot tell us which ideas to generate, or which processes and memories to use. Logic no more explains how we think than grammar explains how we speak; both can tell us whether our sentences are properly formed, but they cannot tell us which sentences to make. Without an intimate connection between our knowledge and our intentions, logic leads to madness, not intelligence. A logical system without a goal will merely generate an endless host of pointless truths like these:

> *A implies A.*
> *P or not P.*
> *A implies A or A or A.*
> *If 4 is 5, then pigs can fly.*

18.2   CHAINS OF REASONING

Here's a rule that's part of ordinary common sense: If A *depends on* B, and B *depends on* C, then—clearly—A *depends on* C. But what do such expressions mean? And why do we make the same kinds of inferences not only for dependency but also for implication and causality?

If A *depends on* B, and, also, B *depends on* C, then A *depends on* C.
If A *implies* B, and, also, B *implies* C, then A *implies* C.
If A *causes* B, and, also, B *causes* C, then A *causes* C.

What do all these different ideas have in common? All lend themselves to being linked into chainlike strings. Whenever we discover such sequences—however long they may be—we regard it as completely natural to compress them into single links, by deleting all but the beginning and the end. This lets us "conclude," for example, that A *depends on*, *implies*, or *causes* C. We do this even with imaginary paths through time and space.

Floor *holds* Table *holds* Saucer *holds* Cup *holds* Tea
Wheel *turns* Shaft *turns* Gear *turns* Shaft *turns* Gear

Sometimes we even chain together different kinds of links:

House *walk to* Garage *drive to* Airport *fly to* Airport
Owls *are* Birds, and Birds *can* Fly. So, Owls *can* Fly.

The chain containing "walk," "drive," and "fly" may appear to use several different kinds of links. But although they differ in regard to *vehicles*, they all refer to paths through space. And in the Owl-Bird example, "are" and "can" seem more different at first, but we can translate them both into a more uniform language by changing "Owls *are* Birds" into "An Owl *is a* Typical-Bird" and "Birds *can* Fly" into "A Typical-Bird *is a* thing-which-can-Fly." Both sentences then share the same type of "*is a*" link, and this allows us to chain them together more easily.

For generations, scientists and philosophers have tried to explain ordinary reasoning in terms of logical principles—with virtually no success. I suspect this enterprise failed because it was looking in the wrong direction: common sense works so well not because it is an approximation of logic; logic is only a small part of our great accumulation of different, useful ways to chain things together. Many thinkers have assumed that *logical necessity* lies at the heart of our reasoning. But for the purposes of psychology, we'd do better to set aside the dubious ideal of faultless deduction and try, instead, to understand how people actually deal with what is *usual* or *typical*. To do this, we often think in terms of causes, similarities, and dependencies. What do all these forms of thinking share? They all use different ways to make chains.

18.3 CHAINING

Why is chaining so important? Because, as we've just seen, it seems to work in many different realms. More than that, it can also work in several ways at once, inside the same world. Consider how, with no apparent mental strain at all, we can first imagine the same kind of arch to be a bridge, a tunnel, or a table—and then we can imagine chains of these, according to quite different views:

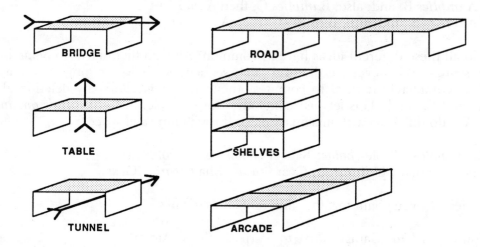

Chaining seems to permeate not only how we reason, but how we think of structures in space and time. We find ourselves involved with chains whenever we imagine or explain. Why does the ability to build mental chains help us solve so many different kinds of problems? Perhaps because all sorts of chains share common properties like these:

> *When chains are stressed, the weakest links break first.*
> *To fix a broken chain, one needs to repair only its broken links.*
> *No part of a chain can be removed if both ends remain fixed.*
> *If pulling A makes B move, there must be a chain connecting A and B.*

Each separate rule seems common sense, at least when we apply it to a solid thing like a bridge, a fence, or a physical chain. But why do chains apply so well to insubstantial "lines of thought"? It is because there's such a good analogy between how chains can break and how reasoning can fail.

18.4 LOGICAL CHAINS

"Logic" is the word we use for certain ways to chain ideas. But I doubt that pure deductive logic plays much of a role in ordinary thinking. Here is one way to contrast logical reasoning and ordinary thinking. Both build chainlike connections between ideas. The difference is that in logic there's no middle ground; a logic link is either there or not. Because of this, a logical argument cannot have any "weakest link."

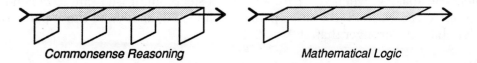

Commonsense Reasoning   *Mathematical Logic*

Logic demands just one support for every link, a single, flawless deduction. Common sense asks, at every step, if all of what we've found so far is in accord with everyday experience. No sensible person ever trusts a long, thin chain of reasoning. In real life, when we listen to an argument, we do not merely check each separate step; we look to see if what has been described so far seems plausible. We look for other evidence beyond the reasons in that argument. Consider how often we speak of reasoning in terms of structural or architectural expressions, as though our arguments were like the towers *Builder* builds:

> *"Your argument is **based** on **weak** evidence."*
> *"You must **support** that with more evidence."*
> *"That argument is **shaky**. It will **collapse**."*

In this way, commonsense reasoning differs from "logical" reasoning. When an ordinary argument seems weak, we may be able to support it with more evidence. But there is no way for a link inside a logic chain to use additional support; if it's not quite right, then it's absolutely wrong. Indeed, this weakness is actually the source of logic's own peculiar strength, because *the less we base our conclusions on, the fewer possibilities can exist for weaknesses in our arguments!* This strategy serves mathematics well—but it doesn't help us much in dealing with uncertainties. We cannot afford to stake our lives on chains that fall apart so easily.

I do not mean to say that there is anything wrong with logic; I only object to the assumption that ordinary reasoning is largely based on it. What, then, *are* the functions of logic? It rarely helps us get a new idea, but it often helps us to detect the weaknesses in old ideas. Sometimes it also helps us clarify our thoughts by refining messy networks into simpler chains. Thus, once we find a way to solve a certain problem, logical analysis can help us find the most essential steps. Then it becomes easier to explain what we've discovered to other people—and, also, we often benefit from explaining our ideas to ourselves. This is because, more often than not, instead of explaining what we actually did, we come up with a new formulation. Paradoxically, the moments in which we think we're being logical and methodical can be just the times at which we're most creative and original.

18.5 STRONG ARGUMENTS

When people disagree, we often say that one side's position seems "stronger" than the other. But what has "strength" to do with reasoning? In logic, arguments are simply either right or wrong, since there is not the slightest room for matters of degree. But in real life, few arguments are ever absolutely sure, so we simply have to learn how various forms of reasoning are likely to go wrong. Then we can use different methods to make our chains of reasoning harder to break. One method is to use several different arguments to prove the same point—putting them "in parallel." By analogy, when you park a car on a steep hill, it isn't safe to depend on brakes alone. No brake can work unless all its parts do—and, unhappily, all those parts form a long and slender chain no stronger than its weakest link.

> -- *Driver's foot presses on brake pedal.*
> --- *Brake pedal forces piston into master cylinder.*
> ---- *This forces brake fluid to flow from cylinder.*
> ----- *Brake fluid flows through tubes to brakes at wheels.*
> ------ *Pistons in brake cylinders apply force to brake shoes.*
> ------- *Brake shoes press on wheel drums, stopping wheels.*

An expert driver also leaves the car in gear and turns the wheels into the curb. Then, though no one of these tricks is perfectly secure, the combination cannot fail unless three things go wrong at once. This whole is stronger than any of its parts.

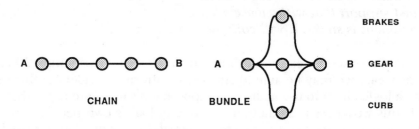

A chain can break with any single injury, but a parallel bundle cannot fail unless every one of its links has been broken. Our car can't roll away unless all three—brake, wheel, and parking gear—go wrong at once. Parallel bundles and serial chains are only the simplest ways to link together various parts. Here are some others.

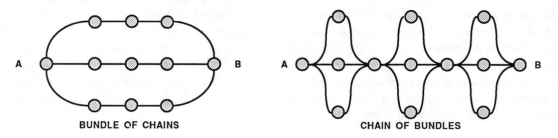

Each serial connection makes a structure weaker, while each parallel connection makes it stronger.

18.6 MAGNITUDE FROM MULTITUDE

We like to think of reasoning as rational—yet we often represent our arguments as fights between adversaries positioned to determine which can wield the greater strength or force. Why do we use such energetic and aggressive images of weakness, strength, defeat, and victory, of boxing in and breaking down an enemy's defense? Why don't we just use cool, clear, faultless reasoning to prove we are right? The answer is that we rarely need to know that anything is absolutely wrong or right; instead, we only want to choose the best of some alternatives.

Here are two different strategies for deciding whether one group of reasons should be considered "stronger" than another. The first strategy tries to compare opposing arguments in terms of magnitudes, by analogy to how two physical forces interact:

> **Strength from Magnitude:** *When two forces work together, they add to form a single larger force. But when two forces oppose each other directly, their strengths subtract.*

Our second strategy is simply to *count* how many different reasons you can find for choosing each alternative:

> **Strength from Multitude:** *The more reasons we can find in favor of a particular decision, the more confidence we can have in it. This is because if some of those reasons turn out to be wrong, other reasons may still remain.*

Whichever strategy we use, we tend to speak of the winning argument as the "stronger" one. But why do we use the same word "strong" for two such different strategies? It is because we use them both for the same purpose: *to reduce the likelihood of failure.* It comes out the same in the end, whether we base a decision on a single "strong" argument—that is, one unlikely to be wrong—or on several weaker arguments, in hopes that they won't fail all at once.

What makes us so prone to formulate our reasoning in terms of conflicting adversaries? It must be partly cultural, but some of it could also be based on inheritance. When we use architectural metaphors that speak of arguments as not supported properly, we could be exploiting structures that evolved within our spatial agencies. Similarly, when we represent our reasoning in terms of battling adversaries, we might be exploiting the use of agencies that first evolved for physical defense.

18.7 WHAT IS A NUMBER?

Why do we find it so hard to explain the meanings of things? Because what something "means" depends upon each different person's state of mind. If so, you might suspect that nothing means exactly the same thing to any two different persons. But if that were the case, where could you start? If every meaning in a person's mind depended on all the other meanings there, wouldn't everything go in circles? And if you couldn't break into those circles, wouldn't it all become too subjective to make good science? No. There is nothing wrong with phenomena in which many things depend on one another. And you don't have to be *in* those circles in order to understand them; you simply have to make good theories about them. It is a pleasant dream to imagine things being defined so perfectly that different people could understand things in exactly the same ways. But that ideal can't be achieved, because in order for two minds to agree *perfectly*, at every level of detail, they'd have to be identical.

The closest we can come to agreeing on meanings is in mathematics, when we talk of things like "Three" and "Five." But even something as impersonal as "Five" never stands isolated in a person's mind but becomes part of a huge network. For example, we sometimes think of "Five" for counting things, as when we recite "One, Two, Three, Four, Five" while taking care 1) to touch each thing only once, and 2) never to touch anything more than once. One way to ensure *that* is to pick up each thing as it's counted and remove it. Another way is to match a group of things to a certain standard set of Five—such as the fingers of your hand—or to that silent stream of syllables spoken in the mind. If, one by one, the things are matched and none are left behind, then there were Five. Another way to think of Five is to imagine some familiar shape—a pentagon, an **X** or **V** or **W**, a star, or even an airplane:

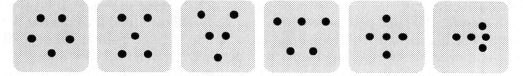

That way, a child might even come to understand a larger number before a smaller one. I actually knew one child who seemed to know Six before she knew Five, because she'd played so much with sets of triangles and hexagons.

Each number meaning works in different problem worlds. To ask which meaning is correct —to count, match, or put into groups—is foolishness: each method helps the others, and all of them together make a mass of skills that grow in power and efficiency. The really useful "meanings" are not the flimsy logic chains of definitions, but the much harder-to-express networks of ways to remember, compare, and change things. A logic chain can break easily, but you get stuck less often when you use a cross-connected meaning-network; then, when any sense of meaning fails, you simply switch to another sense. Consider, for example, how many different Twos a child knows: two hands, two feet, two shoes, two socks, and all their inter-changeabilities. As for Threes, recall the popular children's tale about three bears. The bears themselves are usually perceived as Two and One—Momma and Poppa Bear, plus Baby Bear. But their forbidden porridge bowls are seen as quite another sort of Three: too hot, too cold, and then just right; a compromise between extremes.

18.8 MATHEMATICS MADE HARD

That theory is worthless. It isn't even wrong!
—WOLFGANG PAULI

Scientists and philosophers are always searching for simplicity. They're happiest when each new thing can be defined in terms of things that have already been defined. If we can keep doing this, then everything can be defined in successive layers and levels. This is how mathematicians usually define numbers. They begin by defining Zero—or, rather, they assume that Zero needs no definition. Then they define One as the "successor" of Zero, Two as the successor of One, and so on. But why prefer such slender chains? Why not prefer each thing to be connected to as many other things as possible? The answer is a sort of paradox.

> *As scientists, we like to make our theories as delicate and fragile as possible. We like to arrange things so that if the slightest thing goes wrong, everything will collapse at once!*

Why do scientists use such shaky strategies? So that when anything goes wrong, they'll be the first to notice it. Scientists adore that flimsiness because it helps them find the precious proofs they love, with each next step in perfect mesh with every single previous one. Even when the process fails, it only means that we have made a new discovery! Especially in the world of mathematics, it is just as bad to be nearly right as it is to be totally wrong. In a sense, that's just what mathematics *is*—the quest for absolute consistency.

But that isn't good psychology. In real life, our minds must always tolerate beliefs that later turn out to be wrong. It's also bad the way we let teachers shape our children's mathematics into slender, shaky tower chains instead of robust, cross-connected webs. A chain can break at any link, a tower can topple at the slightest shove. And that's what happens in a mathematics class to a child's mind whose attention turns just for a moment to watch a pretty cloud.

Teachers try to convince their students that equations and formulas are more expressive than ordinary words. But it takes years to become proficient at using the language of mathematics, and until then, formulas and equations are in most respects even less trustworthy than commonsense reasoning. Accordingly, the investment principle works against the mathematics teacher, because even though the potential usefulness of formal mathematics is great, it is also so remote that most children will continue to use only their customary methods in ordinary life, outside of school. It is not enough to tell them, "*Someday you will find this useful,*" or even, "*Learn this and I will love you.*" Unless the new ideas become connected to the rest of the child's world, that knowledge can't be put to work.

The ordinary goals of ordinary citizens are not the same as those of professional mathematicians and philosophers—who like to put things into forms with as few connections as possible. For children know from everyday experience that the more cross-connected their commonsense ideas are, the more useful they're likely to be. Why do so many schoolchildren learn to fear mathematics? Perhaps in part because we try to teach the children those formal definitions, which were designed to lead to meaning-networks as sparse and thin as possible. We shouldn't assume that making careful, narrow definitions will always help children "get things straight." It can also make it easier for them to get things scrambled up. Instead, we ought to help them build more robust networks in their heads.

18.9 ROBUSTNESS AND RECOVERY

Most machines that people build stop working when their parts break down. Isn't it amazing that our minds can keep on functioning while they're making changes in themselves? Indeed, they must, since minds can't simply shut down work when "closed for renovations." But how do we keep functioning while vital parts are modified—or even lost? It's a fact that our brains can keep on working well in spite of injuries in which great multitudes of cells are killed. How could anything be so robust? Here are some possibilities:

Duplication. It is possible to design a machine so that every one of its functions is embodied in several duplicated agents, in different places. Then, if any agent is disabled, one of its duplicates can be made to "take over." A machine based on this duplication-scheme could be surprisingly robust. For example, suppose that every function were duplicated in ten agents. If an accident were to destroy half the agents of that machine, the chance that any particular function would be entirely lost is the same as the chance that ten tossed coins would all come up tails—that is, less than one chance in a thousand. And many regions of the human brain do indeed have several duplicates.

Self-Repair. Many of the body's organs can regenerate—that is, they can replace whichever parts are lost to injury or disease. However, brain cells do not usually share this ability. Consequently, healing cannot be the basis of much of the brain's robustness. This makes one wonder why an organ as vital as the brain has evolved to be less able than other organs to repair or replace its broken parts. Presumably, this is because it simply wouldn't help to replace individual brain-agents—unless the same healing process could also restore all the learned connections among those agents. Since it is those networks that embody what we've learned, merely to replace their separate parts would not restore the functions that were lost.

Distributed Processes. It is possible to build machines in which no function is located in any one specific place. Instead, each function is "spread out" over a range of locations, so that each part's activity contributes a little to each of several different functions. Then the destruction of any small portion will not destroy any function entirely but will only cause small impairments to many different functions.

Accumulation. I'm sure that all of the above methods are employed in our brains. But we also have another source of robustness that offers more advantages. Consider any learning-scheme that begins by using the method of *accumulation*—in which each agent tends to accumulate a family of subagents that can accomplish that agent's goals in several ways. Later, if any of those subagents become impaired, their supervisor will still be able to accomplish *its* job, because other of its subagents will remain to do that job, albeit in different ways. So accumulation—the very simplest kind of learning—provides both robustness and versatility. Our learning-systems can build up centers of diversity in which each agent is equipped with various alternatives. When such a center is damaged, the effects may scarcely begin to show until the system's reserves are nearly exhausted.

WORDS AND IDEAS

I am not yet so lost in lexicography, as to forget that words are the daughters of the earth, and that things are the sons of heaven. Language is only the instrument of science, and words are but the signs of ideas: I wish, however, that the instrument might be less apt to decay, and that signs might be permanent, like the things which they denote.

—SAMUEL JOHNSON

19.1 THE ROOTS OF INTENTION

*The wind blows where it will, and you hear the sound of it, but you
do not know whence it comes or whither it goes; so it is with every
one who is born of the Spirit.*
—St. John

Language builds things in our minds. Yet words themselves can't be the substance of our thoughts. They have no meanings by themselves; they're only special sorts of marks or sounds. If we're to understand how language works, we must discard the usual view that words *denote*, or *represent*, or *designate*; instead, their function is *control*: each word makes various agents change what various other agents do. If we want to understand how language works, we must never forget that our thinking-in-words reveals only a fragment of the mind's activity.

We often seem to think in words. Yet we do this with no conscious sense of where and why those words originate or how they then proceed to influence our further thoughts and what we subsequently do. Our inner monologues and dialogues proceed without any effort, deliberation, or sense of how they're done. Now you might argue that you *do* know what brings those words to mind—in the sense that they are how you "express" your intentions and ideas. But that amounts to the same thing—since your intentions, too, appear to come and go in ways you do not understand. Suppose, for example, that at a certain moment you find you want to leave the room. Then, naturally, you'd look for the door. And this involves two mysteries:

> *What made you want to leave the room?* Was it simply that you became tired of staying in that room? Was it because you remembered something else you had to do? Whatever reasons come to mind, you still must ask what led to *them*. The further back you trace your thoughts, the vaguer seem those causal chains.

> The other side of the mystery is that we are equally ignorant of how we *respond* to our own intentions. Given a desire to leave the room, what led you to the thought of "*door*"? You only know that first you thought, "*It's time to go,*" and then you thought, "*Where is the door?*"

We're all so used to this that we regard it as completely natural. Yet we have barely any sense of why each thought follows the last. What connects the idea of *leaving* with the idea of *door*? Does this result from some direct connection between two partial states of mind, of *leaving* and of *door*? Does it involve some sort of less direct connection, not between those states themselves, but only between some *signals* that somehow *represent* those states? Or is it the product of yet more complex mechanisms?

Our introspective abilities are too weak to answer such questions. The words we think seem to hover in some insubstantial interface wherein we understand neither the origins of the symbol-signs that seem to express our desires nor the destinations wherein they lead to actions and accomplishments. This is why words and images seem so magical: they work without our knowing how or why. At one moment a word can seem enormously meaningful; at the next moment it can seem no more than a sequence of sounds. And this is as it should be. It is the underlying emptiness of words that gives them their potential versatility. The less there is in a treasure chest, the more you'll be able to put in it.

WORDS AND IDEAS

19.2 THE LANGUAGE-AGENCY

The use of language is not confined to its being the medium through which we communicate ideas to one another. . . . Words are the instrument by which we form all our abstractions, by which we fashion and embody our ideas, and by which we are enabled to glide along a series of premises and conclusions with a rapidity so great as to leave in memory no trace of the successive steps of this process; and we remain unconscious of how much we owe to this.
—JOHN L. ROGET

We're normally quite unaware of how our brain-machines enable us to see, or walk, or remember what we want. And we're equally unaware of how we speak or of how we comprehend the words we hear. As far as consciousness can tell, no sooner do we hear a phrase than all its meanings spring to mind—yet we have no conscious sense of how those words produce their effects. Consider that all children learn to speak and understand—yet few adults will ever recognize the regularities of their grammars. For example, all English speakers learn that saying *"big brown dog"* is right, while *"brown big dog"* is somehow wrong. How do we learn which phrases are admissible? No language scientist even knows whether brains must learn this once or twice—first, for knowing what to say, and second, for knowing what to hear. Do we reuse the same machinery for both? Our conscious minds just cannot tell, since consciousness does not reveal how language works.

However, on the other side, language seems to play a role in much of what our consciousness does. I suspect that this is because our language-agency plays special roles in how we think, through having a great deal of control over the memory-systems in other agencies and therefore over the huge accumulations of knowledge they contain. But language is only a part of thought. We sometimes seem to think in words—and sometimes not. What do we "think in" when we aren't using words? And how do the agents that work with words communicate with those that don't? Since no one knows, we'll have to make a theory. We'll start by imagining that the language-system is divided into three regions.

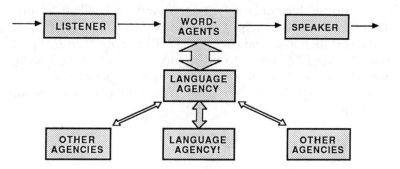

The upper region contains agents that are concerned specifically with words. The lower region includes all the agencies that are affected by words. And in the center lie the agencies involved with how words engage our recollections, expectations, and other kinds of mental processes. There is also one peculiarity: the language-agency seems to have an unusual capacity to control its own memories. Our diagram suggests that this could be because the language-agency can exploit itself as though it were just another agency.

19.3 WORDS AND IDEAS

How does an insubstantial word like "apple" lead you to think of a real thing—an object of a certain size that is red, round, sweet, and has a shiny, thin-peeled skin? How could a plain acoustic sound produce such complex states of mind, involving all those qualities of color, substance, taste, and shape? Presumably, each different quality involves a different agency. But then—in view of all we've said about why different agents can't communicate—how could such varying recipients all "understand" the selfsame messages? Do language-agents have unusual abilities to communicate with different kinds of agencies?

Many people have tried to explain language as though it were separate from the rest of psychology. Indeed, the study of language itself was often divided into smaller subjects, called by traditional names like *syntax*, *grammar*, and *semantics*. But because there was no larger, coherent theory of thinking to which to attach those fragments, they tended to lose contact with one another and with reality. Once we assume that language and thought are different things, we're lost in trying to piece together what was never separate in the first place. This is why, in the following pages, I'll put aside most of the old language theories and return to the questions that led to them:

> *How are words involved with mental processes?*
> *How does language enable people to communicate?*

In the next few sections, we'll introduce two kinds of agents that contribute to the power of words. The first kind, called "polynemes," are involved with our long-term memories. A polyneme is a type of K-line; it sends the same, simple signal to many different agencies: *each of those agencies must learn, for itself, what to do when it receives that signal.* When you hear the word "apple," a certain polyneme is aroused, and the signal from this polyneme will put your *Color* agency into a state that represents redness. The same signal will set your *Shape* agency into a state that represents roundness, and so forth. Thus, the polyneme for "apple" is really very simple; it knows nothing whatever about apples, colors, shapes, or anything else. It is merely a switch that turns on processes in other agencies, each of which has learned to respond in its own way.

Later we'll discuss another type of language-agent that we'll call an "isonome." Each isonome controls a short-term memory in each of many agencies. For example, suppose we had just been talking about a certain apple, and then I said, *"Please put it in this pail."* In this case, you would assume that the word "it" refers to the apple. However, if we had been discussing your left shoe, you would assume "it" referred to that shoe. A word like "it" excites an isonome whose signal has no particular significance by itself, but controls what various agencies do with certain recent memories.

19.4 OBJECTS AND PROPERTIES

What does a word like "apple" mean? This is really many questions in one.

How could hearing the word "apple" make you "imagine" an apple?
How could seeing an apple activate a word-agent for "apple"?
How could thinking about an apple make one think of the word for "apple"?
How could seeing an apple make one wordlessly recall the flavor of an apple?

It's usually impossible to perfectly "define" a word because you cannot capture everything you mean in just a phrase; an apple means a thousand things. However, you can usually say *some* of what you mean by making lists of *properties*. For example, you could say that an "apple" is something round and red and good to eat. But what exactly is a "property"? Again, it's hard to define that idea—but there are several things to say about what properties we like our properties to have.

We like the kinds of properties that do not change capriciously.

The color of your car will stay the same from day to day, and, barring accidents, so will its basic size and shape, as well as the substances of which it is made. Now, suppose you were to paint that car a new color: its shape and size would remain the same. This suggests another thing we like to find in our properties:

The most useful sets of properties are those whose members do not interact too much.

This explains the universal popularity of that particular combination of properties: *size, color, shape,* and *substance*. Because these attributes scarcely interact at all with one another, you can put them together in any combination whatsoever, to make an object that is either large or small, red or green, wooden or glass, and having the shape of a sphere or of a cube. And we derive a wonderful power from representing things in terms of properties that do not interact: *this makes imagination practical.* It lets us anticipate what will happen when we invent new combinations and variations we've never seen before. For example, suppose that a certain object almost works for a certain job—except for being a bit too small; then you can imagine using a "larger" one. In the same way, you can imagine changing the color of a dress or its size, shape, or the fabric of which it's made, without altering any of its other properties.

Why is it so easy to imagine the effects of such changes? First, these properties reflect the nature of reality; when we change an object's color or shape, its other properties are usually left unchanged. However, that doesn't explain why such changes do not interact inside the mind. Why is it so easy to imagine a small brown wooden cube or a long red silk skirt? The simplest explanation is that we represent each of the properties of material, color, size and shape in separate agencies. Then those properties can simultaneously arouse separate partial states of mind at once, in several divisions of the mind. That way, a single word can activate many different kinds of thoughts at once! Thus the word "apple" can set your *Color* agency into a "redness" state, put your *Shape* agency into a "roundness" state—or, really, into a representation of an indented sphere with a stem—and cause your *Taste* and *Size* agencies to react in accord with memories of previous experiences with apples. How does language do such things?

What happens when a single agent sends messages to several different agencies? In many cases, such a message will have a different effect on each of those other agencies. As I mentioned earlier, I'll call such an agent a "polyneme." For example, your word-agent for the word "apple" must be a polyneme because it sets your agencies for color, shape, and size into unrelated states that represent the independent properties of being red, round, and "apple-sized."

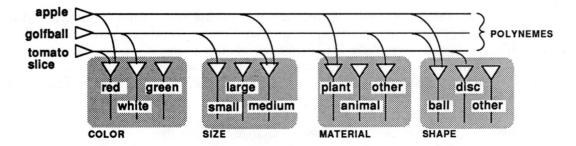

But how *could* the same message come to have such diverse effects on so many agencies, with each effect so specifically appropriate to the idea of "apple"? There is only one explanation: Each of those other agencies must already have *learned* its own response to that same signal. Because polynemes, like politicians, mean different things to different listeners, each listener must learn its own, different way to react to that message. (The prefix *poly-* is to suggest diversity, and the suffix *-neme* is to indicate how this depends on memory.)

> *To understand a polyneme, each agency must learn its own specific and appropriate response. Each agency must have its private dictionary or memory bank to tell it how to respond to every polyneme.*

How could all those agencies learn how to respond to each polyneme? If each polyneme were connected to a K-line in each agency, each of those K-lines would need only to learn what partial state to arouse inside its agency. The drawing below suggests that those K-lines could form little "memorizers" next to the agencies that they affect. Thus, memories are formed and stored close to the places where they are used.

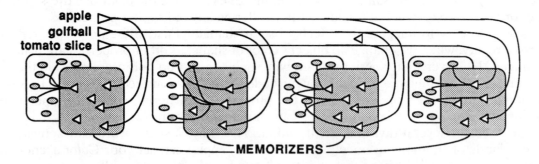

Can any simple scheme like this give rise to all the richness of the meaning of a real language-word? The answer is that *all* ideas about meaning will seem inadequate by themselves, since nothing can mean anything except within some larger context of ideas.

19.6 RECOGNIZERS

When we see an apple, how do we know it as an *apple*? How do we recognize a friend—or even know when we're seeing a person? *How do we recognize things?* The simplest way to recognize something is to verify that it has certain properties. To recognize an apple, in many circumstances, it might suffice to look for something that is red **AND** round **AND** apple-sized. In order to do that, we need a kind of agent that detects when all three conditions occur at once. The simplest form of this would be an agent that becomes active whenever all three of its inputs are active.

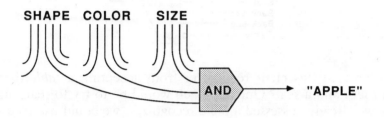

We can use **AND**-agents to do many kinds of recognition, but the idea also has serious limitations. If you tried to recognize a chair that way, you would usually fail, if you insisted on finding *"four legs **AND** a seat **AND** a back."* You scarcely ever see four legs of a chair at the same time, since at least one leg is usually out of view. Besides, if someone's sitting in the chair, you often cannot see the seat at all. In real life, no recognition-scheme will always work if it's based on absolutely perfect evidence. A more judicious scheme would not demand that every feature of a chair be seen; instead, it would only "weigh the evidence" that a chair is present. For example, we could make a chair-agent that becomes active whenever five or more of our six chair features are in sight:

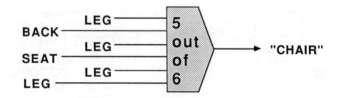

This scheme, too, will make mistakes. It will miss the chair if too many features are out of sight. It will mistake other objects for chairs if those features are present but in the wrong arrangement—for example, if all four legs are attached to the same side of a "seat." Indeed, it usually won't suffice merely to verify that all the required parts are there—one also has to verify their dimensions and relationships; otherwise our recognizer would not distinguish a chair from a couch or even from a bunch of boards and sticks. Failure to verify relationships is the basis of a certain type of nonsense joke:

> *What has eight legs and flies?*
> --- *A string quartet on a foreign tour.*

19.7 WEIGHING EVIDENCE

There are important variations on the theme of "weighing evidence." Our first idea was just to count the bits of evidence in favor of an object's being a chair. But not all bits of evidence are equally valuable, so we can improve our scheme by giving different "weights" to different kinds of evidence.

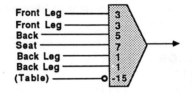

How could we prevent this chair recognizer from accepting a *table* that appeared to be composed of four legs and a seat? One approach would be to try to rearrange the weights to avoid this. But if we already possessed a table recognizer, we could use its output as evidence *against* there being a chair by adding it in with a *negative* weight! How should one decide what number weights to assign to each feature? In 1959, Frank Rosenblatt invented an ingenious evidence-weighing machine called a "Perceptron." It was equipped with a procedure that automatically learned which weights to use from being told by a teacher which of the distinctions it made were unacceptable.

All feature-weighing machines have serious limitations because, although they can measure the presence or absence of various features, they cannot take into account enough of the relations among those features. For example, in the book *Perceptrons*, Seymour Papert and I proved mathematically that no feature-weighing machine can distinguish between the two kinds of patterns drawn below—no matter how cleverly we choose the weights.

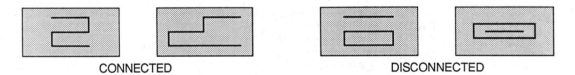

CONNECTED   DISCONNECTED

Both drawings on the left depict patterns that are *connected*—that is, that can be drawn with a single line. The patterns on the right are disconnected in the sense of needing two separate lines. Here is a way to prove that no feature-weighing machine can recognize this sort of distinction. Suppose that you chopped each picture into a heap of tiny pieces. It would be impossible to say which heaps came from connected drawings and which came from disconnected drawings—simply because each heap would contain identical assortments of picture fragments! Every heap would contain exactly four pictures of right-angle turns, two "line endings," and the same total lengths of horizontal and vertical line segments. It is therefore impossible to distinguish one heap from another by "adding up the evidence," because all information about the *relations* between the bits of evidence has been lost.

19.8 GENERALIZING

We're always learning from experience by seeing some examples and then applying them to situations that we've never seen before. A single frightening growl or bark may lead a baby to fear all dogs of similar size—or, even, animals of every kind. How do we make generalizations from fragmentary bits of evidence? A dog of mine was once hit by a car, and it never went down the same street again—but it never stopped chasing cars on other streets.

Philosophers of every period have tried to generalize about how we learn so much from our experiences. They have proposed many theories about this, using names like "abstraction," "induction," "abduction," and so forth. But no one has found a way to make consistently correct generalizations—presumably because no such foolproof scheme exists, and whatever we "learn" may turn out to be wrong. In any case, we humans do not learn in accord with any fixed and constant set of principles; instead, we accumulate societies of learning-schemes that differ both in quality and kind.

We've already seen several ways to generalize. One way is to construct uniframes by formulating descriptions that suppress details we regard as insignificant. A related idea is built into our concept of a "level-band." Yet another scheme is implicit in the concept of a polyneme, which tries to guess the character of things by combining expectations based upon some independent properties. In any case, there is an intimate relationship between how we "represent" what we already know and the generalizations that will seem most plausible. For example, when we first proposed a "recognizer" for a chair, we composed it from the polynemes for several already familiar ideas, namely seats, legs, and backs. We gave these features certain weights.

If we changed the values of those evidence weights, this would produce new recognizer-agents. For example, with a *negative* weight for "back," the new agent would reject chairs but would accept benches, stools, or tables. If all the weights were increased (but the required total were kept the same), the new recognizer would accept a wider class of furniture or furniture with more parts hidden from view—as well as other objects that weren't furniture at all.

Why would there be any substantial likelihood that such variations would produce useful recognizers? That would be unlikely indeed, if we assembled new recognizers by combining old ones selected at random. But there is a much better chance for usefulness if each new recognizer is made by combining signals from agents that have already proven themselves useful in related contexts. As Douglas Hofstadter has explained:

> *Making variations on a theme is the crux of creativity. But it is not some magical, mysterious process that occurs when two indivisible concepts collide; it is a consequence of the divisibility of concepts into already significant subconceptual elements.*

19.9 RECOGNIZING THOUGHTS

How do we recognize our own ideas? At first, that must seem a strange question. But consider two different situations. In the first case, I hold up an apple and ask, *"What is this?"* We've already seen how such a sight could lead to activating polynemes for words like "apple" or "fruit." In the second case, there is no apple on the scene, and I ask instead, *"What do we call those round, red, thin-peeled fruits?"* Yet this time, too, you end up with an apple thought. Isn't it remarkable that one can "recognize" a thing merely from hearing words? What is there in common to our recognizing things in two such different ways? The answer is that inside the brain, these situations really aren't so different. In neither case is there a real apple in the brain. In both cases, some part of mind must recognize what's happening in certain other parts of mind.

Let's pursue this example and imagine that those words have caused three partial states to become active among your agencies. Your *Taste* agency is in the condition corresponding to *apple taste*, your *Physical Structure* agency is representing *thin-peeled*, and your *Substance* agency is in the state that corresponds to *fruit*. Thus, even though there is no apple in sight, this combination would probably activate one of your "apple" polynemes. Let's call them "apple-nemes" for short. How could we make a machine do that? We would simply attach another recognizer to the apple-neme, a recognizer whose inputs come from memories instead of from the sensory world.

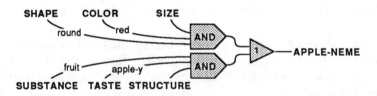

In a limited sense, such an agent could be said to recognize a certain state of mind or—if we dare to use the phrase—a certain combination of ideas. *In this sense, both physical and mental objects could engage similar representations and processes.* As we accumulate these recognizers, each agency will need a second kind of memory—a sort of recognition dictionary of recognizers to recognize its various states.

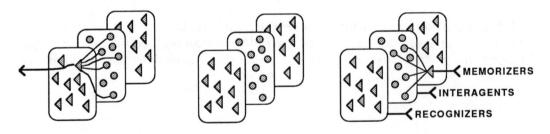

This simple scheme can explain only a little of how we represent ideas, since only certain things can be described by such simple lists of properties. We usually need additional information about constraints and relationships among the parts of things—for example, to represent the knowledge that the wheels of a car must be mounted *underneath* its body. To discover how we might represent such things is becoming a major concern of modern research in both psychology and Artificial Intelligence.

19.10 CLOSING THE RING

Now let's redraw the diagram for the language-agency, but fill in more details from the last few sections.

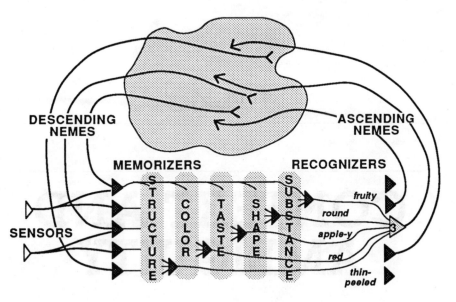

Something amazing happens when you go around a loop like this! Suppose you were to imagine three properties of an apple—for example, its substance, taste, and thin-peeled structure. Then, even if there were no apple on the scene—and even if you had not yet thought of the word "apple"—the recognition-agent on the left will be aroused enough to excite your "apple" polyneme. (This is because I used the number three for the required sum in the apple polyneme's recognizer instead of demanding that all five properties be present.) That agent can then arouse the K-lines in other agencies, like those for color and shape—and thus evoke your memories of other apple properties! In other words, if you start with enough clues to arouse one of your apple-nemes, it will automatically arouse memories of the other properties and qualities of apples and create a more complete impression, "simulus," or hallucination of the experience of seeing, feeling, and even of eating an apple. This way, a simple loop machine can reconstruct a larger whole from clues about only certain of its parts!

Many thinkers have assumed that such abilities lie beyond the reach of all machines. Yet here we see that retrieving the whole from a few of its parts requires no magic leap past logic and necessity, but only simple societies of agents that "recognize" when certain requirements are met. If something is red and round and has the right size and shape for an apple—and nothing else seems wrong—then one will probably think "apple."

This method for arousing complete recollections from incomplete clues—we could call it "reminding"—is powerful but imperfect. Our speaker might have had in mind not an apple, but some other round, red, fruit, such as a tomato or a pomegranate. Any such process leads only to guesses—and frequently these will be wrong. Nonetheless, to think effectively, we often have to turn aside from certainty—to take some chance of being wrong. Our memory systems are powerful *because* they're not constrained to be perfect!

CHAPTER 20

CONTEXT AND AMBIGUITY

Any attempt at a philosophical arrangement under categories of the words of our language must reveal the fact that it is impossible to separate and circumscribe the several groups by absolutely distinct boundaries. Were we to disengage their interwoven ramifications, and seek to confine every word to its main or original meaning, we should find some secondary meaning has become so firmly associated with many words and phrases, that to sever the alliance would be to deprive our language of the richness due to an infinity of natural adaptations.

—John L. Roget

20.1 AMBIGUITY

We often find it hard to "express our thoughts"—to summarize our mental states or put our ideas into words. It is tempting to blame this on the ambiguity of words, but the problem is deeper than that.

Thoughts themselves are ambiguous!

At first, one might complain that that's impossible. *"I'm thinking exactly what I'm thinking; there's no way it could be otherwise. And this has nothing to do with whether I can express it precisely."* But *"what you're thinking now"* is itself inherently ambiguous. If we interpret it to mean the states of *all* your agencies, that would include much that cannot be "expressed" simply because it is not accessible to your language-agency. A more modest interpretation of *"what you're thinking now"* would be a partial indication of the present states of some of your higher-level agencies. But the significance of any agency's state depends on how it is likely to affect the states of other agencies. This implies that in order to "express" your present state of mind, you have to partially anticipate what some of your agencies are about to do. Inevitably, by the time you've managed to express yourself, you're no longer in the state you were before; your thoughts were ambiguous to begin with, and you never *did* succeed in expressing them but merely replaced them with other thoughts.

This is not just a matter of words. The problem is that our states of mind are usually subject to change. The properties of physical things tend to persist when their contexts are changed—but the "significance" of a thought, idea, or partial state of mind depends upon which other thoughts are active at the time and upon what eventually emerges from the conflicts and negotiations among one's agencies. It is an illusion to assume a clear and absolute distinction between "expressing" and "thinking," since expressing is itself an active process that involves simplifying and reconstituting a mental state by detaching it from the more diffuse and variable parts of its context.

The listener, too, must deal with ambiguity. You understand *"I wrote a note to my sister,"* despite the fact that the word "note" could mean a short letter or comment, a banknote, a musical sound, an observation, a distinction, or a notoriety. If all our separate words are ambiguous by themselves, why are sentences so clearly understood? Because the context of each separate word is sharpened by the other words, as well as by the context of the listener's recent past. We can tolerate the ambiguity of words because we are already so competent at coping with the ambiguity of thoughts.

20.2 NEGOTIATING AMBIGUITY

Many common words are ambiguous enough that even simple sentences can be understood in several ways.

> *The astronomer married the star.*

It probably was a movie star—though the listener may have experienced a moment of confusion. The trouble is that the word "star" is linked to different polynemes for a celestial body, a theatrical celebrity, or an object with a certain shape. The momentary confusion comes because the word "astronomer" gives us an initial bias toward the celestial sense of "star." But that inhuman meaning causes conflict in our *marriage-agent*, and this soon leads to another, more consistent interpretation. The problem is harder when a sentence contains two or more ambiguous words.

> *John shot two bucks.*

Alone, the word "shot" could refer either to shooting a gun or, in American slang, to making a bet. By itself, the word "buck" could mean either a dollar or a male deer. These alternatives permit at least *four* conceivable interpretations. Two of them are quite implausible, because people rarely shoot bullets at dollars or bet deer. But the other two are possible, since, unfortunately, people *do* bet dollars and shoot at deer. Without more clues, we have no way to choose between these interpretations. Yet we wouldn't have the slightest doubt that "buck" means dollar if the previous context gave a hint that money or gambling was involved—rather than hunting, forestry, or outdoor life.

How do "contexts" work to clarify such ambiguities? The activity of an *outdoors* polyneme would start by producing a small bias for arousing *deer* rather than *dollar* and *gun* instead of *bet*. Then the "ring-closing" effect would swiftly make that preference sharp. Other polynemes like those for *hunting* and *killing* will soon be engaged and will combine to activate the recognizers for yet other, related polynemes like those for *forest* and *animal*. Soon this will produce a collection of mutually supporting polynemes that establish a single, consistent interpretation.

One might fear that this would lead, instead, to an avalanche that arouses *all* the agents of the mind. That would be less likely to happen if the different possible senses of each word are made to compete, by being assembled into *cross-exclusion* groups. Then, as the polynemes for *deer* and *gun* gain strength, they will weaken and suppress the competing nemes for *money* and for *wagering*—and that will weaken, in turn, the other polynemes that support the alternative context of making bets. The end effect of this will be almost instantaneous. In only a few cycles of the meaning ring, the agents associated with *deer* and *gun* will completely suppress their competitors.

20.3 VISUAL AMBIGUITY

We usually think of "ambiguity" as an aspect of language—but ambiguities are just as common in vision, too. What's that structure shown below? It could be regarded as nine separate blocks, as an arch supported by two other arches, or as a single, complicated, nine-block arch!

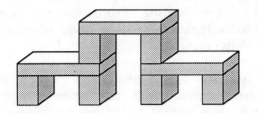

What process makes us able to see that superarch as composed of three little arches rather than of nine separate blocks? How, for that matter, do we recognize those as blocks in the first place, instead of seeing only lines and corners? These "ambiguities" are normally resolved so quickly and quietly that our higher-level agencies have no sense of conflict at all. To be sure, we sometimes have the sense of perceiving the same structure in several ways at once—for example, as both a single complex arch and as three separate simpler arches. But we usually lock in on one particular interpretation.

Sometimes no lower-level information can resolve an ambiguity—as in the case of this example by Oliver Selfridge.

Here, there is no difference whatever between the **H** and the **A,** yet we see them as having distinct identities in their different contexts. Evidently, the "simulus" produced by the visual sense is strongly affected by the state of some language-related agency. Furthermore, just as we can describe the same figure in different ways, we often can describe different figures in the same way. Thus, we recognize all these figures as similar, though no two of them are actually the same:

$$A \lhd \upharpoonright \forall A$$

If we described each of these in terms of the lengths, directions, and locations of their lines, they would all seem very different. But we can make them all seem much the same by describing each of them in the same way, perhaps like this: *"a triangle with two lines extended from one of its vertices."* The point is that what we "see" does not depend only on what reaches our eyes from the outside world. The manner in which we interpret those stimuli depends to a large extent on what is already taking place inside our agencies.

20.4 LOCKING-IN AND WEEDING-OUT

Most language-words are linked to many different polynemes, which correspond to the many "meaning-senses" of each word. To arouse so many polynemes at once would often lead to conflicts as each tries to set one's agencies into different states at the same time. If there are no other contextual clues, some of these conflicts would be resolved in accord with their connection strengths. For example, upon hearing *"The astronomer married the star,"* a playwright would tend to give priority to the theatrical sense of "star," whereas an astronomer would think first of a distant sun, *other things being equal*.

But other things are not usually equal. At every moment a person's mind is already involved with some "context" in which many agents are actively aroused. Because of this, as each new word arouses different polynemes, these will compete to change the states of those agents. Some of those changes will gain support as certain combinations of agents reinforce one another. Others that lose support and are left to stand alone will tend to weaken, and most ambiguities will thus be weeded out. In a few cycles, the entire system will firmly "lock in" on one meaning-sense for each word and firmly suppress the rest.

A computer program that actually worked this way was developed by Jordan Pollack and David Waltz. When applied to the sentence, *"John shot two bucks,"* and supplied with the faintest context clue, the program would indeed usually settle into a single, consistent interpretation. In other words, after a few cycles, the agents ended up in a pattern of mutually supporting activities in which only one sense of each word remained strongly active while all the other meaning-senses were suppressed. Thereafter, whether this "alliance" of word-senses was involved with hunting or with gambling, it became so self-supporting that it could resist any subsequent small signal from outside. In effect, the system had found a stable, unambiguous interpretation of the sentence.

What can be done if such a system settles on a wrong interpretation? Suppose, for example, that an "outdoors" clue had already made the system decide that John was hunting, but later it was told that John was gambling in the woods. Since a single new context clue might not be able to overcome an established alliance of meaning-senses, it might be necessary for some higher-level agency to start the system out afresh. What if the end result of locking-in were unacceptable to other agencies? Simply repeating the process would only lead to making the same mistake again. One way to prevent that would be to record which meaning-senses were adopted in the previous cycle and suppress them temporarily at the start of the next cycle. This would probably produce a new interpretation.

There is no guarantee that this method will always find an interpretation that yields a meaning consistent with all the words of the sentence. Then, if the locking-in process should fail, the listener will be confused. There are other methods that one could attempt, for example, to imagine a new context and then restart the ring-closing process. But no single method will always work. To use the power of language, one must acquire many different ways to understand.

20.5 MICRONEMES

That old idea of classifying things by properties is not entirely satisfactory, because so many kinds of qualities interact in complicated ways. Every situation or condition we experience is influenced or, so to speak, colored, by thousands of contextual shades and hues, just as looking through a tinted glass has faint effects on everything we see.

material: *animate, inanimate; natural, artificial; ideal, actual*
perceptual: *color, texture, taste, sound, temperature*
solidity: *hardness, density, flexibility, strength*
shape: *angularity, curvature, symmetry, verticality*
permanence: *rarity, age, fragility, replaceability*
location: *office, home, vehicle, theater, city, forest, farm*
environment: *indoors, outdoors; public, private*
activity: *hunting, gambling, working, entertaining*
relationship: *cooperation, conflict; negotiating, confronting*
security: *safety, danger; refuge, exposure; escape, defeat*

Some of these conditions and relationships may correspond to language-words, but for most of them we have no words, just as we have no expressions for most flavors and aromas, gestures and intonations, attitudes and dispositions. I'll call them "micronemes"—those inner mental context clues that shade our minds' activities in ways we can rarely express. There is a somewhat different microstructure to each person's thoughts; indeed, their inexpressibility reflects our individuality. Nevertheless, we can clarify our image of the mind by envisioning these unknown influences as embodied in the forms of particular agents. Accordingly, in the next few sections, we'll envision our micronemes as K-lines that reach into many agencies with widespread effects on the arousal and suppression of other agents—*including other micronemes*. These micronemes participate in all those "locking-in" and "weeding-out" processes, so that, for example, the activity of your microneme for "outdoors" makes a small contribution to arousing your "hunting" microneme. While each such effect may be relatively small, the effects of activating many micronemes will usually combine to establish a context within which most words are understood unambiguously.

For example, in one context the word "Boston" might bring to mind some thoughts about the American Revolution. In a different setting, the same word might lead one to think instead of a geographic location. Other contexts might yield thoughts about famous universities, sporting teams, life-styles, accents of speech, or traditional meals. Each of these concepts must be represented by a certain network of agents that are connected, directly or indirectly, to the word-agent for Boston. But hearing or thinking that word by itself is not enough to determine which of those word-sense networks to activate; this must also depend upon other aspects of the present context. Our hypothesis is that this comes about principally through each word-sense agent learning to recognize the activation of certain combinations of micronemes.

Even modest families of micronemes could span vast ranges of contexts. A mere forty *independent* micronemes could specify a trillion different contexts—and we surely have thousands, and perhaps millions of different micronemes.

20.6 THE NEMEIC SPIRAL

Our polynemes and micronemes grow into great branching networks that reach every level of every agency. They approximate the general form of a hierarchy, but one that is riddled with shortcuts, cross-connections, and exceptions. No one could ever comprehend all the details of the connections that develop inside a single human individual; that would amount to grasping how all that person's thoughts and inclinations interact. At best, we can envision only the broadest outlines of such structures:

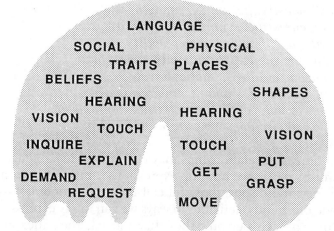

In regions near the agencies for speech, some elements of this network might signify or represent ideas and thoughts we can easily express in words. But because speaking is a social act, we are far less able to express the significance of the nemes involved with agencies that are not directly concerned with communication. This is because those agencies are less constrained by the discipline of public language; accordingly, the nemes inside those agencies can vary more from one person to the next.

In any case, our higher-level agencies are generally unaware of what our lower-level agents do; they supervise and regulate—but scarcely comprehend at all—what happens among their subordinates. For example, a high-level agency might find that a certain subagency is unproductive, because it is responding to too many micronemes—or to too few—and adjust its sensitivity accordingly. Like a B-brain, a controlling agency could make such judgments without comprehending the local meanings of those micronemes. This could also provide a basis for controlling the *levels* of the activities of those other agencies—just as we suggested when we discussed the idea of a "spiral" computation within an agency and its K-line trees. When the work appears to be going well, the supervisor can direct the lower-level processes to "spiral down" toward small details. But when there seem to be too many obstacles, the level of activity would be made to "spiral up" instead, to levels capable of diagnosing and altering an ineffective plan.

20.7 CONNECTIONS

To speak and understand a language well, an ordinary person must learn thousands of words. To learn the proper use of a single word must involve great numbers of connections between the agents for that word and other agents. What could cause such connections to be made, and what might be their physical embodiments?

Any comprehensive theory of the mind must include some ideas about the nature of the connections among agents. Consider that a person can learn to "associate" virtually any combination of ideas or words. Does this require us to assume that it is possible for a given K-line agent to become connected, directly, to any of thousands or millions of other agents? That seems out of the question, in view of what we know about connections in the human brain. Many brain cells have fibers that branch out enough to approach many thousands of other cells—but few of them branch out enough to reach *millions* of other cells, and as far as we know, a mature brain cell can only make new connections to other cells that already lie close to the fibers that branch into or out from it. Furthermore, we do not seem to grow many new brain cells after birth; on the contrary, their number actually decreases. To be sure, brain cells continue to "mature" for several years, and probably their fibers grow extensively. But no one yet knows whether this comes about as a result of learning new connections or whether it must happen first, to make it feasible for those cells to learn new connections.

Even the arrangements of long-distance connections between our brain cells do not permit *direct* connection between arbitrary pairs of agents, for those long connections are generally arranged in relatively orderly bundles, less regular but otherwise resembling the parallel pathways from skin to brain. Fortunately, direct connections are not really necessary, for the same reasons that every telephone in the world can easily be connected to any other telephone without the need for connecting a billion separate wires to each house. Instead, telephone systems make their connections indirectly, by using agencies called "exchanges" that require only moderate numbers of wires. I don't mean to suggest that brains use the sorts of switching-schemes found in telephone systems but only to say that it is not necessary for every K-line agent to directly contact every agent to which it might eventually become linked.

There are several factors that reduce the magnitude of the interconnection problem. First, in order to reproduce the major features of a remembered partial state of mind, it should suffice to activate only a representative sample of its agents. Second, according to our theory of knowledge-trees, most K-lines' connections are indirect to begin with, since they connect only to other, nearby K-line trees. A polyneme, too, need be connected only to a single memorizer agent near each agency. And no K-line needs potential connections to *all* the agents in any agency, since it is enough to make connections only in a certain level-band.

20.8 CONNECTION LINES

Or in the night, imagining some fear,
How easy is a bush supposed a bear.
—WILLIAM SHAKESPEARE

The diagram below depicts a connection-scheme that permits many agents to communicate with one another, yet uses surprisingly few connection wires. It was invented by Calvin E. Mooers in 1946, before the modern era of computers. Here is how we could use just ten wires to enable any of several hundred "transmitting-agents" to activate any of a similar number of "receiving-agents." The trick is to make each transmitting-agent excite not one, but five of those wires, chosen at random from the available ten. Then each receiving-agent is provided with an **AND**-agent connected to recognize the same five-wire combination.

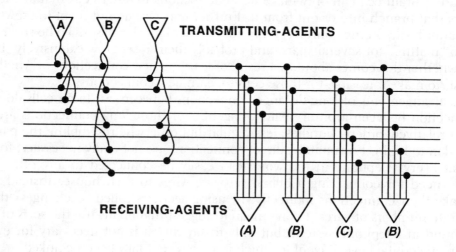

In this example, each receiving-agent is aroused by precisely one transmitting-agent. If we wanted each receiving-agent to respond to several transmitting-agents, we could join together several separate recognizers so that the receiving-agent's input looks like a tree with a recognizer at the tip of each branch. How could those receivers learn which input patterns to recognize? One way would be to use the kind of evidence-weighing machinery we described earlier. Indeed, for brain cells that would seem quite plausible, since a typical brain cell actually has a treelike network to collect its input signals. No one yet knows quite what those networks do, but I wouldn't be surprised if many of them turn out to be simple Perceptron-like learning machines.

The network shown in the diagram above has a serious deficiency: it can transmit only one signal at a time. The problem is that if several transmitting-agents were aroused at once, almost all ten connecting wires would be activated, which would then arouse *all* the receiving-agents and cause an avalanche. However, we can make that problem disappear by providing the system with enough additional connection wires. For example, suppose there were ten thousand connection wires rather than ten, and that each transmitting-agent became attached to about fifty of them. Then, even if one hundred agents were to send their signals all at once, there would be less than one chance in a trillion that this would erroneously activate any particular receiving-agent!

20.9 DISTRIBUTED MEMORY

Let's redraw our connection line–scheme in the form of three layers of agents.

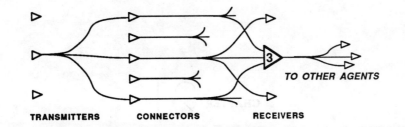

TRANSMITTERS   CONNECTORS   RECEIVERS

The transmitting-agents can simply be K-lines or memorizers, since each of them sends signals to a variety of other agents. The receiving-agents can be simple recognizers, since each of them is aroused only by certain combinations of connector-agents. However, because a typical agent must both arouse other agents and be aroused by other agents, it must tend to branch both at its inputs and at its outputs. So our network can be drawn to look like this:

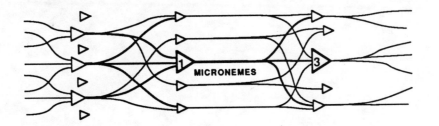

When we represent the agents this way, we see that they can all be simple evidence-weighing agents, only with different "threshold" values. Each recognizer could start out with connections to many connector-agents and then, somehow, *learn* to recognize certain signals by altering its connection weights. Would it be practical to build learning machines according to such a simple scheme? That was the dream of several scientists in the 1950s, but none of their experiments worked well enough to stimulate further work. Recently, a new type of network machine has shown more promise: the so-called Boltzmann machine resembles a Perceptron in having an automatic procedure for learning new connection weights, but it also has some ability to resolve ambiguities by using a variety of "ring-closing" process. The next few years should tell us more about such machines. Perhaps they could provide a basis for memory-systems that work very much the way K-lines do, to reproduce old partial states of mind.

In designing these clever ways to reduce the numbers of connecting wires, most researchers have proposed wiring the connections at random, so that no signal on any particular wire has any significance by itself but represents only a fragment of each of many unrelated activities. This has the mathematical advantage of producing very few accidental interactions. However, this seems to me a bad idea: *it would make it very hard for a transmitting-agency to learn how to exploit a receiving-agency's abilities.* I suspect that when we understand the brain, we'll discover that small groups of connection lines do indeed have local significance—because they will turn out to be the most important agents of nearby levels. The connection lines themselves will constitute our micronemes!

20.9 DISTRIBUTED MEMORY

CHAPTER 21

TRANS-FRAMES

21.1   THE PRONOUNS OF THE MIND

We often say what we want to say in fewer words than might seem possible.

> *"Do you see the table over there?"*   *"Yes."*
> *"Do you see the red block on it?"*   *"Yes."*
> *"Good. Please bring it to me."*

That first "it" saves the speaker from having to say *"that table"* again. The second "it" does the same for *"the red block."* Accordingly, many people consider a pronoun like "it" to be an abbreviation or substitute for another phrase used recently. But when we look at this more carefully, we see that a pronoun need not refer to any phrase at all. For instance, the word *"this"* in the previous sentence was not an abbreviation for any particular phrase. Instead, it served as a signal to you, the reader, to examine more carefully a certain partial state of mind —in this case, a certain theory about pronouns—that I assumed was aroused in your mind by the sentences that preceded it. In other words, pronouns do not signify objects or words; instead, they represent conceptions, ideas, or activities that the speaker assumes are going on inside the listener's mind. But how can the listener tell which one of the activities is signified when there are several possibilities?

> *"Do you remember the ring Jane liked?"*   *"Yes."*
> *"Good. Please buy it and give it to her."*

How do we know that "her" means Jane and that "it" means the ring—and not the other way around? We can tell that "her" is not the ring because English grammar usually restricts the pronoun "her" to apply only to a female person—though it could also mean an animal, a country, or a ship. But you would know that "it" means "the ring" in any case, because your *Buy* agency would not accept the thought of buying Jane, nor would your agency for *Give* accept the thought of giving gifts to rings. If someone said, "Buy Jane and give her to that ring," both *Buy* and *Give* would have such strong conflicts that the problem would ascend to the listener's higher-level agencies, which would react with disbelief.

Our language often uses pronounlike words to refer to mental activities—but we do not do this only in language: it happens in all the other higher-level functions of our minds. Later we'll see how *Find* will find a block—rather than, say, a toy giraffe—even though *Builder* has only said *"find."* The trick is that *Find* will use whichever description happens to be available in its current context. Since it's already working for *Builder*, its subagents will assume that it should find a building-block.

Whenever we talk or think, we use pronounlike devices to exploit whatever mental activities have already been aroused, to interlink the thoughts already active in the mind. To do this, though, we need to have machinery we can use as temporary "handles" for taking hold of, and moving around, those active fragments of mental states. To emphasize the analogy with the pronouns of our languages, I'll call such handles *"pronomes."* The next few sections speculate about how pronomes work.

21.2 PRONOMES

Why are sentences so easy to understand? How do we compress our ideas into strings of words and, later, get them out again? Typically, an English sentence is built around a verb that represents some sort of act, event, or change:

Jack drove from Boston to New York on the turnpike with Mary.

As soon as you hear such a thing, parts of your mind become engaged with these sorts of concerns related to driving:

ROLE	CONCERN	ROLE	CONCERN
ORIGIN	The initial state.	DESTINATION	The final state.
ACTION	What act was done.	ACTOR	Who caused it.
DIFFERENCE	What was changed.	CAUSE	What made the change.
RECIPIENT	Who was affected.	METHOD	How it was done.
MOTIVE	Why it was done.	OBSTACLE	What was the problem.
TRAJECTORY	The selected path.	INSTRUMENT	What tools were used.
OBJECT	What was affected.	VEHICLE	What vehicle was used.
TIME	When it happened.	PLACE	Where it occurred.

These concerns and "roles" seem so important that every language has developed special word-forms or grammatical constructions for them. *How do we know who drove the car?* We know that it's "Jack"—because the *Actor* comes before the verb. *How do we know a car was involved?* Because that is the default *Vehicle* for "drive." *When did all this happen?* In the past —because the verb *drive* has the form *dr-o-ve*. *Where did the journey start and end?* We know that those places are Boston and New York, respectively, because in English the prepositions *from* and *to* precede the *Origin* and *Destination*. But we often use the same prepositions for different kinds of concerns. In the sentence about driving, "from" and "to" refer to places in space. But in the sentence below they refer to intervals of time:

He changed the liquid from water to wine.

The liquid has changed its composition from what it was "at" some previous time. In English we use prepositions like "from," "to," and "at" both for places in space and for moments in time. This is not an accident, since *representing both space and time in similar ways lets us apply the selfsame reasoning skills to both of them.* Thus, many of our language-grammar "rules" embody or reflect some systematic correspondences—and these are among our most powerful ways to think. Many other language-forms have evolved similarly to make it easy for us to formulate, and communicate, our most significant concerns. The next few sections discuss how the "pronomes" we mentioned earlier could be involved in processes we use to make both verbal and nonverbal "chains of reasoning."

21.3 *TRANS-FRAMES*

Whenever we consider an action, such as moving from one place to another, we almost always have particular concerns like these:

Where does the action start? Where does it end?
What instrument is used? What is its purpose or goal?
What are its effects? What difference will it make?

We could represent several of these questions with a simple diagram, which we'll call the *Trans*-frame.

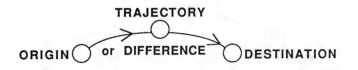

In the early 1970s, Roger Schank developed ways to represent many situations in terms of a relatively few kinds of relations which he called "conceptual dependencies." One of these, called *P-Trans*, represents a physical motion from one place to another. Another, called *M-Trans*, represents the sort of mental transportation involved when John tells Mary his telephone number; some information "moves" from John's memory to Mary's memory. A third type of conceptual dependency, called *A-Trans*, represents what is involved when Mary buys John's house. The house itself doesn't move at all, but its "ownership" is transferred from John's estate to Mary's estate.

But why should we want to represent, in the same way, three such different ideas: *transportation* in space, *transmission* of ideas, and *transfer* of ownership? I suspect that it is for the same reason that our language uses the same word fragment *trans* for all of them: this is one of those pervasive, systematic cross-realm correspondences that enables us to apply the same or similar mental skills to many different realms of thought. For example, suppose you were to drive first from Boston to New York, and then from New York to Washington. Obviously the overall effect would be equivalent to driving from Boston to Washington—but that wouldn't be so "obvious" unless you used a certain kind of mental chaining skill. Similarly, if John told you his phone number, and you then told it to Mary, this would end up much as though John had told Mary directly. And if you first bought John's house and then sold it to Mary, the net result, again, would be as though Mary had bought it directly from John. All three forms of *Trans*-frames can be used in chains! This means that once you learn efficient chain-manipulating skills, you can apply them to many different kinds of situations and actions. Once you know how to do it, constructing mental chains seems as easy as stringing beads. All you have to do is replace each *Trans*-frame's *Destination* with the next one's *Origin*.

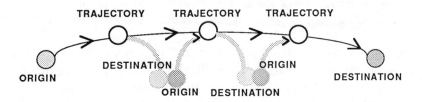

21.4 COMMUNICATION AMONG AGENTS

If agents had huge minds like ours, they could talk the way people do—and *Add* could say, "*Please, Get an apple and Put it in the pail.*" Perhaps our largest agencies can deal with messages like that, but smaller agencies like *Get* cannot interpret such expressions because they're much too specialized to understand complicated wants and needs. Then how could *Get* know *what* to get—in order to find an apple rather than a block, a fork, or a paper doll? To examine this problem, we'll have to make some assumptions about what happens in a listener's mind. For the moment, let's simply assume that the result is to activate a *Builder*-like society with these ingredients:

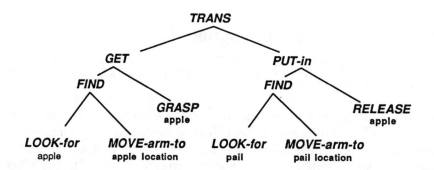

At first sight, it seems as though all these agents are involved with the apple and the pail. But a closer look shows only the low-level agents *Look-for* and *Grasp* are actually concerned with the physical aspects of actual objects; all the others are merely "middle-level managers." For example, the agent *Get* doesn't actually "get" anything; it only turns on *Find* and *Grasp* at the proper time. To be sure, *Look-for* will need some information about what to look for—that is, about an apple's appearance, and *Move* will need information about the apple's actual location. Nevertheless, we'll see that this information can become available to those agents without any need for messages from *Get*. To see how agents can operate without explicit messages, let's compare two ordinary language descriptions of how to put an apple into a pail. Our first script mentions each object explicitly in every line, the way one might speak to a novice.

> *Look for an **apple**.   Move the arm and hand to the **apple's** location.*
> *Prepare the hand to grasp an **apple**-shaped object. Grasp the **apple**.*
> *Now look for the **pail**.   Move the arm and hand to the **pail's** location.*
> *Release the hand's grip on the **apple**.*

Now let's rewrite this description in a style more typical of normal speech.

> *Look for an **apple**.   Move the arm and hand to its location.*
> *Prepare the hand to grasp an object of that shape. Grasp it.*
> *Now look for the **pail**.   Move the arm and hand to its location.*
> *Release the hand's grip.*

This second script uses the words "apple" and "pail" only once. This is how we usually speak; once something has been mentioned, we normally don't use its name again. Instead, whenever possible, we replace a name by a pronoun word. In the next few sections I'll argue that it is not only in language that we replace things with pronounlike tokens; we also do this in many other forms of thought. It could scarcely be otherwise, because in order to make sense to us, our sentences must mirror the structures and methods we use for managing our memories.

21.5 AUTOMATISM

How do higher-level agencies tell lower-level agents what to do? We might expect this problem to be harder for smaller agencies because they can understand so much less. However, smaller agencies also have fewer concerns and hence need fewer instructions. Indeed, the smallest agencies may need scarcely any messages at all. For example, there's little need to tell *Get*, *Put*, or *Find* what to "get," "put," or "find"—since each can exploit the outcome of the *Look-for* agency's activity. But how can *Look-for* know what to look for? We'll see the answer in a trick that makes the problem disappear. In ordinary life this trick is referred to by names like "expectation" or "context."

Whenever someone says a word like "apple," you find yourself peculiarly disposed to "notice" any apple in the present scene. Your eyes tend to turn in the direction of that apple, your arm will prepare itself to reach out in that direction, and your hand will be disposed to form the corresponding grasping shape. This is because many of your agencies have become immersed in the "context" produced by the agents directly involved with whatever subject was mentioned recently. Thus, the polyneme for the word "apple" will arouse certain states of agencies that represent an object's color, shape, and size, and these will automatically affect the *Look-for* agency—simply because *that agency must have been formed in the first place to depend upon the states of our object-description agencies.* Accordingly, we can assume that *Look-for* is part of a larger society that includes connections like these:

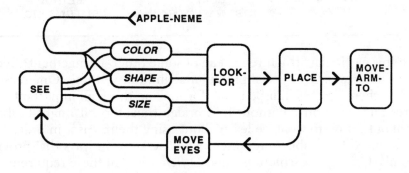

This diagram portrays an automatic "finding machine." Whether an apple was actually seen, imagined, or suggested by naming it, the agents for *Color*, *Shape*, and *Size* will be set into states that correspond to "red, round, and apple-sized." Accordingly, when *Look-for* becomes active, it cannot help but seek an object with those properties. Then, according to our diagram, once such a thing is found, its *location* will automatically be represented by the state of an agency called *Place*—again, because this is the environment within which *Look-for* grew. The same must be true of the agency *Move-arm-to*, which must also have grown in the context of some location-representing agency like *Place*. So when *Move-arm-to* is aroused, it will automatically tend to move the arm and hand toward that location without needing to be told. Thus, such an arrangement of agencies can carry out the entire apple-moving script with virtually no "general-purpose" communication at all.

This could be one explanation of what we call "focus of mental attention." Because the agency that represents locations has a limited capacity, whenever some object is seen or heard —or merely imagined—other agencies that share the same representation of location are likely to be forced to become engaged with the same object. Then this becomes the momentary "it" of one's immediate concern.

21.6 TRANS-FRAME PRONOMES

Our first *Trans*-frame scheme was connected to only four pronomes—*Origin, Destination, Difference,* and *Trajectory.* These are just enough to link together a simple chain of reasoning. But what about all the other roles that things can play—like *Actors, Recipients, Vehicles, Goals, Obstacles,* and *Instruments?* In order to keep track of these, we surely need some other pronomes, too. So now we shall imagine a larger kind of *Trans*-frame scheme that engages, all at once, a much larger constellation of different pronome roles.

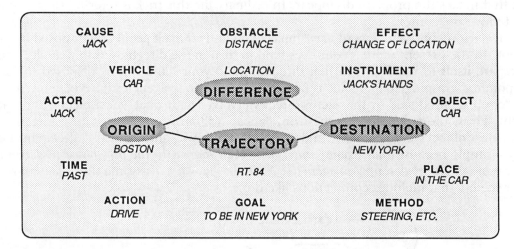

Why propose this particular structure instead of some other arrangement? Because I suspect that *Trans*-like structures have a special prominence in how we think. One reason is that some sort of bridgelike scheme seems indispensable for making those all-important connections between structures and functions. Without that bridge trajectory, it might be hard to connect what we learn about things to what we learn about using them. Also, in order to use chainlike thinking skills, we need to be able to represent what we know in ways that provide connection points for roles like *Origins, Destinations,* and *Differences.* All these requirements suggest the use of bridgelike frames.

One might wonder if we need to use any "standard" representations at all. The answer is that we do, indeed, if only because of the Investment principle. No matter what kind of representations we adopt, there's nothing we could do with them until we also learn effective skills and memory-scripts that work with them. And since such complex skills take time to grow, we'd never have enough time to learn new sets of representations for every different new idea. No powerful skills would ever emerge without *some* reasonably uniform schemes for representing knowledge.

Trans-like representation-schemes have been very useful in Artificial Intelligence research projects. They have been useful, among other things, for making theories of problem solving, for making clever computer programs to simulate "expertise" in various specialties, and for making programs that understand languagelike expressions to limited degrees. In the next few sections we'll see how to use them for making several different kinds of "chains of reasoning."

From every moment to the next, a person's state of mind is involved with various objects, topics, goals, and scripts. When you hear the words of the sentence, *"Put the apple in the pail,"* somehow this causes the subjects of "apple," "pail" and "putting in" to "occupy your mind." Later we'll speculate about how these become assigned to appropriate roles. Here, to make the story short, let's just assume that at a certain point the language-agency interprets the verb "put" to activate a certain *Trans*-frame and to assign the apple-neme to the *Object* pronome of that *Trans*-frame. The automatic "finding machine" described earlier then assigns the apple's location to the *Origin* pronome. Similarly, the pail's location is assigned to the *Destination* pronome. (As for the *Instrument* pronome, this is assigned to the listener's hand by default.) Now each entity "on the listener's mind" is represented by one or another pronome assignment. We're almost done, except that in order to actually perform the imagined action, we need some kind of control process to activate the proper agencies in the proper sequence!

> *Activate **apple-neme**, **Look-for**, and **Move**.*
> *Then activate **Grasp.***

> *Activate **pail-neme**, **Look-for**, and **Move**.*
> *Then activate **Ungrasp.***

This suggests a way to learn a skill. The first few times you try to do something new, you must experiment to find which agents to activate, and at what times, and for how long. Later, you can prepare a script that will do the job more quickly and easily by accumulating memories of which agent-activations were successful, together with memories of which polynemes were assigned to various pronomes at those moments. For example, if you were to "play back" the "*Trans*-script" shown above, your arm would find and put a second apple in that pail—without invoking any higher-level agencies at all! However, this script has a dreadful limitation: it will work only to put apples into pails. What if you later wished to put a block into a box or a spoon into a bowl? We could do that by dividing the process into two scripts: a "pronome-assignment script" and an "action script."

> Assignment Script:
> *Assign the apple-neme to the **Origin** pronome.*
> *Assign the pail-neme to the **Destination** pronome.*

> Action Script:
> *Activate **Origin**. Then turn on **Look-for**, **Move**, and **Grasp**.*
> *Activate **Destination**. Then turn on **Look-for**, **Move**, and **Ungrasp**.*

Now notice that the action script never actually mentions the apple or pail at all but refers only to the pronomes that represent them. Thus the same action script will serve as well for putting a block into a box as for putting an apple into a pail!

21.8 ATTENTION

When several objects move at once, it's hard to keep track of them all. This also seems to be the case in every other realm of thought; the more things we think about, the harder it is to pay attention to them all. We're forced to focus on a few while losing track of all the rest. What causes these phenomena? I'll argue that they're aspects of the processes we use to control our short-term memories. These skills develop over time; an adult can do things with memories that infants cannot do at all, such as remembering details of an action's purpose and trajectory, and how various obstacles were overcome. An infant, though, can barely keep track of what it's holding in one hand and is likely to forget what's in its other hand.

How does memory-control begin? Perhaps our infants first acquire control over a single pronome, which gives them the ability to keep in mind a "temporary polyneme." This amounts to being able to maintain only a single "object of attention"; let's call it **IT**. Now even the ability to keep track of a single **IT** requires the development of certain skills of memory-control, for it takes the normal infant several months to become able to tolerate even a small interruption without losing its previous focus of interest.

One kind of interruption comes, for example, when watching a ball that happens to roll behind a box. To a very young infant, that **IT** will simply disappear from mind. An older infant will remember **IT** and expect the ball soon to reappear; we can see this in the way the older infant's eyes look toward the far side of the box. If the ball does not soon reappear, the older child will actively reach around the box for it, which shows that the child has maintained some sort of representation of **IT**. Another variety of interruption can come from inside the child's own mind, from refocusing on the same object, but at a different level of detail. For example, when a young child concentrates upon a doll's shoe, it may forget its original concern with the doll itself. Later, that concern with the shoe may be replaced, in turn, when the baby becomes occupied with the end of the shoelace.

But what's an **IT**? The ability to focus attention could start with some machinery for keeping track of simple polynemes for object-things. In later stages, an **IT** could represent more complex processes or scripts that keep track of entire *Trans*-actions with their various pronomes for *Objects, Origins, Destinations, Obstacles, Trajectories,* and *Purposes*. Eventually our **IT**s develop into complex systems of machinery that represent the things that are "on one's mind" at the moment. In later life, we become more able to maintain several **IT**s at once. This enables us to construct comparisons, predictions, and imaginary plans, and to begin to construct explanations in terms of chains of causes and reasons.

CHAPTER 22

EXPRESSION

Henri finally came to the conclusion that the book owed its success simply to misunderstandings. Lambert believed he had meant to exalt individualism through collective action, and Lechaume, on the other hand, believed it preached the sacrifice of the individual to collectivism. Everyone emphasized the book's moral character. And yet Henri . . . had thought of a man and of a situation, of a certain relationship between that man's past life and the crisis through which he was passing, and of a great many other things which none of the critics had mentioned. Was it his fault or the readers'? The public, Henri was forced to conclude, had liked a completely different book from the one he believed he was offering them.

—SIMONE DE BEAUVOIR

22.1 PRONOMES AND POLYNEMES

To represent the action *"Put the apple in the pail,"* the *Origin* pronome must be assigned to an apple-neme, and the *Destination* pronome to a polyneme for pail. However, at another time, another process might need the *Origin* to represent a block and the *Destination* to represent a tower top. Each pronome must be assigned to different things at different times, and only for long enough to complete the task of the moment. In other words, *a pronome is a type of short-term memory.*

This suggests a simple way to embody the idea of a pronome: each pronome is simply a "temporary K-line." The basic difference, then, between a pronome and a K-line is that a pronome's connections are temporary rather than permanent. We can "assign" a pronome by temporarily connecting it to whichever currently active agents it reaches. Then, when we "activate" that pronome again, those same agents will be aroused. To make the *Origin* pronome represent an apple, first activate an apple-neme; this will arouse certain agents. Next, quickly "assign" the *Origin* pronome. Those agents will then become attached to that pronome and presumably remain attached until the pronome is reassigned.

If we compare pronomes and polynemes from this point of view, we see that they are closely related.

> *Polynemes are permanent K-lines. They are long-term memories.*
> *Pronomes are temporary K-lines. They are short-term memories.*

It is not yet known today how brains form long-term memories. One hypothesis would be that we don't really have temporary K-lines at all, but that after a pronome's K-line is used, it becomes permanent, and the pronome machinery gets connected to another, previously un-used K-line. However this works, we know little about it except that it requires a substantial amount of time to form a permanent memory—a time on the order of half an hour. If there is any serious disturbance in that interval, no memory will be formed. There also is some evidence that we can form new long-term memories at rates on the order of no more than perhaps one every few seconds, but this is very imprecise because we have no good definition of what we mean by separate memories. In any case, this seems to suggest that we might have several hundred such processes going on at once.

Why does the process take so long? Perhaps because it simply takes that long to synthesize chemicals used to make permanent connection bridges between agents. Perhaps most of that time is consumed in *searching* for an unused K-line agent, particularly for one that already has the required potential connections. Or perhaps the required connections could emerge from "distributed memories" like those we mentioned briefly in section 20.9.

22.2 ISONOMES

We introduced the concept of a polyneme to explain how an agent could communicate with many other kinds of agencies. In order for a polyneme to work, each of its recipients must learn its own way to react. Now we've seen a second way, for a pronome is also an agent that can interact with many other agencies. The difference is that a pronome has essentially the *same* effect on each of its recipients—namely, to activate or to assign a certain short-term memory-unit. I'll introduce a new word—"isonome"—for any agent that has this sort of uniform effect on many agencies.

>**An isonome has a similar, built-in effect on each of its recipients.**
> *It thus applies the same idea to many different things at once.*
>**A polyneme has different, learned effects on each of its recipients.**
> *It thus connects the same thing to many different ideas.*

Why should isonomes exist at all? Because our agencies have common genetic origins, they tend also to be architecturally similar. So they'll tend to lie in roughly parallel formations like the pages of a book, operate in generally similar ways, and have similar memory-control processes. Then any agent whose connections tend to run straight *through* the pages of that book from cover to cover will tend to have similar effects on all of them.

Both isonomes and polynemes are involved with memories—but polynemes are essentially the memories themselves, while isonomes *control* how memories are used. Pronomes are a particular type of isonome; there must also be "interruption isonomes" that work similarly but manage memories on larger scales—for example, for storing away the several pronome memories of an entire *Trans*-frame all at once. (We'll see how something like this must be done whenever we encounter a grammar word like "who" or "which.") Yet other types of isonomes must be involved whenever an agent is used to control the level-band of activity in another agency without concern for all the fine details of what happens inside that agency. So the power of polynemes stems from how they learn to arouse many different processes at once, while isonomes draw their power from exploiting abilities that are already common to many agencies.

22.3 DE-SPECIALIZING

Soon after learning how to put an apple into a pail, a child will discover that it now can put the apple into a box or put an onion into the pail. What magic tricks allow us to "de-specialize" whatever skills we learn? We've already seen one way to do this simply by replacing certain polynemes with less specific isonomes. For example, our first apple-into-pail procedure was so specialized that it could be used only to put apples into pails—because it is based on using specific polynemes for those objects. However, the second script just as easily puts onions into pails or umbrellas into suitcases, because it engages no polynemes at all, but only the *Origin* and *Destination* pronomes. This script is more versatile because those pronomes can be assigned to *anything*! Learning to think in terms of isonomes must be a crucial step in many types of mental growth.

None of our many chaining tricks would have much use if each were permanently tied to one specific polyneme like "owl" or "car" or "cup" or "gear." However, once we learn to build our process scripts with isonomes, each can be applied to many kinds of reasoning—to logic, cause, dependency, and all the rest. But changing polynemes to isonomes will not always work. What could keep a child from trying to apply the script that works on *"put the apple in the block"* to *"put the ocean in the cup"*? To prevent such absurdities, our script must also place appropriate constraints on the *Origin* and *Destination*—for example, to ensure that the *Destination* must represent a *container* large enough to hold the *Origin* thing, and that the container be open toward the top. If all this seems too obvious to say, just watch a baby's first attempts to put an object in a pail or pick up food with a spoon or a fork. It takes many weeks or months of work to bring such skills to the point of usefulness. If we generalize too recklessly by changing all our polynemes to isonomes, few of our generalizations will actually work.

What we call "generalizing" is not any single process or concept, but a *functional* term for the huge societies of different methods we use to extend the powers of our skills. No single policy will work for all domains of thought, and each refinement of technique will affect the quality of the generalizations we make. Converting polynemes to isonomes may be a potentially powerful skill, but it must be adapted to different realms. Once we accumulate enough examples of how a new script fails and succeeds in several situations, we can try to build a uniframe to embody good constraints. But no matter which policy we adopt, we must always expect some exceptions. You cannot carry birds in pails, no matter how well they fit inside. Premature generalizations could lead to such large accumulations of constraints, censors, and exceptions that it would be better to retain the original polynemes.

22.4 LEARNING AND TEACHING

One frustration every teacher knows arises when a child learns a subject well enough to pass a test, yet never puts that skill to use on problems met in "real life." It doesn't often help to scold but it sometimes helps to explain, through examples, how to apply the concept to other contexts. Why do some children seem to do this for themselves, automatically and spontaneously, while others seem to have to learn essentially the same thing over and over in different domains? Why are some children better than others at "transfer of learning" from one domain to another? It doesn't explain anything to say that those children are "smarter," "brighter," or "more intelligent." Such vaguely defined capacities vary greatly even among different parts of the same mind.

The power of what we learn depends on how we represent it in our minds. We've seen how the same experience can lead to learning different action scripts by replacing certain polynemes with isonomes. Certain of those versions will apply only to specific situations, others will apply to many more situations, and yet others will be so general and vague as to lead only to confusion. Some children learn to represent knowledge in versatile ways; others end up with accumulations of inflexible, single-purpose procedures or with almost useless generalities. How do children acquire their "representation skills" in the first place? An educational environment can lead a child to build large, complicated processes from smaller ones by laying out sequences of steps. Good teachers know what size to make each step and can often suggest analogies to help the child's mind to use what it already knows for building larger scripts and processes. By making each step small enough, we can keep the child from getting lost in unfamiliar worlds of meaningless alternatives; then the child will remain able to use previous skills to test and modify the growing new structures. But when a new fragment of knowledge or process constitutes too abrupt a break from the past, then none of the child's old recognizers and action scripts will apply to it; the child will get stuck, and "transfer of learning" won't occur. Why are some children better than others at "teaching themselves" to make changes inside their minds?

Each child learns, from time to time, various *better ways to learn*—but no one understands how this is done. We tend to speak about "intelligence" because we find it virtually impossible to understand how this is done from watching only what the child *does*. The problem is that one can't observe a child's strategies for "learning how to learn"—because those strategies are twice removed from what we can see. It is hard enough to guess the character of the A-brain systems that directly cause those actions. Think how much more difficult it would be for an observer to imagine the multilayer teacher-learner structures that must have worked inside the child to train the A-brain agencies! And that observer has no way at all to guess what crucial "lucky accidents" may have led those hidden B-brains to persistent concerns with finding better ways to learn. Perhaps our educational research should be less concerned with teaching children to acquire particular skills and more concerned with how we learn to learn.

22.5 INFERENCE

Linking structures together into chains is one of our most useful kinds of reasoning. Suppose you learned that *"John gave the kite to Mary"* and then *"Mary gave the kite to Jack."* You could then conclude that the kite went from John to Jack. How do we draw such conclusions? Some people think we use "logic" for this. A simpler theory is that we do it by fitting together *Trans*-frames into chains. Suppose you see two frames like these:

All A's are B's   and   All B's are C's.

Then just combine the first *Origin* with the second *Destination* to make this new "deduction-frame":

All A's are C's.

To do this sort of "reasoning," we have to use our isonomes to rearrange our short-term memories. But this requires more than simple chaining. For example, all older children can infer that Tweety can fly from *"Tweety is a bird"* and *"All birds can fly."* To do this, though, one has to deal with a disparity: the first **B** is "a bird" while the second **B** is "all birds." To be able to make such chains would be virtually useless if we could do it only when both pronome assignments were absolutely identical. Over the years, children improve their abilities to decide when two different structures are similar enough to justify making chain-links. This often requires us to recall and apply other types of knowledge at appropriate level-bands of detail.

Children take many years to learn effective ways to use their pronomes and isonomes. The youngest ones can neither rearrange their representations of physical scenes nor make the kinds of inference we're discussing here. To think like adults, we must develop and learn to use memory-controlling processes that manipulate several sets of pronome values at the same time. Just such a process was concealed in our simple script for *"Put the apple in the pail"*—which first appears to be merely a matter of assigning "apple" to the *Origin* and "pail" to the *Destination*. But you can't *Put* something until you *Get* it, so this must actually involve *two Trans*-frame operations. The first is for moving your hand to the apple, and the second is for moving the apple to the pail. During the transition, your pronomes have to change their roles since the apple's location is the *Destination* of the first *Trans*, but then becomes the *Origin* of the second *Trans*. No matter that this seems too obvious to state; *some* mental process has to switch that pronome's role.

By learning to manipulate our isonomes, we become able to combine mental representations into structures that resemble bridges, chains, and towers. Our language-agencies learn to express these in the form of compound sentences, by using conjunctive grammar words like *"and," "because,"* or *"or."* But language is not the only realm in which we learn to "conceptualize"—that is, to treat our mental processes almost as though they were object-things. After you solve a hard problem, you may find yourself representing the steps you took as if they were the parts of a physical structure. Doing this can enable you to reassemble them into other forms that achieve the same results with much more speed and much less thought.

22.6 EXPRESSION

Language lets us treat our thoughts as though they were much like ordinary things. Suppose you meet someone who is trying to solve a problem. You ask what's happening. *"I'm thinking,"* you are told. *"I can see that,"* you say, *"but what are you thinking about?"* *"Well, I was looking for a way to solve this problem, and I think I've just found one."* We speak as though ideas resemble building-blocks that one can find and grasp!

Why do we "thing-ify" our thoughts? One reason is that this enables us to reapply the wonderful machines our brains contain for understanding worldly things. Another thing it does is help us organize our expeditions in the mental world, much as we find our ways through space. Consider how the strategies we use to "find" ideas resemble the strategies we use for finding real things: *Look in the places they used to be or where they're usually found—but don't keep looking again and again in the same place.* Indeed, for many centuries our memory-training arts have been dominated by two techniques. One is based on similarities of sounds, exploiting the capacities of our language-agencies to make connections between words. The other method is based on imagining the items we want to remember as placed in some familiar space, such as a road or room one knows particularly well. This way, we can apply our thing-location skills to keeping track of our ideas.

Our ability to treat ideas as though they were objects goes together with our abilities to reuse our brain-machinery over and over again. Whenever an agency becomes overburdened by a large and complicated structure, we may be able to treat that structure as a simple, single unit by thing-ifying—or, as we usually say, *"conceptualizing"*—it. Then, once we replace a larger structure by representing it with a compact symbol-sign, that overloaded agency may be able to continue its work. This way, we can build grand structures of ideas—much as we can build great towers from smaller parts.

I suspect that, *as they're represented in the mind,* there's little difference between a physical object and an idea. Worldly things are useful to us because they are "substantial"—that is, because their properties are relatively permanent. Now we don't usually think of ideas as substantial, because they don't have the usual properties of worldly things—such as color, shape, and weight. Yet "good ideas" must also have substantiality, albeit of a different sort:

> *No conception or idea could have much use unless it could remain unchanged— and stay in some kind of mental "place"—for long enough for us to find it when we need it. Nor could we ever achieve a goal unless it could persist for long enough. In short, no mind can work without some stable states or memories.*

This may sound as though I'm speaking metaphorically, since a mental "place" is not exactly like a worldly place. But then, when you *think* of a place you know, that thought itself is not a worldly place, but only a linkage of memories and processes inside your mind. This wonderful capacity—to think about thoughts as though they were things—is also what enables us to contemplate the products of our thoughts. Without that ability to reflect, we would have no general intelligence—however large our repertoire of special-purpose skills might grow. Of course this same capacity enables us to think such empty thoughts as *"This statement is about itself,"* which is true but useless, or *"This statement is not about itself,"* which is false and useless, or *"This statement is false,"* which is downright paradoxical. Yet the benefit of being able to conceptualize is surely worth the risk that we may sometimes be nonsensical.

22.7 CAUSES AND CLAUSES

For virtually every change we see, we tend to seek some cause. And when we find no cause on the scene, we'll postulate that one exists, even though we might be wrong. We do this so consistently that I wouldn't be surprised to find that brains have built-in tendencies to try to represent all situations in certain special ways:

> THINGS. *Whatever we may see or touch, we represent the scene in terms of separate object-things. We do the same for representing processes and mental states. In languages, these object-symbols tend to correspond to nouns.*

> DIFFERENCES. *Whenever we discern a change or just compare two different things, we represent this as a difference thing. In languages, these often correspond to verbs.*

> CAUSES. *Whenever we conceive of an action, change, or difference, we try to assign a cause to it—that is, some other person, process, or thing that we can hold to be responsible for it. In languages, causes often take the forms of things.*

> CLAUSES. *Whatever structures we conceive are dealt with like single things. In languages, this corresponds to treating an entire phrase as though it were a single word.*

In English, almost every sentence form demands some sort of *Actor* noun—and I think this reflects the need to find a motive or a cause. Consider how we place that *it* in "*Soon it will begin to rain.*" We're always chopping complex situations into artificially clear-cut chunks which we perceive as separate things. Then we notice various differences and relationships among those parts and assign them to various parts of speech. We string our words into clauses and our clauses into chains, often interrupting one by inserting fragments of others inside it, yet proceeding as though there were no interruptions at all. It has been alleged that the construction of such structures is unique to the grammar-machinery of language, but I suspect that languages evolved those forms because of mechanisms deeper in the grain of how we think. For example, when we talked about visual ambiguity, we saw that our vision-systems are highly proficient at representing structures that interrupt one another. This suggests that both our visual and linguistic abilities to deal with "interruptions" could be based on similar methods with which we "manage" what is represented in our short-term memories.

In any case, our brains appear to make us seek to represent dependencies. Whatever happens, where or when, we're prone to wonder who or what's responsible. This leads us to discover explanations that we might not otherwise imagine, and that helps us predict and control not only what happens in the world, but also what happens in our minds. But what if those same tendencies should lead us to imagine things and causes that do not exist? Then we'll invent false gods and superstitions and see their hand in every chance coincidence. Indeed, perhaps that strange word "I"—as used in "*I just had a good idea*"—reflects the selfsame tendency. If you're compelled to find some cause that causes everything you do—why, then, that something needs a name. You call it "*me.*" I call it "*you.*"

22.8 INTERRUPTIONS

What enables us to tolerate an interruption and then return to our previous thoughts? This must engage the agents that control our short-term memories. It is important also to recognize that many interruptions come not only from outside, but also from inside the mind. For example, all but the simplest discourses make interruptions in the trains of thought they start. Consider this sentence:

> *The thief who took the moon moved it to Paris.*

We can regard this as expressing one thought that is interrupted by another. The principal intention of the speaker is to express this *Trans*-frame:

> *The thief   moved   the moon   (from?)   to Paris.*
> **Actor   Trans   Object   Origin   Destination**

The speaker, realizing that the listener may not know who the thief was, interrupts the main sentence with a "relative clause"—"*who took the moon*"—to further describe that *Actor* thief. As it happens, this interrupting clause also has the form of a *Trans*-frame—so now the language-agency must deal with two such frames at once.

> *Who   took   the moon   (from?)   (to?)*
> **Actor   Trans   Object   Origin   Destination**

English tends to use certain *wh* words, like "which" and "who," to interrupt a listener's language-agency and cause its short-term memories to temporarily store away some of their present pronome assignments. This provides the language-agency with more capacity to understand the interrupting phrase. In the case of the moon sentence, the word "who" instructs the listener to prepare to elaborate the description of the *Actor* thief. Once this is done, the language-agency can "re-member" its previous state in the process of understanding the main sentence. We can often tell when to use an interruption process even though the initial *wh* word is missing; however, this doesn't always work so well:

> *The cotton clothing is made of is grown in the south.*

This sentence is confusing because the reader tends to treat the word "cotton" in "*cotton clothing*" as an adjective that modifies "clothing," when the writer meant it as a noun. The same sentence is easier to understand when set in a larger context:

> *Where do people grow the cotton that is used to make clothing?*
> *--- The cotton clothing is made of is grown in the south.*

The first sentence activates the noun sense of "cotton" and asks a *question* about that subject. Now a question is really a sort of command: it makes the reader focus attention on a certain subject. Here, it prepares the reader to add more structure to the representation of the cotton noun, so there is less need for an explicit interruption signal. Still, it is very curious how rarely we bother to use any signal at all for marking the *end* of an interrupting phrase. We never say a word that means "un-who." Evidently, we're usually ready to assume that the interrupting phrase is complete.

22.9 PRONOUNS AND REFERENCES

We sometimes think of words like *"who"* or *"it"* as *"pronouns"*—that is, signals that represent or substitute for other nouns or phrases. But as we've seen, pronouns don't refer to words so much as to partial states that are active in the listener's mind. In order to "refer" to such an activity, the listener must assign it to some short-term memory-unit—that is, to some pronome. However, communication will fail unless the listener can correctly guess which pronome the speaker wishes to assign to that activity. This can be a problem when there is more than one available choice. For example, consider the pronoun *"it"* in the following sentence:

> *The thief who took the moon moved it to Paris.*

How does the listener understand that *it* must mean the moon? English grammar constrains the choice: *it* cannot be assigned to the thief because *it* cannot refer to a person at all. (We apologize to all the small children we have called "it" in this book.) But grammar alone can't determine the choice, since *"it"* could also mean the sun, as it does in this little dialogue:

> *Good grief; what's happened to the sun?*
> *Oh, that! The thief who took the moon moved it to Paris.*

The way *"it"* works is not so much grammatical as psychological. The expression *"moved it"* causes the listener's language-agencies to seek a pronome that represents something that could have been moved. This could be either the sun or the moon. But the previous question, *What happened to the sun?* has already prepared the listener to expect to hear about an action whose *Object* pronome represents the sun, just as our earlier question about cotton made the listener anticipate an answer concerning that topic. Furthermore, the new phrase, *"The thief . . . moved it,"* fulfills this expectation by activating a *Trans*-frame whose *Actor* and *Action* pronomes already have assignments; this frame requires only an *Object* to be complete. So the word *"it"* is perfectly suited to fill the role of *"sun"* in that unassigned *Object* slot.

What does "expectation" mean? At each point in a dialogue, both parties are already involved with various concerns and desires. These establish contexts in which each new word, description, or representation, however ambiguous, gets merged with whichever short-term memory best matches it. Why do we make such assignments so quickly, instead of waiting until all the ambiguities are resolved? That is a practical matter. Our language-agencies must dispose of each phrase as soon as possible, so that they can apply their full capacities to deal with what comes afterward. If something in the conversation does not match anything that came before, the listener must activate a new memory-unit. This tends to slow the process down, because it consumes our limited supplies of short-term memory and makes subsequent matching more difficult. If the listener cannot make suitable assignments quickly enough, the conversation will seem incoherent and communication will break down.

Eloquent speakers avoid this by designing each new expression to be easily attached to structures already active in the listener; otherwise the listener is entitled to complain that the language isn't clear. A speaker can also indicate which subjects have *not* been mentioned yet, to spare the listener from struggling to make a nonexistent match; we use expressions like *"by the way"* to tell the listener *not* to attach what comes next to any presently active pronome. To do such things, the speaker must anticipate some of what is happening inside the listener's mind. The next section describes a way of doing this—by using the speaker's own mind as a model and assuming that the listener will be similar.

22.10 VERBAL EXPRESSION

How easily people can communicate. We listen and speak without the slightest sense of what's involved! One of us expresses an idea, the other understands it, and neither thinks anything complicated has happened; it seems as natural to talk as it is to walk. Yet both simplicities are illusions. To walk, you must engage a vast array of agencies to move your body down the street. To talk, you must engage a vast array of agencies to build new structures in another person's mind. But *how* do you know just what to say to affect that other person's agencies?

Let's suppose that Mary wants to tell Jack something. This means there is a certain structure **p** somewhere inside the network of Mary's agencies—and that Mary's language-agency must construct a similar structure inside Jack's mind. To do this, Mary will need to speak words that will activate appropriate activities inside Jack's agencies, then correctly link them together. How can she do that? Here is what we'll call the "re-duplication" theory of how we formulate what we say:

> *Mary proceeds, step by step, to construct a new version of* **p**—*call it* **q**—*inside her own mind.* In doing this, she will apply various memory-control operations to activate certain isonomes and polynemes.

> *As Mary performs each internal operation, her speech-agency selects certain corresponding verbal expressions—and these cause similar operations to occur inside Jack.* As a result, Jack builds a structure similar to **q**.

> *To be able to do that, Mary must have learned at least one expressive technique that corresponds to each frequently used mental operation.* And Jack must have learned to recognize those expressive techniques—we'll call them *grammar-tactics*—and to use them to activate some corresponding isonomes and polynemes.

To build her new version of **p**, Mary could employ a goal-achieving scheme: she keeps comparing **p** with the latest version of **q**, and whenever she senses a significant difference, she applies some operation to **q** that removes or reduces the difference. For example, if Mary notices that **p** has an *Origin* pronome where **q** lacks one, her memory-control system will focus on **p**'s *Origin*. In this case, if **p** itself is a motion frame, the usual speech-tactic is to use the word "*from*." Next she must describe the substructure attached to **p**'s *Origin* pronome. If this were a simple polyneme like "Boston," Mary's speech-agency could simply pronounce the corresponding word. But if that pronome is assigned to some more complicated structure, such as an entire frame, Mary's language-agency must interrupt itself to copy *that*. This is expressed, as we have seen, by using words like "who" or "which." In any case, Mary continues this difference-duplication process until she senses no significant discrepancies between **q** and **p**. Of course, what Mary finds significant depends on what she "wants to say."

This "re-duplication" theory of speech describes only the first stages of how we use language. In later stages, the mental operations we use to construct **q** are not always immediately applied to pronouncing words. Instead, we learn techniques for storing sequences of grammar-tactics temporarily; this makes it possible to modify and rearrange our words and sentences before we say them. Learning these arts takes a long time: most children need a decade or more to complete their language-systems and many keep learning, throughout their lives, to sense new sorts of discrepancies and discover ways to express them.

22.11   CREATIVE EXPRESSION

There is a wonderful capacity that comes along with the ability to "express" ideas. Whatever we may want to say, we probably won't say exactly *that*. But in exchange, there is a chance of saying something else that is both good and new! After all, the "thing we want to say"—the structure **p** we're trying to describe—is not always a definite, fixed structure that our language-agents can easily read and copy. If **p** exists at all, it's likely to be a rapidly changing network involving several agencies. If so, then the language-agency may only be able to make guesses and hypotheses about **p** and try to confirm or refute them by performing experiments. Even if **p** were well defined in the first place, this very process is liable to change it, so that the final version **q** won't be the same as the original structure **p**. Sometimes we call this process "thinking in words."

In other words, whether or not what you "meant" to say actually existed before you spoke, your language-agencies are likely either to *reformulate* what did exist or create something new and different from anything you had before. Whenever you try to express with words any complicated mental state, you're forced to oversimplify—and that can cause both loss and gain. On the losing side, no word description of a mental state can ever be complete; some nuances are always lost. But in exchange, when you're forced to separate the essences from accidents, you gain the opportunity to make reformulations. For example, when stuck on a problem, you may "say to yourself" things like *"Now, let's see—just what was I really trying to accomplish?"* Then, since your language-agency knows so little about the actual state of those other agencies, it must answer such questions by making theories about them, and these may well leave you in a state that is simpler, clearer, and better suited to solving your problem.

When we try to explain what we think we know, we're likely to end up with something new. All teachers know how often we understand something for the first time only after trying to explain it to someone else. Our abilities to make language descriptions can engage all our other abilities to think and to solve problems. If speaking involves thinking, then one must ask, *"How much of ordinary thought involves the use of words?"* Surely many of our most effective thinking methods scarcely engage our language-agencies at all. Perhaps we turn to words only when other methods fail. But then the use of language can open entirely new worlds of thought. This is because once we can represent things in terms of strings of words, it becomes possible to use them in a boundless variety of ways to change and rearrange what happens in our other agencies. Of course, we never realize we're doing this; instead we refer to such activities by names like *paraphrase* or *change of emphasis*, as though we weren't changing what we're trying to describe. The crucial thing is that during the moments in which those word-strings are detached from their "meanings," they are no longer subject to the constraints and limitations of other agencies, and the language-systems can do what they want with them. Then we can transmit, from one person's brain to another, the strings of words our grammar-tactics produce, and every individual can gain access to the most successful formulations that others can articulate. This is what we call culture—the conceptual treasures our communities accumulate through history.

CHAPTER 23

COMPARISONS

What, in effect, are the conditions for the construction of formal thought? The child must not only apply operations to objects—in other words, mentally execute possible actions on them—he must also "reflect" those operations in the absence of the objects which are replaced by pure propositions. Thus "reflection" is thought raised to the second power. Concrete thinking is the representation of a possible action, and formal thinking is the representation of a representation of possible action. . . . It is not surprising, therefore, that the system of concrete operations must be completed during the last years of childhood before it can be "reflected" by formal operations. In terms of their function, formal operations do not differ from concrete operations except that they are applied to hypotheses or propositions [whose logic is] an abstract translation of the system of "inference" that governs concrete operations.

—Jean Piaget

23.1 A WORLD OF DIFFERENCES

Much of ordinary thought is based on recognizing differences. This is because it is generally useless to do anything that has no discernible effect. To ask if something is significant is virtually to ask, *"What difference does it make?"* Indeed, whenever we talk about "cause and effect" we're referring to imaginary links that connect the differences we sense. What, indeed, are goals themselves, but ways in which we represent the kinds of changes we might like to make?

It is interesting how many familiar mental activities can be represented in terms of the differences between situations. Suppose you have in mind two situations **A** and **Z**, and **D** is your description of the difference between them. Suppose also that you are thinking of applying a certain procedure **P** to the first situation, **A**. There are several kinds of thinking you might do.

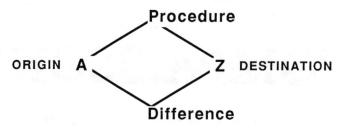

PREDICTING. *To the extent that you can predict how various **P**'s will affect **A**, you can avoid the expense and risk of actually performing those actions.*
EXPECTING. *If you expect **P** to produce **Z** but it actually produces **Y**, then you can try to explain what went wrong in terms of difference between **Y** and **Z**.*
EXPLAINING. *If actions like **P** usually lead to differences of type **D**, then when you observe such a **D**, you can suspect that it was caused by something like **P**.*
WANTING. *If you are in situation **A**, but wish to be in a situation more like **Z**, it may help to remember ways to remove or reduce differences like **D**.*
ESCAPING, ATTACKING, and DEFENDING. *If **P** causes a disturbing type of difference **D**, we can try to improve matters by finding some action that counteracts or opposes **P**.*
ABSTRACTING. *In many forms of thought, the differences we notice between objects at each level become the "objects" of our higher-level thoughts.*

Not only are differences important by themselves; more often than we realize, we think about *differences between differences*. For example, the "height" of a physical object is really a difference between the locations of its top and its bottom. And this means that the higher-level agents in our Societies-of-More must actually deal with *differences between differences*. For example, the agent *Taller* has to react to the difference between two heights—but as we've just seen, a height is already a difference between two locations!

The ability to consider differences between differences is important because it lies at the heart of our abilities to solve new problems. This is because these "second-order-differences" are what we use to remind ourselves of other problems we already know how to solve. Sometimes this is called "reasoning by analogy" and is considered to be an exotic or unusual way to solve problems. But in my view, it's our most ordinary way of doing things.

23.2 DIFFERENCES AND DUPLICATES

It is important for us to be able to notice differences. But this seemingly innocent requirement poses a problem whose importance has never been recognized in psychology. To see the difficulty, let's return to the subject of mental rearrangements. Let's first assume that the problem is to compare two room-arrangement descriptions represented in two different agencies: agency **A** represents a room that contains a couch and a chair; agency **Z** represents the same room, but with the couch and chair exchanged.

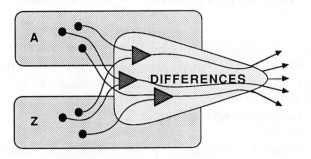

Now if both agencies are to represent furniture arrangements in ways that some third agency **D** can compare, then the "difference-detecting" agency **D** must receive two sets of inputs that match almost perfectly. Otherwise, every other, irrelevant difference between the outputs of **A** and **Z** would appear to **D** to be differences in those rooms—and **D** would perceive so many spurious differences that the real ones would be indiscernible!

> **The Duplication Problem.** *The states of two different agencies cannot be compared unless those agencies themselves are virtually identical.*

But this is only the tip of the iceberg, for it is not enough that the descriptions to be compared emerge from two almost identical agencies. Those agencies must, in turn, receive inputs of near identical character. And for *that* to come about, each of their subagencies must also fulfill that same constraint. The only way to meet all these conditions is for both agencies—*and all the subagencies upon which they depend*—to be identical. *Unless we find another way, we'll need an endless host of duplicated brains!*

This duplication problem comes up all the time. What happens when you hear that *Mary bought John's house?* Must you have separate agencies to keep both *Mary* and *John* in your mind at once? Even that would not suffice, for unless both person-representing agencies had similar connections to all your other agencies, those two representations of "persons" would not have similar implications. The same kind of problem must arise when you compare your present situation with some recollection or experience—that is, when you compare how you react to the two different partial states of mind. But to compare those two reactions, what kind of simultaneous machinery would be needed to maintain both momentary personalities? How could a single mind hold room for two—one person old, the other new?

Fortunately, there is a way to get around the duplication problem entirely. Let's take a cue from how a perfume makes a strong impression first, but then appears to fade away, or how, when you put your hand in water that is very hot or very cold, the sensation is intense at first— but will soon disappear almost entirely. As we say, we "get used to" these sensations. Why? Because *our senses react mainly to how things change in time*. This is true even for the sensors in our eyes—though normally we're unaware of it because our eyes are always moving imperceptibly. Most of the sensory agents that inform our brains about the world are sensitive *only* to various sorts of time changes—and that, surely, is also true of most of the agents *inside* the brain.

> *Any agent that is sensitive to changes in time can also be used to detect differences. For whenever we expose such an agent, first to a situation A and then to a situation B, any output from that agent will signify some difference between A and B.*

This suggests a way to solve the duplication problem. Since most agents can be made to serve as difference-agents, *we can compare two descriptions simply by presenting them to the same agency at different times*. This is easily done if that agency is equipped with a pair of high-speed, temporary K-line memories. Then we need only load the two descriptions into those memories and compare them by activating first one and then the other.

> *Store the first description in pronome **p**.*
> *Store the second description in pronome **q**.*
> *Activate **p** and **q** in rapid succession.*
> *Then any changes in the agents' outputs represent differences between A and B!*

We can use this trick to implement the scheme we described for escaping from a topless-arch. Suppose that **p** *describes the present situation and* **q** *describes a box that permits no escape.* Each *Move* agent is designed to detect the appearance of a wall. If we simply "blink" from the present situation to the box frame, one of these agents will announce the appearance of any box wall that was not already apparent in the present situation. Thus, automatically, this scheme will find all the directions that are not closed off. If the outputs of the *Move* agents were connected to cause you to move in the corresponding direction, this agency would lead you to escape!

The method of time blinking can also be used to simplify our difference-engine scheme for composing verbal expressions, since now the speaker can maintain both **p** and **q** inside the selfsame agency. If not for this, each speaker would need what would amount to a duplicate society of mind in order to simulate the listener's state. Although the method of time blinking is powerful and efficient, it has some limitations; for example, it cannot directly recognize relations among more than two things at a time. I suspect that people share this limitation, too —and this may be why we have relatively few language-forms, like "between" and "middle," for expressing three-way comparisons and relationships.

23.4 THE MEANINGS OF MORE

Let's return just one time more to all the things that *More* can mean. Each usage has a different sense—*more powerful; more meaningful*—and each such meaning must be learned. In other words, each use of *More* involves a connection with an agent for some adjective. But *More* must also engage some systematic use of isonomes, since all the different meanings share a certain common character.

> *When we hear the word "more," we become disposed to make comparisons.*

This suggests that *More* engages both an accumulation of different meanings and also some systematic, isonomelike effect. Indeed, *More* could exploit our time-blinking mechanism, which already uses isonomes to make comparisons. To do that, *More* would have to activate a memory-control process that "blinks" whichever pronomes have been assigned to the things to be compared. Then their differences would be computed automatically.

More needs two additional ingredients. We'd never ask, all by itself, "*Which one is more, an apple or a pear?*"—because our general-purpose comparison script would generate difference-descriptions in too many agencies. We also need to know which *kind* of difference is of concern at the moment. So we scarcely ever say "more" by itself but usually attach some modifier—*more red*, say, or *more expensive*. Of course, if our focus of concern is already clear from the context—for example, if it is clear that we want to know whether apples or pears are *more expensive*—then we need not express this explicitly.

Finally, it is one thing to find a difference, but another to know whether to call it "more" or "less." It may seem self-evident that *Taller* corresponds to "more," and *Thinner* corresponds to "less"—yet that is something we once had to learn. This is the other ingredient of *More*: we need another polyneme to say which sorts of differences should be considered positive. In English we sometimes encode such preferences as choices between pairs of adjectives like "large" and "small," but we do not have pairs of words for concepts such as "triangular" or "red," presumably because we do not think of them as having "natural" opposites. Instead, we can use word-pairs like "more red" and "less triangular." We can even modify the words themselves; we often say "redder" or "rounder"—but for some reason we never say "triangularer."

How does one answer a question like "*Which is bigger, a large mouse or a small elephant?*" We can't compare two descriptions until we engage enough knowledge to construct suitable representations of them. One way to compare mouse and elephant would be to envision another entity of intermediate size. A suitcase would be suitable for this, since it could hold the largest mouse but not the smallest elephant. How do you find such standards for comparison? That can take considerable time, during which you have to search your memories for structures that can serve as links for longer chains of comparisons. As life proceeds, each person's concept of *More* grows more and more elaborate. When it comes to notions like *more similar, more interesting*, or "*more difficult*," there seems no limit to the complexity of what a word like "more" can represent.

23.5 FOREIGN ACCENTS

It is not unusual for an adult to learn a second language with nearly perfect mastery of grammar and vocabulary. But once past adolescence, most people never manage to imitate the new language's *pronunciation* perfectly, no matter how long and hard they work at it. In other words, they speak with "foreign accents." Even when another speaker tries to help with *"Say it like this, not that,"* the learner is unable to learn what changes to make. Most people who change countries in their later teens *never* learn to speak the way the natives do.

Why do adults find it so hard to learn how to pronounce new word sounds? Some like to say that this reflects a general decline in the learning capacities of older people, but that appears to be a myth. Instead, I suspect this particular disability is caused, more or less directly, by a genetically programmed mechanism that disables our ability to learn to make new connections in or between the agents we use to represent speech sounds. There is evidence that our brains use different machinery for recognizing language sounds than for recognizing other sorts of sounds, particularly for the little speech-sound units that language scientists call "phonemes." Most human languages use less than a hundred phonemes.

Why should we be able to learn many different speech sounds before the age of puberty but find it so much harder to learn new ones afterward? I suspect that this link to puberty is no coincidence. Instead, one or more of the genetically controlled mechanisms that brings on sexual maturity also acts to reduce the capacities of these particular agencies to learn to recognize and make new sounds! But why did this peculiar disability evolve? What evolutionary survival advantage would favor individuals whose genes *reduce*, after that age, this particular ability to learn? Consider this hypothesis:

> The onset of the childbearing age is the biological moment when a person's social role changes from learner to teacher. The "evolutionary purpose" of suppressing speech-sound learning may simply serve to prevent the parent from learning the child's speech—thus making the child learn the adult's speech instead!

Wouldn't parents *want* to teach the children their language anyway? Not necessarily. In the short run, a parent is usually more concerned with communication than with instruction. Accordingly, if we found it easier to imitate our children's sounds, that's what we'd do. *But if parents were inclined and able to learn to speak the ways their children do, those children would lose both incentive and opportunity to learn to speak like adults, and—if every child acquired a different set of language sounds—no common, public language would ever have evolved in the first place!* If this is right, puberty-linked genes for suppressing speech-sound learning may have formed fairly early in the evolution of human languages. No one knows when that occurred, but if biologists could find and date the genes for this, we could obtain a clue about the time of language's unknown origin, perhaps within the last half million years.

FRAMES

The justification of Napoleon's statement—if, indeed, he ever made it—that those who form a picture of everything are unfit to command, is to be found in the first of these defects. A commander who approaches a battle with a picture before him of how such and such a fight went on such and such an occasion, will find, two minutes after the forces have joined, that something has gone awry. Then his picture is destroyed. He has nothing in reserve except another individual picture, and this also will not serve him for long. Or it may be that when his first pictured forecast is found to be inapplicable, he has so multifarious and pressing a collection of pictures that equally he is at a loss what practical adjustment to make. Too great individuality of past reference may be very nearly as embarrassing as no individuality of past reference at all. To serve adequately the demands of a constantly changing environment, we have not only to pick items out of their general setting, but we must know what parts of them may flow and alter without disturbing their general significance and functions.

—F. C. Bartlett

24.1 THE SPEED OF THOUGHT

*For there exists a great chasm between those, on the one side,
who relate everything to a single central vision, one system more or
less coherent or articulate, in terms of which they understand,
think and feel—a single, universal, organizing principle in terms
of which alone all that they are and say has significance—and, on
the other side, those who pursue many ends, often unrelated and
even contradictory, connected, if at all, only in some* de facto *way,
for some psychological or physiological cause, related by no moral
or aesthetic principle. . . .*

—ISAIAH BERLIN

When we enter a room, we seem to see the entire scene at a glance. But, really, it takes time to see—to apprehend all the details and see if they confirm our expectations and beliefs. Our first impressions often have to be revised. Still, one wonders how so many kinds of visual cues can lead so quickly to consistent views. What could explain the blinding speed of sight?

The secret is that sight is intertwined with memory. When face to face with someone you've just met, you seem to react almost instantly—*but not as much to what you see as to what that sight "reminds" you of.* The moment you sense the presence of a *person*, a whole world of assumptions are aroused that are usually true about people in general. At the same time, certain superficial cues remind you of particular people you've already met. Unconsciously, then, you will assume that this stranger must also resemble *them*, not only in appearance but in other traits as well. No amount of self-discipline can keep those superficial similarities from provoking assumptions that may then affect your judgments and decisions. When we disapprove of this, we complain about stereotypes—and when we sympathize with it, we speak of sensitivity and empathy.

It's much the same with language, too. If someone said, *"It's raining frogs,"* your mind would swiftly fill with thoughts about the origins of those frogs, about what happens to them when they hit the ground, about what could have caused that peculiar plague, and about whether or not the announcer had gone mad. Yet the stimulus for all of this is just three words. How do our minds conceive such complex scenes from such sparse cues? The additional details must come from memories and reasoning.

Most older theories in psychology could not account for how a mind could do such things—because, I think, those theories were based on ideas about "chunks" of memory that were either much too small or much too large. Some of those theories tried to explain appearances only in terms of low-level "cues," while other theories tried to deal with entire scenes at once. None of those theories ever got very far. The next few sections describe what seems to be a useful compromise; at least it has led to some better results in some projects concerned with Artificial Intelligence. Our idea is that each perceptual experience activates some structures that we'll call *frames*—structures we've acquired in the course of previous experience. We all remember millions of frames, each representing some stereotyped situation like meeting a certain kind of person, being in a certain kind of room, or attending a certain kind of party.

24.2 FRAMES OF MIND

A *frame* is a sort of skeleton, somewhat like an application form with many blanks or slots to be filled. We'll call these blanks its *terminals*; we use them as connection points to which we can attach other kinds of information. For example, a frame that represents a "chair" might have some terminals to represent a seat, a back, and legs, while a frame to represent a "person" would have some terminals for a body and head and arms and legs. To represent a *particular* chair or person, we simply fill in the terminals of the corresponding frame with structures that represent, in more detail, particular features of the back, seat, and legs of that particular person or chair. As we'll see, virtually any kind of agent can be attached to a frame-terminal. It can be a K-line, polyneme, isonome, memory-control script, or, best of all, another frame.

In principle, we could use frames without attaching their terminals to anything. Normally, though, the terminals come with other agents already attached—and these are what we called "default assignments" when we first talked about level-bands. If one of your person-frames is active, and you actually see some arms and legs, their descriptions will be assigned to the proper terminals. However, if certain parts cannot be seen, perhaps because they're out of view, the missing information will be filled in by default. We use default assumptions all the time: that's how, when you see someone wearing shoes, you "know" that there are feet in them. From where do those assumptions come? I'll argue that

Default assumptions fill our frames to represent what's typical.

As soon as you hear a word like "person," "frog," or "chair," you assume the details of some "typical" sort of person, frog, or chair. You do this not only with language, but with vision, too. For example, when someone is seated across the table from you, you may be unable to see any part of that person's chair. Still, this situation will probably activate a sitting-frame. But a sitting-frame will surely have a terminal for what to sit upon, and that will be assigned, by default, to some stereotypical chair. Then, though there is no chair in sight, a chair-frame will be supplied by default.

Default assignments are of huge significance because they help us represent our previous experience. We use them for reasoning, recognizing, generalizing, predicting what may happen next, and knowing what we ought to try when expectations aren't met. Our frames affect our every thought and everything we do.

Frames are drawn from past experience and rarely fit new situations perfectly. We therefore have to learn how to adapt our frames to each particular experience. What if a given situation closely matches several different frames at once? Some such conflicts could be resolved by the "locking-in" negotiations we described earlier; then only the frames that manage to suppress their competitors can influence one's other agencies. But the other frames could lurk offstage, awaiting opportunities to intervene.

24.3 HOW *TRANS*-FRAMES WORK

In order to be more concrete, let's make a little theory of how a frame might actually work. Consider, for example, a *Trans*-frame that is filled in to represent this sentence:

Jack drove from Boston to New York on the turnpike with Mary.

Whenever this particular frame is active, if you wonder about the *Destination* of that trip, you'll almost instantly think of New York. This suggests that the polyneme for New York must be aroused by the coincidence of two mental events, namely, the arousal of this particular travel-frame and the arousal of the pronome for *Destination*. Now how could a brain-agent recognize such a coincidence? Simple: we need only assume that the polyneme for New York is attached to an **AND**-agent with two inputs; one of them represents the arousal of the travel-frame itself, and the other represents the arousal of the *Destination* pronome. Accordingly, each terminal of our frame could simply be an **AND**-agent with two inputs.

According to this simple scheme, a frame could consist of little more than a collection of **AND**-agents, one for each of the frame's pronome terminals! Then the entire frame for the New York trip would look like this:

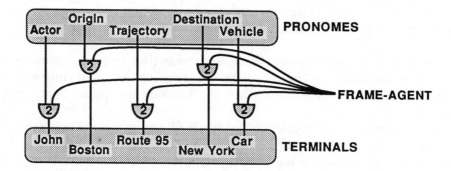

When a frame-agent is activated—either by seeing, hearing, or imagining something—this supplies each of those **AND**-agents with one of these two inputs. The second input is provided by some pronome which can thereby activate whichever agent or frame is presently assigned to that terminal. If several pronomes are active at the same time, all the corresponding agents will be activated, too. When the frame above is active, the pronome for *Origin* will activate the K-line for Boston, and the pronome for *Vehicle* will activate the K-line for car.

How could such a frame be made to *learn* which polynemes should fill its terminals? We could begin with each terminal initially connected to a virgin K-line; then each terminal will represent whatever the corresponding K-line learns. Notice that to build frames this way, we need only connect **AND**-agents to K-lines that can in turn be constructed from little more than simple **AND**-type agents. One of the great surprises of modern computer science was the discovery that so much can be done with so few kinds of ingredients.

24.4 DEFAULT ASSUMPTIONS

*Only by art can we get outside ourselves; instead of seeing only one
world, our own, we see it under multiple forms, and as many as
there are original artists, just so many worlds
have we at our disposal.*
—MARCEL PROUST

When someone says, "John threw a ball," you probably unconsciously assume a certain set of features and qualities of the ball, like color, size, and weight. These are your assumptions by default, the kind we talked about when we first introduced the idea of level-bands. Your assumptions about that ball might be derived from some ball you owned long ago—or, possibly, your newest one. It is our theory that such optional details are usually attached too weakly to hold against the sharp insistence of reality, so that other stimuli will find them easy to detach or otherwise adapt. This is why default assumptions make weak images, and why we aren't too amazed when they turn out wrong. It is no surprise that frames share so many properties of K-lines, since the terminals of frames themselves will lie in level-bands near the K-lines whose fringes represent our expectations and default assumptions.

But why use default assumptions at all, instead of simply seeing what's really there? *Because unless we make assumptions, the world would simply make no sense.* It would be as useless to perceive how things "actually look" as it would be to watch the random dots on untuned television screens. What really matters is being able to see what things *look like*. This is why our brains need special machinery for representing what we see in terms of distinct "objects." The very idea of an object embodies making many assumptions that "go without saying"—for example, that it has substance and boundaries, that it existed before we saw it, and that it will remain afterward—in short, that it will act like other typical objects. Thus, though we never see every side of an object at once, we always assume that its unseen sides exist. I suspect that the larger part of what we know–or think we know–is represented by default assumptions, because there is so little that we know with perfect certainty.

We use default assumptions in personal relations, too. Why do so many people give such credence to astrology, to classify friends by the months of their births? Perhaps it seems a forward step, to class all persons into just twelve types—to those who once supposed that there were less. And how does the writer's craft evoke such lifelike characters? It's ridiculous to think that people could be well portrayed in so few words. Instead, our story writers use phrases that activate great networks of assumptions that already lie in the minds of their readers. It takes great skill to create those illusions—to activate unknown processes in unknown readers' minds and to shape them to one's purposes. Indeed, in doing so, a writer can make things clearer than reality. For although words are merely catalysts for starting mental processes, so, too, are real things: we can't sense what they really are, only what they remind us of. As Proust went on to say:

Each reader reads only what is already inside himself. A book is only a sort of optical instrument which the writer offers to let the reader discover in himself what he would not have found without the aid of the book.

24.5 NONVERBAL REASONING

Even when you were very young, if someone had told you that most Snarks are green—and, also, that every Boojum is a Snark—you would have been able to conclude that most Boojums are green. What could have led you to that conclusion? Presumably, you answer questions about the properties of Boojums by attaching your polyneme for Snark to whichever of your memory-units currently represents a Boojum. Accordingly, you assume that the color property of a Boojum is green by using your usual way of recalling the properties of things you know— activating their polynemes to set your various agencies into the corresponding states. *In other words, we do this kind of reasoning by manipulating our memories to replace particular things by typical things.* I mention all this because it is often assumed that adults are better than children are at what is often called *abstract* or *logical* reasoning. This idea is unfair both to adult and child because logical thinking is so much simpler—and less effective—than common-sense thinking. Actually, what appears to be a matter of "logic" is usually not logical at all and frequently turns out to be wrong. In the case above, you would have been wrong because Boojums are albino Snarks.

The situation is different when you happen to know more about a particular example. For example, suppose you first had learned that penguins cannot fly and then learned that penguins are a kind of bird. When you discover that, should you replace all of your penguin properties with those of your "generic" bird? Clearly not, since then you'd lose your hard-earned penguin facts. To deal with this effectively, children must develop complex skills, not merely to replace one representation with another, but to compare two representations and then move around inside them, making different changes at different levels. These intricate skills involve the use of isonomes that control the level-band of the activities inside our agencies.

In any case, to reason well, our memory-control agencies must learn to "move" our memories around as though those memories were building-blocks. Conceivably, those agencies have to learn such skills before we can learn to build with blocks in the outside world of object-things. Unfortunately, we know very little about how such processes work. Indeed, we're virtually unaware that they even exist, because these kinds of "commonsense" inferences and assumptions come to mind without the slightest conscious effort or activity. Perhaps this unawareness is a consequence of the speed with which those skills employ the very same short-term memory-units that might otherwise be used to record those agents' own recent activities.

24.6 DIRECTION-NEMES

When you think about an object in a certain *place*, many different processes go on inside your mind. Some of your agencies know the visual direction in which that object lies, others can direct your hand to reach toward it, and yet other agencies anticipate how it would feel if it touched your skin. It is one thing to know that a block has flat sides and right angles, another to be able to recognize a block by sight, and yet another to be able to shape your hand to grasp that shape or recognize it from feeling it in your grasp. How do so many different agencies communicate about places and shapes?

No one yet knows how shapes and places are represented in the brain. The agencies that do such things have been evolving since animals first began to move. Some of those agencies must be involved with postures of the arm and hand, others must represent what we discover from the images inside our eyes, and yet others must represent the relations between our bodies and the objects that surround us.

How can we use so many different kinds of information at once? In the following sections I'll propose a new hypothesis to deal with this: that many agencies inside our brains use frames whose terminals are controlled by *interaction-square arrays*. Only now we'll use those square-arrays not to represent the interactions of different causes, but to describe the relations between closely related locations. For example, thinking of the appearance of a certain place or object would involve arousing a squarelike family of frames, each of which in turn represents a detailed view of the corresponding portion of that scene. If we actually use such processes, this could explain some psychological phenomena.

> If you were walking through a circular tube, you could scarcely keep from thinking in terms of bottom and top and sides—however vaguely their boundaries are defined. Without a way to represent the scene in terms of familiar parts, you'd have no well-established thinking skills to apply to it.

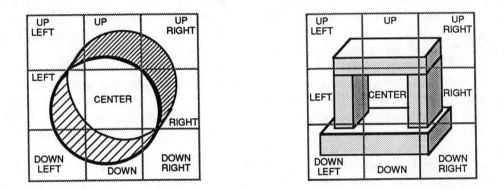

The diagram is meant to suggest that we represent directions and places by attaching them to a special set of pronome-like agents that we shall call "direction-nemes." Later we'll see how these might be involved in surprisingly many realms of thought.

24.7 PICTURE-FRAMES

Whenever we see a thing so utterly new that it resembles nothing we've ever seen before, this means that none of our prelearned frames will fit it very well. But this rarely happens to adults. For example, we have each accumulated enough room-frames to represent most rooms we're likely to see, such as kitchens, bedrooms, offices, factories, and concert halls; one of these will usually match whichever place we happen to be. Besides, we can almost always use a less specific frame that fits most any room at all—a frame with terminals that correspond to nothing more than ceiling, floor, and walls. Then each of those six surfaces could be represented, in turn, by a subframe that has terminals for several vaguely defined regions. To be specific, let's employ our direction-neme idea and divide each surface—for ceiling, floor, and each of the walls—into zones that correspond to the nine regions of an interaction-square. A typical wall might then be represented in this fashion:

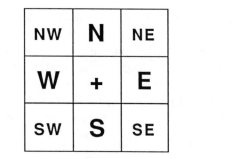

In spite of its simplicity, we can represent quite a lot of information with this scheme. It provides enough structure to recall later that *there was a window toward the left, some shelves high to the right side of the wall, and a table to the right.* If this does not seem precise enough, the fact is that we usually don't remember things so exactly, except when they attract special attention; normally it would be enough to know only roughly where that television was, and we could assume by default that it was supported by the tabletop. It takes surprisingly few such observations to enable us to tell, later, whether much has been changed.

Given more time, one could keep noticing more details and include them by attaching additional subframes. This would overcome the limitations of starting with so few terminals. For example, one might notice that the window is closer to the shelf than to the television set, and closer to the ceiling line than to either the shelf or the television set. And if the outline of the desk and television set reminds you of a goatlike animal, your representation can include that fact.

Suppose you had first assumed this to be a living room but later recognized the table to be a *kitchen* table? Must you undo all the work you've done, to activate a different kitchen-frame and start all over again? No, because later we'll see a convenient way to switch over to another frame—while still retaining what was learned so far. The trick will be to make all our frames for different rooms *share the same terminals*, so that when we interchange those frames, the information stored in them remains.

24.8 HOW PICTURE-FRAMES WORK

Now that we've seen how "picture-frames" could represent memories of spatial arrangements, let's ask how we actually build such frames. We'll use the same technique that we used to build *Trans*-frames, except for one small change. To make a picture-frame, we'll simply replace the pronomes of our *Trans*-frame scheme by a set of nine direction-nemes! The diagram below also includes an agent to serve for turning on the frame itself.

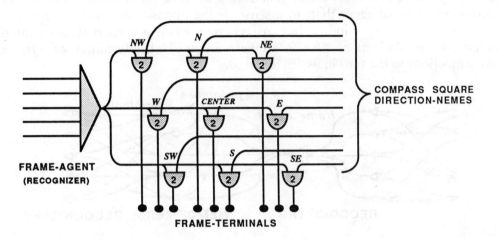

To apply the picture-frame idea to how our vision-systems work, imagine that you're looking at some real-world scene. Your eyes move in various directions, controlled in some way by direction-nemes. Now suppose that every time you move your eyes, the same direction-nemes also activate the K-lines attached to the corresponding terminals of a certain vision-frame. Suppose, also, that those K-lines are ready to form new memories. *Then each time you look in a different direction, your vision-system will describe what you see—and the corresponding K-line will record what you see when you look in that direction!*

Now suppose that the same frame is activated at some later date—but this time by means of memory and not from looking at some scene. Then, as any of your agencies conceives of looking in a certain direction, the thought itself will involve the activation of the corresponding direction-neme; then, before you have a chance to think of anything else, the corresponding K-line will be aroused. This creates a most remarkable effect:

> *Whichever way your "mind's eye" looks, you'll seem to see the corresponding aspect of the scene. You will experience an almost perfect "simulus" of being there!*

How "real" could such a recollection seem? In principle, it could even seem as real as vision itself—since it could make you seem to sense not only how an object looks, but also how it tastes and feels. Shortly we'll see how this could yield not merely the sense of seeing a scene, but also the sense of being able to move around inside it.

24.9 RECOGNIZERS AND MEMORIZERS

How do frames become activated? This amounts to asking how we recognize familiar situations or things. There is no limit to how complicated such a question can become, since there are no natural boundary lines between recognizing, remembering, and all the rest of how we think. For questions like this, with no place to start, we have to construct some boundary lines from our own imagination.

We'll simply assume that every frame is activated by some set of recognizers. We can regard a recognizer as a type of agent that, in a sense, is the opposite of a K-line—since instead of arousing a certain state of mind, it has to recognize when a certain state of mind occurs. Accordingly, the recognizers of a frame are very much like the terminals of a frame, except that the connections to the terminals are reversed.

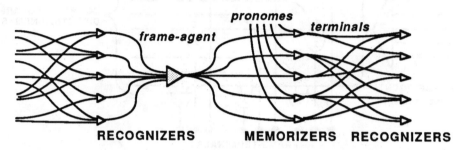

This suggests that not only frames but agencies in general might be organized in the form of agents sandwiched between recognizers and memorizers.

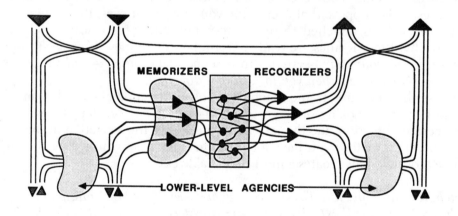

This sketch of how our agencies are organized is oversimplified. Each agent, be it a frame, a K-line, or whatever, must have some machinery for learning when it should become active— and that may require more than simply recognizing the presence of certain features. For example, to recognize an object as a car, it isn't enough that it include some assortment of parts like body, wheels, and license plate; the frame must also recognize that those parts are in suitable relationships—that the wheels be properly attached to the body, for example. Workers in the field of Artificial Intelligence have experimented with a variety of ways to make frame-recognizers, but the field is still in its infancy. The recognizers of our higher-level agencies might have to include mechanisms as complex as difference-engines in order to match their relational descriptions to actual situations.

CHAPTER 25

FRAME-ARRAYS

The device of images has several defects that are the price of its peculiar excellences. Two of these are perhaps the most important: the image, and particularly the visual image, is apt to go farther in the direction of the individualization of situations than is biologically useful; and the principles of the combination of images have their own peculiarities and result in constructions which are relatively wild, jerky and irregular, compared with the straightforward unwinding of a habit, or with the somewhat orderly march of thought.

—F. C. BARTLETT

25.1 ONE FRAME AT A TIME?

Each of the drawings below can be seen in at least two different ways.

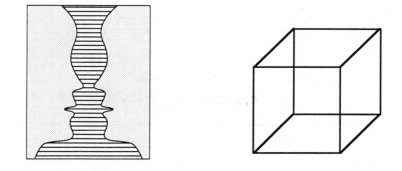

The drawing on the left might represent either a single candlestick or two people facing each other. The drawing on the right looks like a cube—but first it looks like a cube as seen from above and then, suddenly, it looks like a cube as seen from below. Why does each drawing seem to change its character from time to time? Why can't we see both forms at once? Because, it seems, our agencies can tolerate just one interpretation at a time.

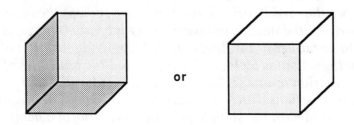

or

We must ask certain questions here. First, *what enables us to see those pictures as composed of the features we call by names like edges, lines, corners, and areas?* Our vision-systems seem virtually compelled to group the outputs of our sensors into entities like these. Next, *what enables us to see those features as grouped together to form larger objects?* Apparently, our vision-systems again are virtually compelled to represent each of those features, be it a corner, edge, or area, as belonging to one and only one larger object at a time. I won't discuss those questions in this book except to suggest a general hypothesis:

> *Our vision-systems are born equipped, on each of several different levels, with some sort of "locking-in" machinery that at every moment permits each "part," at each level, to be assigned to one and only one "whole" at the next level.*

We should also ask, *how do we recognize those objects as examples of familiar things like* **faces, cubes,** *or* **candlesticks?** And again we'll make the similar hypothesis that our memory-frame machinery also uses "locking-in" machinery that permits each "object" to be attached only to one frame at a time. The end result is that in every region of the picture, the frames must compete with one another to account for each feature.

When we first discussed how *Builder* works, we assumed that it employed a vision-agent, *See*, to locate the various blocks it needs. However, we never discussed how *See* itself might work. A person simply "looks and sees"—but that's more complicated than it seems. For instance, even a simple cube looks different from each point of view, since as you move, the images it makes inside your eye keep changing in both shape and size.

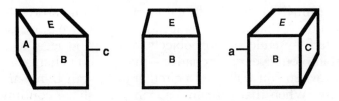

How strange and dangerous moving would be if every step made everything seem wholly new! But that's not how it seems to us. When we move to the right, so that **A** becomes invisible, we remember what we learned when we saw it, and it still seems part of what we're seeing now. How can this be? Here is a theory of why things seem to stay the same, even when what we see of them keeps changing as we move around.

Frame-Arrays. *When we move, our vision-systems switch among a family of different frames that all use the same terminals.*

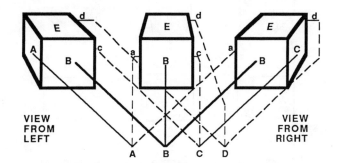

I'll use the term "frame-arrays" for these groups of frames that share the same terminals. When you represent a thing's details with a frame-array, you can continue to move around yet "keep in mind" all that you've observed from those different viewpoints, even though you've never seen them all at once. This gives us the wonderful ability to conceive of all of an object's different views as aspects of a single thing.

I do not mean to suggest that every time you see a new object you build a brand-new frame-array for it. First, you try to match what you see to the frame-arrays in the memories you have accumulated and refined over periods of many years. How do frame-arrays originate? I would assume that this underlying pattern—of families of frames that all share common terminals—is built into the architectures of major sections of the brain. But although that pattern is "built in," developing the skills for using it involves each child in more than a decade of predestined learning.

25.3 THE STATIONARY WORLD

What makes objects seem to stay in place no matter how the viewer moves? To common sense this is no mystery: *it's simply that we're seeing all the time and keeping contact with the world.* However, I suspect that if we had to start seeing all over again from every moment to the next, we'd scarcely be able to see at all. This is because our higher-level agents don't "see" the outputs of the sensors in our eyes at all. *Instead, they "watch" the states of middle-level agencies that don't change state so frequently.* What keeps those "inner models" of the world from changing all the time? This is the function of our frame-arrays: to store what we learn about the world at terminals that stay unchanged when we move our heads and bodies around. This explains a wonderful pseudoparadox: objects in the world seem to change only when the pictures they project into our eyes *don't* change—that is, don't change according to our expectations. For example, when you walk past a circular dish, your frame-arrays expect that circle to turn into an ellipse. When that actually happens, the shape continues to "look" circular. However, should that expected change fail to occur, the shape will seem to change of its own accord.

How, then, do we automatically compensate for changes of view? The system could work just as we described in section 24.8: *by using the same direction-nemes both to control our own motions and to select frames from our frame-arrays.* For example, you might use several frames to represent an image of a cube, arranged in a network like this:

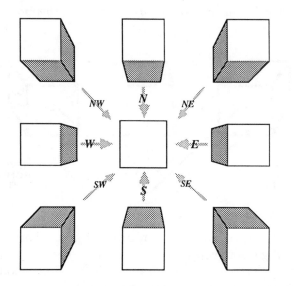

When you activate your "move east" direction-neme in order to make your body move in that direction, the same signal will also make this frame-array replace the middle frame with the one to its left. This will compensate for your change of viewpoint and determine what you "expect" to see—and the scene will appear to remain stationary! Michael Crichton has suggested that when you move inside such a space, you must unconsciously be registering the distortions of the shape, the moving walls and corners. Only you don't interpret these as changes in the room itself, but use them as more accurate cues to orient yourself in the space.

You can bypass this entire system by gently pushing the side of your eye with your finger; then the world will indeed appear to move, because your frame-arrays do not receive the corresponding direction signals!

25.4 THE SENSE OF CONTINUITY

> *And any object being removed from our eyes, though the*
> *impression it made in us remains; yet other objects more present*
> *succeeding, and working on us, the Imagination of the past is*
> *obscured, and made weak; as the voyce of a man is in the noyse of*
> *the day. From whence it followeth, that the longer the time is,*
> *after the sight, or Sense of any object, the weaker is the*
> *Imagination. For the continuall change of mans body, destroys the*
> *parts which in sense were moved; So that distance of time, and of*
> *place, hath one and the same effect in us.*
> —HOBBES

Imagine what these frame-arrays can do! They let us "visualize" imaginary scenes, such as what might happen when we move, because the frames for what we can expect to see are filled in automatically. Not only that, but by using other processes to fill in all those terminals, we can "imagine" scenes and views of things we've never seen before. Still, many people find it hard to consider the thought that mental images could be based on anything as crude as frame-arrays. The world of our experience seems so perfectly continuous. Could such smooth thoughts emerge from sudden frame-to-frame jumps? If the mind kept jerking from one frame to another, wouldn't what we experience seem equally abrupt? Yet we rarely feel our minds change frames, any more than we perceive a visual scene as composed of disconnected spots of light. Why *do* we have the sense that things proceed in smooth, continuous ways? Is it because, as some mystics think, our minds are part of some flowing stream? I think it's just the opposite: our sense of constant, steady change emerges from the parts of mind that manage to insulate themselves against the continuous flow of time!

In other words, our sense of smooth progression from one mental state to another emerges not from the nature of that progression itself, but from the descriptions we use to represent it. Nothing can *seem* jerky except what is *represented* as jerky. Paradoxically, our sense of continuity comes from our marvelous *insensitivity* to most kinds of changes rather than from any genuine perceptiveness. Existence seems continuous to us not because we continually experience what is happening in the present, but because we hold to our memories of how things were in the recent past. Without those short-term memories, all would seem entirely new at every instant, and we would have no sense at all of continuity or, for that matter, of existence.

One might suppose that it would be wonderful to possess a faculty of "continual awareness." But such an affliction would be worse than useless, because the more frequently our higher-level agencies change their representations of reality, the harder it is for them to find significance in what they sense. The power of consciousness comes not from ceaseless change of state, but from having enough stability to discern significant changes in our surroundings. To "notice" change requires the ability to resist it. In order to sense what persists through time, one must be able to examine and compare descriptions from the recent past. We notice change in spite of change, not because of it.

*Our sense of constant contact with the world is not a genuine experience; instead, it is a form of **immanence illusion.** We have the sense of actuality when every question asked of our visual-systems is answered so swiftly that it seems as though those answers were already there. And that's what frame-arrays provide us with: once any frame fills its terminals, the terminals of the other frames in its array are also filled. When every change of view engages frames whose terminals are already filled, albeit only by default, then sight seems instantaneous.*

25.5 EXPECTATIONS

But in the common way of taking the view of any opake object,
that part of its surface, which fronts the eye, is apt to occupy the
mind alone, and the opposite, nay even every other part of it
whatever, is left unthought of at that time: and the least motion
we make to reconnoitre any other side of the object, confounds our
first idea, for want of the connexion of the two ideas.
—WILLIAM HOGARTH

Imagine that you turned around and suddenly faced an absolutely unexpected scene. You'd be as shocked as though the world had changed before your eyes because so many of your expectations were not met. When we look about a familiar place, we know roughly what to expect. But what does *expect* mean?

Whenever we become familiar with some particular environment like an office, home, or outdoor place, we represent it with a frame-array whose terminals have already been filled. Then, for each direction of motion inside that environment, our vision-systems activate the corresponding frames of that array. We also activate the corresponding frames even when we merely *consider* or imagine a certain body motion—and this amounts to "knowing what to expect." In general, each frame of a spatial frame-array is controlled by some direction-neme. However, in surroundings that are either especially familiar or whose relationships we do not understand, we may learn to use more specific stimuli instead of using direction-nemes to switch the frames. For example, when you approach a familiar door, the frame for the room that you expect to find behind that door might be activated, not by your direction of motion, but by your recognition of that particular door. This could explain how a person can reside in the same home for decades, yet never learn which of its rooms share common walls.

In any case, all this is oversimplified. Many of our frame-arrays must require more than nine direction views; they need machinery to modify the sizes and shapes of their objects; they must be adapted to three dimensions; and they must be able to represent what happens at intermediate moments during motion from one view to another. Furthermore, the control of frame selection cannot depend on a single, simple set of direction-nemes, for we must also compensate for the motions of our eyes, neck, body, and legs. Indeed, a major portion of our brain-machinery is involved with such calculations and corrections, and it takes a long time to learn to use all that machinery. The psychologist Piaget found that it takes ten years or more for children to refine their abilities to imagine how the same scene will appear from different viewpoints.

This was the basis of Hogarth's complaint. The artist felt that many painters and sculptors *never* learned enough about spatial transformations. He felt that mental imagery is an acquired skill, and he scolded artists who gave too little time to "perfecting the ideas they have in their minds about the objects in nature." Accordingly, Hogarth worked out ways to train people to better predict how viewpoints change appearances.

[He who undertakes the acquisition of] *perfect ideas of the distances, bearings, and oppositions of several material points and lines in even the most irregular figures, will gradually arrive at the knack of recalling them into his mind when the objects themselves are not before him—and will be of infinite service to those who invent and draw from fancy, as well as to enable those to be more correct who draw from the life.*

25.6 THE FRAME IDEA

*Questions arise from a point of view—from something that helps
to structure what is problematical, what is worth asking, and what
constitutes an answer (or progress). It is not that the view
determines reality, only what we accept from reality and how we
structure it. I am realist enough to believe that in the long run
reality gets its own chance to accept or reject our various views.*
—ALLEN NEWELL

I first conceived the idea of frames in the early 1970s, while working on making a robot that could see, and I described the theory in a 1974 essay entitled "A Framework for Representing Knowledge." The essay influenced the next decade of research on Artificial Intelligence, despite the fact that most readers complained that its explanations were too vague. In retrospect, it seems those explanations were at just the right level-bands of detail to meet the needs of that time, which is why the essay had the effect it did. If the theory had been any vaguer, it would have been ignored, but if it had been described in more detail, other scientists might have "tested" it, instead of contributing their own ideas. Then they might have found my proposals were inadequate. Instead, many versions were suggested by other people, and "frame-based" programming became popular.

Two students in particular, Scott Fahlman and Ira Goldstein, claimed to understand what I had meant—and then explained many details I hadn't imagined at all. Another student, Terry Winograd, worked on making a robot that understood a certain class of English-language sentences; this led to important theories about the relation between grammar and its effect upon a listener. Then, since that robot's task was building towers of children's blocks, Winograd also worked out many details of how to make a *Builder*. You can see how his theories affected this book. Yet another student, Eugene Charniak, worked on the problem of how young children understand the stories they read. He spent at least a solid year thinking about one such story, which had to do with bringing a kite to a birthday party. Shortly, you'll see the influence Charniak had on this book.

All along, I had felt that the frame idea itself was rather obvious and perhaps implicit in the earlier work of psychologists like Bartlett. I considered the more important concept in the 1974 essay to be the idea of a frame-system—renamed "frame-array" in this book. I was surprised that the frame idea became popular while the frame-array idea did not. The neme concept emerged in 1977 (under the term "C-lines"); the K-line idea crystallized in 1979. The concept of pronomes was in my unconscious mind for several years but did not crystallize until, while writing this book, I realized how to reformulate several of Roger Schank's early ideas into the form of *Trans*-frames. The scheme proposed in this book, in which the frame-terminals are controlled by bundles of nemes or isonomes, did not emerge until a full decade after the original concept of a frame-array.

Many questions remain about how frames might work. For example, it should be possible to recognize several different things at once by using different frames in parallel. But how can we see many faces in a crowd at once, or bricks in a wall, or chairs in a room? Do we make many copies of the same frame? I suspect that's impractical. Instead, perhaps we match each frame only to one example at a time—and simply assume that the same frame also applies to every other visible object that shares some characteristic features with the object under attention.

CHAPTER 26

LANGUAGE-FRAMES

Thinking . . . is possible only when a way has been found of breaking up the "massed" influence of past stimuli and situations, only when a device has already been discovered for conquering the sequential tyranny of past reactions. But though it is a later and a higher development, it does not supersede the method of images. It has its own drawbacks. Contrasted with imaging it loses something of vivacity, of vividness, of variety. Its prevailing instruments are words, and, not only because these are social, but also because in use they are necessarily strung out in sequence, they drop into habit reactions even more readily than images do. [With thinking] *we run greater and greater risk of being caught up in generalities that may have little to do with actual concrete experience.*

—F. C. BARTLETT

26.1 UNDERSTANDING WORDS

What happens when a child reads a story that begins like this?

> *Mary was invited to Jack's party.*
> *She wondered if he would like a kite.*

If you asked what that kite was for, most people would answer that it must be a birthday present for Jack. How amazing it is that every normal person can make such complicated inferences so rapidly—considering that the idea of a gift was never mentioned at all! Could any machine do such remarkable things? Consider all the other assumptions and conclusions that almost everyone will make:

> *The "party" is a birthday party.*
> *Jack and Mary are children.*
> *"She" is Mary.*
> *"He" is Jack.*
> *She is considering giving Jack a kite.*
> *She wonders if he would like the kite.*

We call these understandings "common sense." They're made so swiftly that they're often ready in our minds before a sentence is complete! But how is this done? In order to realize that the kite is a present, one has to use such knowledge as that parties involve presents, that presents for children are usually toys, and that kites are appropriate toys to be given as presents. None of this is mentioned in the story itself. How do we bring together all that scattered knowledge so quickly? Here's what I think must happen. Somehow the words *"Mary was invited to Jack's party"* arouses a "party-invitation" frame in the reader's mind—and attached to the terminals of that frame are certain memories of various concerns. *Who is the host? Who will attend? What present should I bring? What clothing shall I wear?* Each of those concerns, in turn, is represented by a frame to whose terminals are already attached, as default assignments, the most usual solutions to that particular kind of problem.

Such knowledge comes from previous experience. I was raised in a culture in which an invitation to a party carries the obligation to arrive well dressed and to bring a birthday present. Accordingly, when I read or hear that Mary was invited to a party, I attribute to Mary the same sorts of subjective reactions and concerns that I would have in such a situation. Therefore, although the story never mentions clothes or gifts at all, to expect their possible involvement seems only simple common sense. But though it is common, it is not simple. The next few sections speculate about how story understanding works.

26.2 UNDERSTANDING STORIES

We'll now see how frames can help to explain how we understand that children's tale. How do we know that the kite is a present for Jack—when neither sentence mentioned this?

Mary was invited to Jack's party.
She wondered if he would like a kite.

After the first sentence activates a party-invitation frame, the reader's mind remains engaged with that frame's concerns—including the question of what type of birthday gift to bring. If this concern is represented by some subframe, what are the concerns of that subframe? That present must be something that will please the party host. "Toy" would be a good default for it, since that's the most usual kind of gift for a child.

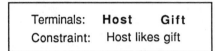

Terminals:	**Host**	**Gift**
Constraint:	Host likes gift	

The first sentence suggests using *Jack* to fill in the [Host] terminal.

Host	**Gift**
Jack	likes	toy

Mary was invited to Jack's party.

The second sentence suggests using *kite* for *toy* and *He* for [Host].

Host	**Gift**
He	likes	kite

She wondered if he would like a kite.

Since "Jack" is a "he" and a "kite" is a "toy," these two frames will merge perfectly—provided that the reader's frame for boy assumes that Jack is likely to enjoy kites. Then our two sentences combine perfectly to fill the present frame's terminals, and our problem is solved!

Terminals:	**Host**	**Gift**
Constraint:	Jack likes	kite

What makes a story comprehensible? What gives it coherency? The secret lies in how each phrase and sentence stirs frames into activity or helps already active ones to fill their terminals. When the first sentence of our story mentions a party, various frames are excited—and these are still active in the reader's mind when the next sentence is read. The ground is prepared for understanding the second sentence because so many agents are already ready to recognize possible references to presents, clothes, and other matters that might be related to birthday parties.

26.3 SENTENCE-FRAMES

*People do not usually state all the parts of a given thought that
they are trying to communicate because the speaker tries to be brief
and leaves out assumed or unessential information. The
conceptual processor searches for a given type of information in a
sentence or a larger unit of discourse that will fill the needed slot.*

—ROGER SCHANK

We've barely started to see what minds must do to comprehend the simplest children's tales. Let's look again at the beginning of our party story.

> *Mary was invited to Jack's party.*

How marvelous that sentence is! How much it says in just six words! Two characters are introduced and quickly cast in clear-cut roles. We learn that there will be a party soon, with Jack as the host and Mary a guest—provided she accepts the invitation. We also learn that this setting is established in the past.

Those six short words tell even more. We can expect the story to focus on Mary's activities rather than Jack's—because "Mary" is the first word that attracts our attention. But to accomplish that, the narrator had to use a clever grammar-tactic. Normally, an English-language sentence begins with a phrase that describes the *Actor* responsible for some action, and we usually represent this with a simple *Trans*-frame.

> JACK   INVIT—ed   MARY
> *Donor*   *action verb*   *Recipient*

In this "active verb" form of sentence-frame, the verb is sandwiched between two nouns; the first describes a *Donor* and the second describes a *Recipient*. However, if our storyteller actually used the active form of sentence-frame, it would tend to mislead the listener into expecting Jack to be the central character of the story—if only because he is mentioned first. Fortunately, English grammar provides an alternative sentence-frame in which the *Recipient* is mentioned first—and which never mentions the *Donor* at all!

> MARY   was   INVIT—ed
> *Recipient*   *was*   *verb—ed.*

How does the understanding listener detect this "passive verb" sentence-frame? Some language-agent has to notice the way the verb is sandwiched between "*was*" and "*-ed.*" As soon as this special subframe is recognized, the language-agency will reassign the first noun, *Mary*, not to the *Donor* terminal, but to the *Recipient* terminal—and thus Mary is represented as receiving the invitation. Why don't we need to say who the donor is? Because in this case the listener can assume it by default. Specifically, the expression "*Jack's party*" evokes a "party-invitation frame," and in such situations it is typical for the host—or the host's parents—to invite the party guests. By thus arousing familiar frames, we can say a great deal in a very few words.

26.4 A PARTY-FRAME

par•ty *n. A gathering for social entertainment, or the
entertainment itself, often of a specific nature. . . .*
—*Webster's Unabridged Dictionary*

Dictionary definitions never say enough. Every child knows that a party is more than just a gathering assembled to celebrate someone's birthday. But no brief definition can describe the complicated customs, rules, and regulations that typical communities prescribe for such ceremonies. When I was a child, a birthday party could be expected to include at least the elements of the following script:

ARRIVAL.
    Greeting. Be well dressed.
GIFT.
    Give birthday present to host or guest of honor.
GAMES.
    Activities like blindfold competitions.
DECOR.
    Balloons. Favors. Crepe-paper decorations.
PARTY-MEAL.
    Hotdogs, candies, ice-cream, etc.
CAKE.
    With candles to represent the host's age.
CEREMONY.
    Host tries to extinguish candles with single breath (to make a wish).
SONG.
    All guests sing birthday song and eat cake.

This is merely an outline, for every item leads to other conditions and requirements. The birthday present has to please the host, of course, but there are other strong constraints on it as well. It ought to be brand new, of good quality, and it should not be ostentatiously extravagant. It ought to be suitably "party-wrapped"—that is, covered with a certain kind of color-printed wrapping paper and tied with colored ribbon. There are also constraints on other items in the script. The birthday cake should be covered with a sweet sugar frosting. In my childhood, the ice cream usually consisted of three colored stripes of different flavors: vanilla, strawberry, and chocolate. Because I did not like the strawberry flavor, my personal party script included the extra steps of finding another child willing to make a trade.

To all their young participants, such parties unfold exactly as a party should, with all these queer complexities. We take our social customs for granted, as though they were natural phenomena. Few guests or hosts will ever wonder why their parties have those explicit forms or ask about their origins. As far as any child can tell, that's just how parties ought to go; they always did and always will. And so it is with almost everything we know.

26.5 STORY-FRAMES

*It's the way in which all tales have been told from Homer and
Sophocles down through Kipling, Hemingway, Bradbury,
Sturgeon, McCaffrey, Zelazny, whoever. Once upon a time, you
say, there lived so-and-so in such and such a place, and while he
was minding his own business the following absolutely astonishing
thing happened to him. And so you begin; and they gather close
about you, for they cannot choose but to hear.*
—ROBERT SILVERBERG

We take it for granted that anyone can understand a story. But every kind of narrative demands some "listening skills." Even the best storytellers find it hard to entertain children, who are prone to interrupt with questions that make perfect sense by themselves but drift away from the story's theme. *"Where does Mary live?" "Does she have a dog?"* To listen well, a child must acquire potent forms of self-control.

The storyteller, too, must work to fix the focus of the listener's mind. If you were speaking about something else and suddenly, completely out of context, remarked, *"Mary was invited to Jack's party,"* an unprepared listener might wonder, *"Mary who?"* and look to see if you were addressing someone else. But you can first prepare the listener by saying, *"Would you like to hear a story?"* or simply, *"Once upon a time . . ."* What is the function of such a phrase? It has a very specific effect: to set the listener into a normal and familiar state of expecting to hear a certain type of narrative—a story. In the English tradition, stories typically begin by specifying the *time*—if only vaguely, by saying *"long ago."* I'm told that in Japan most stories start with saying *where* as well—if only by some empty phrase like *"in a certain time and place."* The biblical book of Job begins with, *"There was a man in the land of Uz . . ."*

Most stories start with just enough to set the scene. Then they introduce some characters, with hints about their principal concerns. Next, the storyteller gives some clues about some "main event" or problem to be solved. From that point on, the listener has a general idea of what comes next: there will be more development of the problem; then it will be resolved, somehow; and then the story will end, perhaps by giving some practical or moral advice. In any case, those magic story-starting words arouse, in knowing listeners' minds, great hosts of expectation-frames to help the listeners anticipate which terminals to fill.

Terminal or Concern	Assignment	Indicated by
Time Setting?	*The past*	*Past tense of verb*
Place Setting?	*Jack's home*	*Destination*
Protagonist?	*Mary is heroine*	*Syntactic emphasis*
Central Concern?	*Mary's subjective reaction*	*Default assumption*
Antagonist?		*Not yet mentioned*

Beyond arousing all these specific expectations, *"once upon a time"* plays one more crucial role: it says that what comes after it is fictional or, in any case, far too remote to activate much personal concern. Instead, it tells the listener to disregard the normal sympathies one should feel when real persons meet the monstrous destinies so usual in children's tales: to be turned into toads, imprisoned in stones, or devoured by terrible dragon beasts.

26.6 SENTENCE AND NONSENSE

Part of what a sentence means depends upon its separate words, and part depends on how those words are arranged.

Round squares steal honestly.
Honestly steal squares round.

What makes these seem so different in character, when both use the very same words? I'll argue that this is because your language-agency, immediately upon hearing the first word-string, knows exactly what to do with it because it fits a well-established sentence-frame. The second string fits no familiar form at all. But how do we fit those sentence-frames? We'll come to that presently, but for the moment, let's simply assume that our young listener has somehow come to classify words into various types, like nouns, adjectives, verbs, and adverbs. (We'll ignore the fact that children go through other stages before they use words as adults do.) Then our first string of words has this form:

Adjective   Noun   Verb   Adverb

Now we'll suppose our listener has learned a specific recognition-frame that is activated on hearing this string of particular types of words. This frame then executes a special process script that makes the following assignments to the terminals of a *Trans*-frame. The neme for "steal" is assigned to the *Trans*-frame's *Action* terminal, while the neme for "squares" is attached to the *Actor* terminal. The frame then activates scripts that modify the action "steal" by applying to it the neme for "honestly" and modify the object "squares" by applying to it the neme for "round." Up to this point, everything works smoothly: the language-agency has found a use for every word. We have special names for the strings of words that we process with such fluency: we call them "phrases" or "sentences."

A word-string seems "grammatical" if all its words fit quickly and easily into frames that connect suitably to one another.

However, at this point some serious conflicts start to appear within some other agencies because of certain incompatibilities. The frame for "steal" requires its *Actor* to be *animate*. A square can't steal, because it's not alive! Besides, the frame for "steal" expects an act that's reprehensible, and that clashes with the modifier for "honestly." If that weren't bad enough, our agency for describing shape can't tolerate the polynemes for "round" and "square" when both are activated at the same time. It doesn't matter that our sentence is grammatical: so much turmoil is set up that most of its meaning cancels out and we regard it as "nonsense." But it is important to recognize that the distinction between sense and nonsense is only partly a matter of grammar, for consider what happens when you hear these three words:

thief -- -- careless -- -- prison --

Although these do not establish any single well-formed grammar-frame, they activate some word-sense nemes that skip past all our grammar-forms to fit a familiar story-frame, a moral tale about a thief who's caught and reaps a just reward. Ungrammatical expressions can frequently be meaningful when they lead to clear and stable mental states. Grammar is the servant of language, not the master.

26.7 FRAMES FOR NOUNS

At various points in their development, most children seem suddenly to comprehend new kinds of sentences. Thus, once they learn to deal with single adjectives, some children quickly learn to deal with longer strings like these:

Dogs bark. Big dogs bark. Big shaggy dogs bark. Big black shaggy dogs bark.

If this were done by using word-string sentence-frames, it would require a separate frame for each different number of adjectives. Another scheme would not use any frames at all but have the language-agency convert each adjective, as it arrives, into some corresponding neme. And yet another scheme to handle this (still popular among some grammar theorists) would have each successive adjective arouse a new subframe inside the previous one. However, when we look more closely at how people use adjectives, we find that these strings are not simple at all. Compare the two phrases below:

The wooden three heavy brown big first boxes . . .
The first three big brown heavy wooden boxes . . .

Our language-agents scarcely know what to do with that first string of words because it doesn't fit the patterns we normally use for describing things. This suggests that we use framelike structures for describing nouns as well as verbs—that is, for describing things as well as actions. To fill the terminals of those frames, we expect their ingredients to arrive in a more or less definite order. We find it hard to understand a group of English adjectives unless they are arranged roughly as shown below.

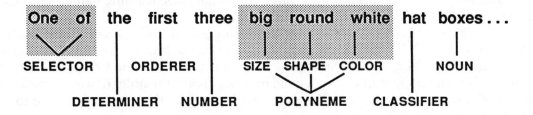

Whenever a language community can agree on forms like these, expression becomes easier. Then every individual can learn, once and for all, where to put—and where to find—the answers to questions most frequently asked. In English one learns to say *"green box,"* while in French one says *"box green."* It doesn't matter which order is used—as long as everyone agrees to do it the same way. But what *are* the *"questions most frequently asked"*—the ones we build into our language-forms? The answer to this is likely to be somewhat circular, since the language culture in which we're raised will probably affect the kinds of questions that will seem most natural to ask. Still, there could be useful clues in features that are common to many different languages.

Many scientists have asked, indeed, why so many human languages use similar structures such as nouns, adjectives, verbs, clauses, and sentences. It is likely that some of these reflect what is genetically built into our language-agencies. But it seems to me even more likely that most of these nearly universal language-forms scarcely depend on language at all—but reflect how descriptions are formed in other agencies. The most common forms of phrases could arise not so much from the architecture of the language-agencies as from the machinery used by other agencies for representing objects, actions, differences, and purposes—as suggested in section 22.7—and from how those other agencies manipulate their memories. In short, the ways we *think* must have a strong and universal influence on how we speak—if only through its influence on the sorts of things we'll want to say.

26.8 FRAMES FOR VERBS

We've seen how a four-word sentence such as *"Round squares steal honestly"* could be made to fit a certain four-terminal frame. But what about a sentence like *"The thief who took the moon moved it to Paris"*? It would be dreadful if we had to learn a new and special ten-word frame for each particular type of ten-word string! Clearly we don't do any such thing. Instead, we use the pronoun "who" to make the listener find and fill a second frame. This suggests a multistage theory. In the earliest stages of learning to speak, we simply fill the terminals of word-string frames with nemes for words. Then, later, we learn to fill those terminals with *other filled-in language-frames*. For example, we can describe our moon sentence as based on a top-level *Trans*-frame for "move" whose *Actor* terminal contains a second *Trans*-frame for "took":

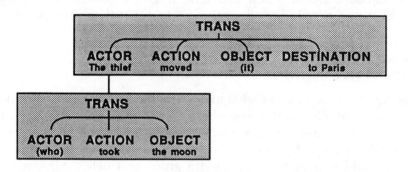

Using frames this way simplifies the job of learning to speak by reducing the number of different kinds of frames we have to learn. But it makes language learning harder, too, because we have to learn to work with several frames at once.

How do we know which terminals to fill with which words? It isn't so hard to deal with *"red, round, thin-peeled fruit,"* since each such property involves a different agency. But that won't work for *"Mary loves Jack,"* since *"Jack loves Mary"* has the very same words, and only their order indicates their different roles. Each child must learn how the order of words affects which terminal each phrase should fill. As it happens, English applies the same policy both to *"Mary loves Jack"* and to our moon sentence:

> Assign the *Actor* pronome to the phrase *before* the verb.
> Assign the *Object* pronome to the phrase *after* the verb.

The policies for assigning phrases to pronomes vary from one language to another. The word order for *Actor* and *Object* is less constrained in Latin than in English, because in Latin those roles can be specified by altering the nouns themselves. In both languages we often indicate which words should be assigned to other pronome roles by using specific prepositions like "for," "by," and "with." In many cases, different verb types use the same prepositions to indicate the use of different pronomes. At first such usages may seem to be arbitrary, but they frequently encode important systematic metaphors; in section 21.2 we saw how "from" and "to" are used to make analogies between space and time. How did our language-forms evolve? We have no record of their earliest forms, but they surely were affected at every stage by the kinds of questions and problems that seemed important at the time. The features of present-day languages may still contain some clues about our ancestors' concerns.

26.9 LANGUAGE AND VISION

Some language scholars seem to think that what we do in language is unique, in the filling of frames with other frames to open up a universe of complicated structure-forms. But consider how frequently we do similarly complex things in understanding visual scenes. The language-agency must be able to interrupt itself, while handling one phrase, to work on parts of another phrase, and this involves some complex short-term memory skills. But in vision, too, there must be similar processes involved in breaking scenes apart and representing them as composed of objects and relationships. The picture below suggests how similar such processes may be. In language, the problem is to recognize that the two words "took" and "out" both belong to the same verb phrase, although they are separated in time. In vision, the problem is to recognize the two regions of a tabletop as being parts of the same object, although they are separated in space.

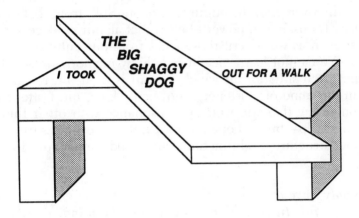

Notice also that we cannot see the tops of the blocks that serve as legs—and yet we do not have the slightest doubt about where they end. Similarly, the ends of language phrases are frequently unmarked—yet again we can tell where they end. In *"The thief who took the moon moved it to Paris,"* the word "who" marks the beginning of a new frame, but there is no special word to indicate the ending of that phrase. Why don't we wrongly assign *"the moon"* to the *Actor* of the spurious phrase, *"The moon moved it to Paris?"* It is because we first heard *". . . who took the moon,"* which caused *"the moon"* to get attached to the *Object* pronome of the *Trans*-frame for *"took"*—so now it's not available to serve as *Actor* in the frame for *"moved."* *"The thief"* is still available to play that role. I don't mean to suggest we can never assign the same phrase to two different roles, only that good speakers choose their forms so that this doesn't happen by accident.

Did our capacity to deal with phraselike structures evolve first in language or in vision? Among our ancestors, vision greatly antedates language, so if these abilities are related, our language-agencies themselves more likely evolved from variants of genes that first evolved in shaping the architecture of our vision-systems. Today we have no way to verify such a conjecture, but future geneticists may become able to trace the ancestry of many such relationships by examining the genes that generate the corresponding brain structures.

26.10 LEARNING LANGUAGE

*Language is itself the collective art of expression, a summary of
thousands upon thousands of individual intuitions. The
individual goes lost in the collective creation, but his personal
expression has left some trace in a certain give and flexibility that
are inherent in all collective works of the human spirit.*
—Edward Sapir

The vocabulary of a language—the words themselves—is the product of a project that spans the history of a culture and can involve millions of person years of work. Every sense of every word records some intellectual discovery that now outlives the myriad other, less distinguished thoughts that never earned a name.

Each person invents some new ideas, but most of these will die when their owners do, except for those that make their way into the culture's lexicon. Still, from that ever-growing reservoir we each inherit many thousands of powerful ideas that all our predecessors found. Yet it is no paradox to say that even as we inherit those ideas from our culture, we each must reinvent them for ourselves. We cannot learn meanings only by memorizing definitions: we must also "understand" them. Each situation in which a word is used must suggest some mixture of materials already in the mind of a listener, who then, alone, must attempt to assemble these ingredients into something that will work in consonance with other things already learned. Definitions sometimes help—but still one must separate the essences from the accidents of the context, link together structures and functions, and build connections to the other things one knows.

*A word can only serve to indicate that someone else may have a valuable idea—
that is, some useful structure to be built inside the mind. Each new word only
plants a seed: to make it grow, a listener's mind must find a way to build inside
itself some structure that appears to work like the one in the mind from which it
was "learned."*

Along with the words, we also have to learn the grammar-tactics for using them. Most children start by using only one or two words at a time. Then, over the next two or three years, they learn to speak in sentences. It usually takes a full decade to learn most of the conventions of adult speech, but we often see relatively sudden advances over concentrated periods of time. How do children learn such complicated skills so quickly? Some language theorists have suggested that children learn to use grammar so readily that our brains must be born with built-in grammar-machinery. However, we've seen that our visual-systems solve many similar problems in even earlier years—and we've also seen that when they learn to play with spoons and pails, children must learn yet other languagelike skills for managing the *Origins, Destinations, Recipients*, and *Instruments* of their actions. Thus, many sections of our brains appear to demonstrate capacities for rearranging pronome roles even before we learn to speak. If so, perhaps we ought not to wonder so much about how children learn to speak so readily. Instead, we ought to wonder why it takes so long, when they already do so many similar things inside their heads.

How do we choose the words we speak, and how do we understand what others say? Earlier, I suggested that in the course of learning language we accumulate various processes and tactics that enable us to partially reproduce our own mental operations in other speakers. These processes affect our choices of words, the forms we select for phrases and sentences, and the styles in which we frame our narratives. There have been many attempts to study how children learn language, but psychologists do not yet have coherent theories about the underlying processes. For example, we do not yet even know whether we learn each bit of grammar only once—or whether we have to learn it twice, for speaking and for understanding what other people say.

We know so little about such matters that we can scarcely even speculate about the nature of those early language-learning steps. Perhaps the process starts with some agents that can enable a child to make various vocal sounds in response to specific internal states. These agents then become involved with built-in "predestined" learning processes that lead to limited abilities to imitate other sounds the child hears by using feedback from its ears. Later stages might then engage new layers of agents that connect word-sound agents to whichever polynemes are most frequently attached to certain pronomes in the language-agency. Once a suitable variety of such processes are established, more layers of frame- and memory-controlling agents could learn to support more complex language skills.

Let's try to imagine what kind of process could produce a language phrase that "expresses" a description of an object. Suppose, for example, that you want to draw attention to a certain very big box. To imagine such a thing in the first place, you might first have to activate your polyneme for "box" and then arouse some other isonomes and polynemes that modify the state of your *Size* agency. To express *very big box* might thus require grammar-tactics that express three mental operations:

> ---"**box**" *expresses the arousal of the box polyneme;*
> ---"**big**" *expresses a process that selects the **Size** agency;*
> ---"**very**" *expresses an isonome that adjusts the sensitivities of agents in whichever agency was selected.*

I do not mean to suggest that a child's earliest three-word noun phrases must be based upon such complicated processes; more likely they begin with simpler sequence scripts. Eventually, though, more complex systems intervene to replace the simple scripts by intricate kinds of frame-arrays that enable the child to make more complex rearrangements of what becomes attached to its expression-frames. Then, as the language-agency acquires more isonome-controlling skills, the child can learn to use pronouns like "it" or "she" to express other structures that are already attached to suitable pronomes. Also, as we develop skills for building chains and trees from other frames, the language-agency can learn to use corresponding grammar-tactics to express those chains—stringing together phrases and sentences with conjunction words like "and" and "but." Similarly, as we improve our methods for controlling memories and managing interruptions, we can learn to combine those skills with clause-interrupting forms like "who" and "which." There seems scarcely any limit to the complexity of our social inventions for expressing mental processes, and it takes most children many years to master all the language arts their ancestors evolved.

26.12 COHERENT DISCOURSE

*Words . . . can indicate the qualitative and relational features of a
situation in their **general** aspect just as directly as, and perhaps
even more satisfactorily than, they can describe its particular
individuality. This is, in fact, what gives to language its intimate
relation to thought processes. For thinking, in the proper
psychological sense, is never the mere reinstatement of some
suitable past situation produced by a crossing of interests, but is
the utilization of the past in solution of difficulties
set by the present. . . .*
— F. C. BARTLETT

Every discourse works on several scales. Each word you hear can change your state in a way that depends upon all the structures you have built while listening to the words that came before. Most of those structures are themselves mere transient things, which persist for only a few moments before you rearrange some of their parts and perhaps discard the rest entirely. Thus, a car might first appear as the subject of a sentence, then become a mere vehicle or instrument in the next sentence; finally, the whole scenario might be used merely to modify a personal trait of some actor in a larger scene. As a discourse proceeds, details on each scale become absorbed into larger-scale representation networks whose outlines become increasingly remote from the individual words that were used to construct them.

It would be wonderful to have a compact, self-contained theory that explains all our language-forms. But that ideal cannot be realized because words are merely the external signs of very complex processes, and there is no clear boundary between *language* and all the rest of what we call *thinking*. To be sure, the boundaries of words themselves are relatively clear, and when they have multiple meanings, our grammar-tactics can often help us to assign the proper senses to various terminals and other structures. These tactics include all sorts of inflections, prepositions, word orderings, and signals that indicate how to include one phrase inside another. We also combine words into larger expressions that range in vagueness of boundaries from compact clichés like "hot dog" to diffuse signals that are scarcely linked to specific words at all; these include our hard-to-describe nuances of phrasing, rhythm, intonation, and shifts of style and flow.

We're normally quite unaware of how our grammar-tactics constrain us in our choices of words. We're often somewhat more aware of other language-tactics we use to guide our listeners' minds—to change the focus from one theme to another, to adjust the levels of detail, to shift between foreground and setting. We learn to use phrases like *"by the way"* to change the topic of concern, to say *"for example"* to shift to a finer level of detail, to say *"but"* to modify an expectation or to interrupt the usual flow, or to say *"in any case"* or *"in spite of that"* to indicate the end of an interruption or elaboration.

But even all this is only a small part of language. To understand what people say, we also exploit our vast stores of common knowledge, not only about how specific words are related to the subjects of concern, but also about how to express and discuss those subjects. Every human community evolves a great array of discourse-forms to shape its stories, explanations, conversations, discussions, and styles of argument. Just as we learn grammar-forms for fitting words to sentence-frames, we also build up stocks of "plots" to organize our story-tales, and standard personalities to fill the roles of their protagonists—and every child must learn these forms.

CHAPTER 27

CENSORS AND JOKES

A man at the dinner table dipped his hands in the mayonnaise and then ran them through his hair. When his neighbor looked astonished, the man apologized: "I'm so sorry. I thought it was spinach."

—SIGMUND FREUD

Our reader must be anxious to know what finally became of Mary and that kite. Here is more of that story.

Mary was invited to Jack's party. She wondered if he would like a kite.
Jane said, "Jack already has a kite. He will make you take it back."

What does the pronoun "it" mean here? Clearly Jane is speaking not of the kite that already belongs to Jack, but of the new kite that Mary is thinking of giving to him. But what leads the listener to assume that this is what the storyteller meant? There are many issues here besides the question of which kite is involved. How do we know "it" refers to a kite at all? Does "take it back" mean take it back from Jack or to return it to the store? For the sake of simplicity, let's put aside the other possibilities and assume that *"it"* must mean a kite. But in order to decide *which* kite is meant, we still must understand the larger phrase *"take it back."* This phrase must refer to some structure already in the listener's mind; the narrator expects the listener to find the appropriate structure by activating an appropriate fragment of commonsense knowledge about giving and receiving birthday presents. But since every listener knows so many things, what sorts of processes could activate the appropriate knowledge without taking too much time? In 1974 Eugene Charniak, a graduate student at MIT, asked how each phrase of this story works to prepare the reader to comprehend the subsequent phrases. He suggested that whenever we hear about a particular event, specific recognition-agents are thereby aroused. These then proceed actively to watch and wait for other related types of events. (Because these recognition-agents lurk silently, to intervene only in certain circumstances, they are sometimes called *"demons."*) For example, whenever a story contains the slightest hint that someone may have purchased a gift, specific demons might be aroused that watch for events like these:

If there is evidence that the recipient rejects the gift, look for signs of it being returned. If you see evidence of a gift being returned, look for signs that the recipient rejected it.

Charniak's thesis raised many questions. How easy should it be to activate demons? How long should they then remain active? If too few demons are aroused, we'll be slow to understand what's happening. But if too many become active, we'll get confused by false alarms. There are no simple solutions to these problems, and what we call "understanding" is a huge accumulation of skills. You might understand certain parts of a story by using separate, isolated demons; you might comprehend other aspects of that same story by using larger-scale processes that try to match the sequence of events to various remembered scripts; yet other understandings might depend upon which agents are aroused by various micronemes. How much of the fascination in telling a story, or in listening to one, comes from the manipulations of our demons' expectations?

27.2 SUPPRESSORS

It would be wonderful never to make mistakes. One way would be to always have such perfect thoughts that none of them is ever wrong. But such perfection can't be reached. Instead we try, as best we can, to recognize our bad ideas before they do much harm. We can thus imagine two poles of self-improvement. On one side we try to stretch the range of the ideas we generate: this leads to more ideas, but also to more mistakes. On the other side, we try to learn not to repeat mistakes we've made before. All communities evolve some prohibitions and taboos to tell their members what they shouldn't do. That, too, must happen in our minds: we accumulate memories to tell ourselves what we shouldn't *think*.

But how could we make an agent to prevent us from doing something that, in the past, has led to bad or ineffectual results? Ideally, that agent would keep us from even *thinking* that bad idea again. But that seems almost paradoxical, like telling someone, *"Don't think about a monkey!"* Yet there is a way to accomplish this. To see how it works, imagine the sequence of mental states that led to a certain mistake:

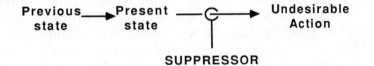

We could prevent the undesired action from taking place by introducing an agent that recognizes the state which, in the past, preceded the undesired action.

> **Suppressor-agents** *wait until you get a certain "bad idea." Then they prevent your taking the corresponding action, and make you wait until you think of some alternative. If a suppressor could speak, it would say, "Stop thinking that!"*

Suppressors could indeed prevent us from repeating actions that we've learned are bad. But it is inefficient to wait until we actually reach undesirable states, then have to "backtrack." It would be more efficient to anticipate such lines of thought so that we never reach those states at all. In the next section we'll see how to do this by using agents called censors.

> **Censor-agents** *need not wait until a certain bad idea occurs; instead, they intercept the states of mind that usually **precede** that thought. If a censor could speak, it would say, "Don't even begin to think that!"*

Though censors were conceived of long ago by Sigmund Freud, they're scarcely mentioned in present-day psychology. I suspect that this is a serious oversight and that censors play fundamental roles in how we learn and how we think. Perhaps the trouble is that our censors work too well. For, naturally, it is easier for psychologists to study only what someone *does*— instead of what someone *doesn't do*.

27.3 CENSORS

To see what suppressors and censors have to do, we must consider not only the mental states that actually occur, but others that might occur under slightly different circumstances.

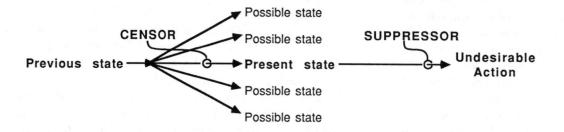

Suppressors work by interceding to prevent actions just before they would be performed. This leads to a certain loss of time, because nothing can be done until acceptable alternatives can be found. Censors avoid this waste of time by interceding earlier. Instead of waiting until an action is about to occur, and then shutting it off, a censor operates earlier, when there still remains time to select alternatives. Then, instead of *blocking* the course of thought, the censor can merely *deflect* it into an acceptable direction. Accordingly, no time is lost.

Clearly, censors can be more efficient than suppressors, but we have to pay a price for this. The farther back we go in time, the larger the variety of ways to reach each unwanted state of mind. Accordingly, to prevent a particular mental state from occurring, an early-acting censor must learn to recognize *all* the states of mind that might precede it. *Thus, each censor may, in time, require a substantial memory bank.* For all we know, each person accumulates millions of censor memories, to avoid the thought-patterns found to be ineffectual or harmful.

Why not move farther back in time, to deflect those undesired actions even earlier? Then intercepting agents could have even larger effects with smaller efforts and, by selecting good paths early enough, we could solve complex problems without making any mistakes at all. Unfortunately, this cannot be accomplished only by using censors. This is because as we extend a censor's range back into time, the amount of inhibitory memory that would be needed (in order to prevent turns in every possible wrong direction) would grow exponentially. To solve a complex problem, it is not enough to know what might go wrong. One also needs some *positive* plan.

As I mentioned before, it is easier to notice what your mind *does* than to notice what it *doesn't* do, and this means that we can't use introspection to perceive the work of these inhibitory agencies. I suspect that this effect has seriously distorted our conceptions of psychology and that once we recognize the importance of censors and other forms of "negative recognizers," we'll find that they constitute large portions of our minds.

Sometimes, though, our censors and suppressors must themselves be suppressed. In order to sketch out long-range plans, for example, we must adopt a style of thought that clears the mind of trivia and sets minor obstacles aside. But that could be very hard to do if too many censors remained on the scene; they'd make us shy away from strategies that aren't guaranteed to work, and tear apart our sketchy plans before we can start to accomplish them.

27.4 EXCEPTIONS TO LOGIC

The thought behind I strove to join
Unto the thought before.
But sequence ravelled out of reach
Like balls upon the floor.
— EMILY DICKINSON

We spend our lives at learning things, yet always find exceptions and mistakes. Certainty seems always out of reach. This means that we have to take some risks to keep from being paralyzed by cowardice. But to keep from having accidents, we must accumulate two complementary types of knowledge:

We search for "islands of consistency" within which ordinary reasoning seems safe.
We work also to find and mark the unsafe boundaries of those domains.

In civilized communities, appointed guardians post signs to warn about sharp turns, thin ice, and animals that bite. And so do our philosophers, when they report to us their paradoxical discoveries—those tales of the Liar who admits to lying and the Barber who shaves all the people who do not shave themselves. These valuable lessons teach us which thoughts we shouldn't think; they are the intellectual counterparts to Freud's emotion censors. It is interesting how frequently we find paradoxical nonsense to be funny, and when we come to the section on jokes, we'll see why this is so. When we look closely, we find that most jokes are concerned with taboos, injuries, and other ways of coming to harm—and logical absurdities can also lead to harm.

We tell our children not to cross the road unless they are sure no car is coming. But what do we mean by "sure"? No one can ever really "prove" that no car is coming, since there is no way to rule out the possibility that some mad scientist has found a way to make cars invisible. In ordinary life we have to deal with "usual" instead of "true." All we can really ask a child to do is *"look both ways before you cross."* In the real world, it makes no sense to ask for absolute certainty.

Unfortunately there are no simple, foolproof ways to get around the inconsistencies of common sense. Accordingly, we each must learn specific ways to keep from various mistakes. Why can't we do that logically? The answer is that perfect logic rarely works. One difficulty is finding foolproof rules for reasoning. But the more serious problem is that of finding foolproof bases for our arguments. It is virtually impossible to state any facts about the real world that actually are always true. We observed this when we discussed *"Birds can Fly."* This statement applies to typical birds, but not to birds imprisoned in small cages, chained with leg irons, or under the influence of high-gravity fields. Similarly, when you're told, "Rover is a dog," you'll assume that Rover has a tail, since your frame for a typical dog has a terminal for a tail. But should you learn that Rover lacks a tail, your mind won't self-destruct; instead, you'll change your Rover-frame—but still expect most other dogs to keep their tails.

Exceptions are a fact of life because few "facts" are always true. Logic fails because it tries to find exceptions to this rule.

27.5 JOKES

Two villagers decided to go bird hunting. They packed their guns and set out with their dog into the fields. Near evening, with no success at all, one said to the other, "We must be doing something wrong." "Yes," agreed the friend, "perhaps we're not throwing the dog high enough."

Why do jokes have such peculiar psychological effects? In 1905, Sigmund Freud published a book explaining that we form censors in our minds as barriers against forbidden thoughts. Most jokes, he said, are stories designed to fool the censors. A joke's power comes from a description that fits two different frames at once. The first meaning must be transparent and innocent, while the second meaning is disguised and reprehensible. The censors recognize only the innocent meaning because they are too simple-minded to penetrate the forbidden meaning's disguise. Then, once that first interpretation is firmly planted in the mind, a final turn of word or phrase suddenly replaces it with the other one. The censored thought has been slipped through; a prohibited wish has been enjoyed.

Freud suggested that children construct censors in response to prohibitions by their parents or peers. This explains why so many jokes involve taboos concerning cruelty, sexuality, and other subjects that human communities typically link to guilt, disgust, or shame. But it troubled Freud that this theory did not account for the "nonsense jokes" people seem to enjoy so much. The trouble was that these seemed unrelated to social prohibitions. He could not explain why people find humor in the idea of "*a knife that has lost both its blade and its handle.*"

Freud considered several explanations to account for pointless nonsense jokes but concluded that none of those theories was good enough. One theory was that people tell nonsense jokes for the pleasure of arousing the expectation of a real joke and then frustrating the listener. Another theory was that senselessness reflects "a wish to return to carefree childhood, when one was permitted to think without any compulsion to be logical, and to put words together without sense, for the simpler pleasures of rhythm or rhyme." Freud put it this way:

> *Little by little the child is forbidden this enjoyment, till there remain only significant combinations of words. But attempts still emerge to disregard restrictions which were learned.*

In yet a third theory, Freud conjectured that humor is a way to ward off suffering—as when, in desperate situations, we make jokes as though the world were nothing but a game. Freud suggested that this is when the superego tries to comfort the childlike ego by rejecting all reality; but he was uneasy about this idea because such kindliness conflicted with his image of the superego's usual stern, strict character.

Despite Freud's complicated doubts, I'll argue that he was right all along. Once we recognize that *ordinary* thinking, too, requires censors to suppress ineffectual mental processes, then all the different-seeming forms of jokes will seem more similar. Absurd results of reasoning must be tabooed as thoroughly as social mistakes and inanities! And that's why stupid thoughts can seem as humorous as antisocial ones.

27.6 HUMOR AND CENSORSHIP

People often wonder if a computer could ever have a sense of humor. This question seems natural to those who think of humor as a pleasant but unnecessary luxury. But I'll argue quite the opposite—that humor has a practical and possibly essential function in how we learn.

> *When we learn in a serious context, the result is to change connections among ordinary agents. But when we learn in a humorous context, the principal result is to change the connections that involve our censors and suppressors.*

In other words, my theory is that humor is involved with how our censors learn; it is mainly involved with "negative" thinking, though people rarely realize this. Why use such a distinct and peculiar medium as humor for this purpose? Because we must make a sharp distinction between our positive, action-oriented memories and the negative, inhibitory memories embodied in our censors.

> *Positive memory-agents must learn which mental states **are** desirable.*
> *Negative memory-agents must learn which mental states **are** undesirable.*

Because these two types of learning required different processes, it was natural to evolve social signals to communicate that distinction. When people do things that we regard as good, we speak to them in encouraging tones—and this switches on their positive learning machinery. However, when people do things we consider stupid or wrong, we then complain in scornful tones or laugh derisively; this switches on their negative learning machinery. I suspect that scolding and laughing have somewhat different effects: scolding tends to produce suppressors, but laughing tends to produce censors. Accordingly, the effect of derisive humor is somewhat more likely to disrupt our present activity. This is because the process of constructing a censor deprives us of the use of our temporary memories, which must be frozen to maintain the records of our recent states of mind.

> *Suppressors merely need to learn which mental states **are** undesirable.*
> *Censors must remember and learn which mental states **were** undesirable.*

To see why humor is so often concerned with prohibition, consider that our most productive forms of thought are just the ones most subject to mistakes. We can make fewer errors by confining ourselves to cautious, "logical" reasoning, but we'll also discover fewer new ideas. More can be gained by using metaphors and analogies, even though they are often defective and misleading. I think this is why so many jokes are based on recognizing inappropriate comparisons. Why, by the way, do we so rarely recognize the negative character of humor itself? Perhaps it has a funny side effect: while shutting off those censored thoughts, our censors also shut off thoughts about themselves—and make themselves invisible.

This solves Freud's problem about nonsense jokes. The taboos that grow within social communities can be learned only from other people. But when it comes to intellectual mistakes, a child needs no helpful friend to scold it when a tower falls, when it puts a spoon in its ear, or thinks a thought that sets its mind into a fruitless and confusing loop. In other words, we can detect many of our own intellectual failures all by ourselves. Freud's theory of jokes was based on the idea that censors suppress thoughts that would be considered "naughty" by those to whom we are attached. He must simply have overlooked the fact that ineffectual reasoning is equally "naughty"—and therefore equally "funny"—in the sense that it, too, ought to be suppressed. There is no need for our censors to distinguish between social incompetence and intellectual stupidity.

27.7 LAUGHTER

What would a Martian visitor think to see a human being laugh? It must look truly horrible: the sight of furious gestures, flailing limbs, and thorax heaving in frenzied contortions. The air is torn with dreadful sounds as though, all at once, that person wheezes, barks, and chokes to death. The face contorts in grimaces that mix smiles and yawns with snarls and frowns. What could cause such a frightful seizure? Our theory suggests a simple answer:

The function of laughing is to disrupt another person's reasoning!

To see and hear a person laugh creates such chaos in the mind that you can't proceed along your present train of thought. Derision makes you feel ridiculous; it prevents you from "being serious." What happens then? Our theory has a second part:

Laughter focuses attention on the present state of mind!

Laughter seems to freeze one's present state of mind in its tracks and hold it up to ridicule. All further reasoning is disrupted, and only the joke-thought remains in sharp focus. What is the function of this petrifying effect?

By preventing you from "taking seriously" your present thought, and thus proceeding to develop it, laughter gives you time to build a censor against that state of mind.

In order to construct or improve a censor, you must retain your records of the recent states of mind that made you think the censored thought. This takes some time, during which your short-term memories are fully occupied—and that will naturally disrupt whichever other processes might try to change those memories.

How could all this have evolved? Like smiling, laughter has a curious ambiguity, combining elements of affection and conciliation with elements of rejection and aggression. Perhaps all these ancestral means of social communication became fused to compose a single, absolutely irresistible way to make another person cease an activity regarded as objectionable or ridiculous. If so, it is no accident that so many jokes mix elements of pleasure, cruelty, sexuality, aggression, and absurdity. Humor must have grown along with our abilities to criticize ourselves, starting with simple internal suppressors that evolved into more sophisticated censors. Perhaps they then split off into B-brain layers that became increasingly able to predict and manipulate what the older A-brains were about to do. At this point, our ancestors must have started to experience what humanists call "conscience." For the first time, animals could start to reflect upon their own mental activities and evaluate their purposes, plans, and goals. This endowed us with great new mental powers but, at the same time, exposed us to new and different kinds of conceptual mistakes and inefficiencies.

Our humor-agencies become internalized in adult life as we learn to produce the same effects entirely inside our own minds. We no longer need the ridicule of those other people, once we can make ourselves ashamed by laughing silently at our own mistakes.

27.8 GOOD HUMOR

Some readers might object that the censor-learning theory of jokes is too narrow to be an explanation of humor in general. What of all the other roles that humor plays in occasions of enjoyment and companionship? Our answer is the same as usual: we can't expect any single, simple theory to explain adult psychology. To ask how humor works in a grown-up person is to ask how *everything* works in a grown-up person, since humor gets involved with so many other things. I didn't mean to suggest that *every* aspect of humor is involved in making censors learn. When humor evolved, as when any other mechanism develops in biology, it must have been built upon other mechanisms that already existed, and embodied mixtures of those other functions. Just as the voice is used for many social purposes, the mechanisms involved in humor are also used for other effects that are less involved with memory. In later life the effect of "functional autonomy" can make it hard to recognize the original function not only of humor, but of many other aspects of adult psychology. To understand how feelings work, we need to understand both their evolutionary and their individual histories.

We've seen how important it is for us to learn about mistakes. To keep from making old mistakes ourselves, we learn about them from our families and friends. But a peculiar problem arises when we tell another person that something is wrong, for if this is interpreted as an expression of disapproval and rejection, it can evoke a sense of pain and loss—and lead to withdrawal and avoidance. Accordingly, to point out mistakes to someone whose loyalty and love we want to keep, we must adopt some pleasant or conciliatory form. Thus humor has evolved its graciously disarming ways to do its basically distasteful job! You don't want the recipient to "kill the messenger who brings bad news"—especially when you're the messenger.

Many people seem genuinely surprised when shown that humor is so concerned with unpleasant, painful, and disgusting subjects. In a certain sense, there's really nothing humorous about most jokes—except, perhaps, in the skill and subtlety with which their dreadful content is disguised; frequently, the thought itself is little more than *"See what happened to somebody else; now, aren't you glad it wasn't you?"* In this sense most jokes are not actually frivolous at all but reflect the most serious of concerns. Why, by the way, are jokes usually less funny when heard again? Because the censors learn some more each time and prepare to act more quickly and effectively.

Why, then, do certain kinds of jokes, particularly those about forbidden sexual subjects, seem to remain persistently funny to so many people? Why do *those* censors remain unchanged for so long? Here we can reuse our explanation of the prolonged persistence of attachment, infatuation, sexuality, and mourning-grief; because these areas relate to self-ideals, their memories, once formed, are slow to change. Thus the peculiar robustness of sexual humor may mean only that the censors of human sexuality are among the "slow learners" of the mind, like retarded children. In fact, we could argue that they literally *are* retarded children—that is, they are among the frozen remnants of our earlier selves.

CHAPTER 28

THE MIND AND THE WORLD

FLUX

Each event is quite unique.
Nothing ever happens twice.
What occurs will not recur.
There can be no second time.

Even gear teeth will have changes
by the time they mesh again.
Though they seem to stay the same,
hard things slowly wear away.

As for softer things, they move,
varying in shape and place
and in memory and hope
twenty-seven thousand days.

Still I keep a single name
labeling a twinkling sea
though it is ten billion waves
that are constituting me.

—THEODORE MELNECHUK

28.1 THE MYTH OF MENTAL ENERGY

Why do angry people act as though some measure of aggression must be spent and, when no proper object lies in reach, strike out and damage harmless things? It almost seems as though our feelings can accumulate like fluids bottled up inside. In earlier times, some scientists identified these quantities with substances like bile and blood. No one believes *those* theories now—yet still we often speak of having mental energy and momentum or of succumbing to depletion or inertia. Do "mental quantities" really exist within the mind? If so, how are they made and stored, brought forth and then spent? And what are their relations to the quantities and magnitudes we read about in technical books? The answer is that words like "energy" and "force" are not used with much precision in everyday psychology. They still have the connotations that they carried several centuries ago, when they referred to commonsense ideas about vitality. Then, "energy" referred to vigor of action and expression, and "force" referred to the binding strength of a commitment or to the fighting strength of an army.

Modern scientists use a concept of energy that, though narrower and more precise, not only explains more perfectly why engines stop when they run out of fuel, but also applies to our bodies as well: each of the cells of which we're made, including those inside the brain, requires some chemical energy in the form of food and oxygen. Accordingly, the body as a whole can do only a limited amount of physical work before it needs another meal. Now many people naively assume that our higher-level mental processes have similar requirements and that they need some second form of fuel—a mythical form of *mental energy*—to keep from becoming bored or mentally exhausted. And yet that simply isn't true! If each of *Builder*'s agents has physical energy enough to do its work, then *Builder*—as an agency—needs nothing more to do *its* work. *Builder*, after all, is but a name for a certain assembly of agents. It can't require *anything* its separate agents do not need.

> *Machines and brains require ordinary energy to do their jobs—and need no other, mental forms of energy. Causality is quite enough to keep them working toward their goals.*

But if our higher-level processes require no extra quantities like fuels or energies, what makes it *seem* to us as though they do? Why do so many people talk about their "levels of mental or emotional energy"? Why do tedious and boring occupations make us feel "run down"? We all experience so many such phenomena that we cannot help thinking our minds depend on many kinds of "mental quantities"—yet scientists apparently have shown that no such quantities exist. How can we explain this? It is not enough to say, simply, that these phenomena are illusions; we must understand why the illusions appear and, if possible, determine what functions they serve. The next few sections show how various illusions of mental force and energy evolve as convenient ways for mental agencies to regulate their transactions, much as many human communities have discovered how to use money.

28.2 MAGNITUDE AND MARKETPLACE

How can a hurt be canceled by a kiss? How can an insult "add" to injury? Why do we so often speak as though our wishes and desires were like forces, which increase one another's effects when they're aligned but cancel out when they're opposed? I'll argue that this is because, at every moment of our lives, we're forced to choose between alternatives we can't compare. Suppose, for example, that you must choose between two homes: one of them offers a mountain view; the other is closer to where you work. It really is a strange idea that two such unrelated things as nearness to work and beautiful scenery could be compared at all. But a person could instead assess that pleasant, restful view as worth a certain amount of travel time. Instead of comparing the items themselves, you could simply compare how much time they seem to be worth.

> *We turn to using quantities when we can't compare the qualities of things.*

This way, for better or for worse, we often assign some magnitude or price to each alternative. That tactic helps to simplify our lives so much that virtually every social community works out its own communal measure-schemes—let's call them *currencies*—that let its people work and trade in harmony, even though each individual has somewhat different personal goals. The establishment of a currency can foster both competition and cooperation by providing us with peaceful ways to divide and apportion the things we have to share.

But who can set prices on things like time or measure the values of comfort and love? What makes our mental marketplaces work so well when emotional states seem so hard to compare? One reason is that no matter how different those mental conditions seem, they must all compete for certain limited resources—such as space, time, and energy—and these, to a rather large extent, are virtually interchangeable. For example, you'd end up with essentially the same result whether you measure things in terms of food or time—because it takes time to find food, and each amount of food helps you survive for some amount of time. Thus the value we place on each commodity constrains, to some extent, the values we'll assign to many other kinds of goods. Because there are so many such constraints, once a community sets up a currency, that currency takes on a life of its own, and soon we start to treat our "wealth" as though it were a genuine commodity, a real substance that we can use, save, lend, or waste.

In a similar way, a group of agencies inside the brain could exploit some "amount" to keep account of their transactions with one another. Indeed agencies need such techniques even more than people do, because they are less able to appreciate each other's concerns. But if agents had to "pay their way," what might they use for currency? One family of agents might evolve ways to exploit their common access to some chemical that is available in limited quantities; another family of agents might contrive to use a quantity that doesn't actually exist at all, but whose amount is simply "computed." I suspect that what we call the pleasure of success may be, in effect, the currency of some such scheme. To the extent that success is interchangeable with time or food or energy, it's useful to treat pleasure as equivalent to wealth.

28.3 QUANTITY AND QUALITY

We've scarcely mentioned at all inside this book the kinds of quantities that could be "measured"—though surely brain cells use them all the time. For example, it seems quite likely that many of our agents employ quantitative schemes for summarizing evidence or establishing the strengths of connections. But I have said little about such matters because I suspect that such matters play diminished roles, the more we move toward higher-level operations of the mind. This is because whenever we're forced to compare magnitudes, we have to pay a heavy price: it tends to terminate what we call "thinking."

> *Whenever we turn to measurements, we forfeit some uses of intellect. Currencies and magnitudes help us make comparisons only by concealing the differences among what they purport to represent.*

By their nature, quantitative descriptions are so one-dimensional and featureless that they cannot help but conceal the structures that give rise to them. This is inescapable, since any act that makes two different things comparable must do it by deflecting our attention from their differences. Numbers themselves are the greatest masters of disguise because they perfectly conceal all traces of their origins. Add five and eight to make thirteen, and tell that answer to a friend: thirteen will be all your friend can know, since no amount of ingenious thought can ever show that it came from adding five and eight! It's much the same inside the head: quantitative judgments help us make decisions only by keeping us from thinking too much about the actual evidence.

No matter that such judgments have faults; you often have no choice but to choose. This happens when you can't stay where you are and must turn either right or left. Somewhere in some agencies, alternatives must be compared—and sometimes one can find no way except by using currencies. Then, various agents in your brain may turn to whatever quantities—chemical, electrical, or whatever—that happen to be available. Any substance or quantity whose availability is limited can be made to serve as a currency. But when we make our theories about how such systems work, we simply must remember not to make the easy mistake of confusing those quantities with their adopted functions and thus, for example, believing that certain drugs are inherently "stimulating" or "depressing," or that certain foodstuffs are *inherently* more "natural," or more "healthy." Most of the properties of a currency are not inherent—but merely conventional.

In any case, we should never assume that the quality or character of a thought process depends directly on the *nature* of the circumstances that evoke it. There is no quality of "sweetness" inherent in sugar itself, which is a mere chemical. Its quality of sweetness is, in effect, a currency involved with certain agencies that are connected to sensors that detect the presence of sugar. Those agencies evolved that way because whenever we have hunger goals, it pays to recognize the taste of sugar as a "sign of success"—simply because sugar itself supplies energy, is easy to detect, and usually indicates the presence of other edible sources of nutrition. Similarly, inside our brains, many agencies have come to influence one another by controlling the amounts of various chemicals in much the way that many kinds of human transactions have come to use substances like candy, coins, or bags of salt—or banknotes backed by promises.

28.4 MIND OVER MATTER

It seems completely natural to us that we should feel pain when we're injured or hunger when we're deprived of food. Such feelings seem to us to be inherent in those predicaments. Then why doesn't a car feel pain when its tire is punctured or feel hungry when its fuel runs low? The answer is that pain and hunger are *not* inherent in being injured or starved: *such feelings must be "engineered."* These physical circumstances do not directly produce the states of mind they arouse; on the contrary, this depends upon intricate networks of agencies and nerve-bundles that took millions of years to evolve. We have no conscious sense of that machinery. When your skin is touched, it seems as though it were your *skin* that feels—and not your brain—because you're unaware of everything that happens in between.

In order for hunger to keep us fed, it must engage some agency that gives priority to food-acquiring goals. But unless such signals came before our fuel reserves were entirely gone, they'd arrive too late to have any use. This is why feeling hungry or tired is not the same as being genuinely starved or exhausted. To serve as useful "warning signs," feelings like pain and hunger must be engineered not simply to indicate dangerous conditions, but to *anticipate* them and warn us *before* too much damage is done.

But what about the feelings of depression and discouragement we get when stuck at boring jobs or with problems we cannot solve? Such feelings resemble those that accompany physical fatigue, but they do not signify genuine depletions because they often easily respond to changes of context, interest, and schedule. Nevertheless, the similarity would be no accident, for probably those feelings arise because our higher-level brain centers have evolved connections that exploit our ancient fuel-exhaustion warning systems. After all, the unproductive use of time is virtually equivalent to wasting hard-earned energy!

Now what about those incidents in which some person seems to go beyond what we supposed were the normal bounds of endurance, strength, or tolerance of pain? We like to believe this demonstrates that the force of will can overrule the physical laws that govern the world. But a person's ability to persist in circumstances we hadn't thought were tolerable need not indicate anything supernatural. Since our feelings of pain, depression, exhaustion, and discouragement are themselves mere products of our minds' activities—and ones that are engineered to warn us *before* we reach our ultimate limits—we need no extraordinary power of mind over matter to overcome them. It is merely a matter of finding ways to rearrange our priorities.

In any case, what hurts—and even what is "felt" at all—may, in the end, be more dependent upon culture than biology. Ask anyone who runs a marathon, or ask your favorite Amazon.

28.5 THE MIND AND THE WORLD

We spend our lives in several realms. The first is the ordinary physical world of "objects" that exist in space and time. Objects obey simple laws. When any object moves or changes, we can usually account for it in terms of other objects pushing it, or else of gravity or wind. We also live in a *social realm* of persons, families, and companies; those entities appear to be ruled by quite different kinds of causes and laws. Whenever a person moves or changes, we look for signs of intentions, ambitions, infatuations, promises, threats, and the like—none of which could affect a brick. We also live in a *psychological* realm—inhabited by entities we call by names like "meanings," "ideas," and "memories." These, too, appear to obey different rules.

The causes in the physical realm seem terribly different from those that work in the social and psychological realms—so different that they seem to belong to different worlds.

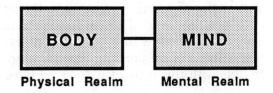

In some respects our bodies act exactly like ordinary objects: they have shapes we can see and touch, and they have locations that change when we're dropped or pushed. Yet in other ways, our bodies act quite differently from other things, and this appears to be because of minds. But what on earth are minds? For ages people have wondered about the relationship between the mind and body; some philosophers became so desperate as to suggest that only the mental world is real and the real world is merely an illusion. (That idea just makes the problem worse, because it can't even explain why there *seems* to be a physical world.) Most thinkers have ended up with images that portray two different kinds of worlds, one of matter and one of mind, somehow connected by mysterious threads of spiritual causality, somewhat like the films and tendrils formed when sticky stuff is pulled apart. Certain modern physicists have even speculated that these connections are somehow involved with the "uncertainty principle" in physics, perhaps because that problem also confounds their usual conceptions of causality. I see no merit in such ideas because as far as I'm concerned, the so-called problem of body and mind does not hold any mystery:

Minds are simply what brains do.

Whenever we speak about a mind, we're speaking of the processes that carry our brains from state to state. And this is what makes minds appear to us so separate from their physical embodiments: it is because concerns about minds are really concerns with *relationships between states*—and this has virtually nothing to do with the natures of the states themselves. This is why we can understand how a society of agents like *Builder* will work without knowing the physical constitution of its agents: what happens depends only on how each agent changes its state in response to its previous state and those of the other agents that connect to it. Other than that, it does not matter in the least what are the individual agents' colors, sizes, shapes, or any other properties that we could sense. So naturally minds seem detached from physical existence. It doesn't matter what agents *are*; it only matters what they *do*—and what they are connected to.

28.6 MINDS AND MACHINES

Why does a mind seem so unlike any other kind of thing? First, as we just said, minds *aren't* things—at least they share none of the usual properties of things, like colors, sizes, shapes, or weights. Minds lie beyond the reach of the senses of sound, touch, sight, smell, and taste. Yet though minds aren't things at all, they certainly have vital links to the things we call brains. What is the nature of those bonds? Are minds peculiar entities, possessed alone by brains like ours? Or could those qualities of minds be shared, to various degrees, by everything? Now, when we said, *"Minds are simply what brains do,"* that should have made us ask as well, *"Does every other kind of process also have a corresponding kind of mind?"* This could lead to an argument. One side might insist that this is merely a matter of degree, because people have well-developed minds while bricks or stones have almost none. Another side might try to draw a sharper boundary, arguing that only people can have minds—and, maybe, certain animals. Which side is right? This isn't a matter of wrong or right, since the issue is not about a fact, but only about when it's wise to use a certain word. Those who wish to reserve the label "mind" for only certain processes are obliged to specify which processes deserve that name. Those who claim that every kind of process has a corresponding type of mind are obliged to classify all minds and processes. The trouble with this is that we don't yet have adequate ways to classify processes.

Why are processes so hard to classify? In earlier times, we could usually judge machines and processes by how they transformed raw materials into finished products. But it makes no sense to speak of brains as though they manufacture thoughts the way factories make cars. The difference is that brains use *processes that change themselves*—and this means we cannot separate such processes from the products they produce. In particular, brains make memories, which change the ways we'll subsequently think. *The principal activities of brains are making changes in themselves.* Because the whole idea of self-modifying processes is new to our experience, we cannot yet trust our commonsense judgments about such matters.

As for brain science, no one ever before tried to study machines with billions of working parts. That would be difficult enough, even if we knew exactly how every part worked, and our present-day technology does not yet allow us to study the brain cells of higher animals while they're actually working and learning. This is partly because those cells are extremely small and sensitive to injury, and partly because they are so crowded together that we have not yet been able to map out their interconnections.

These problems will all be solved once we have better instruments and better theories. In the meantime, the hardest problems we have to face do not come from philosophical questions about whether brains are machines or not. There is not the slightest reason to doubt that brains are anything other than machines with enormous numbers of parts that work in perfect accord with physical laws. As far as anyone can tell, our minds are merely complex processes. The serious problems come from our having had so little experience with machines of such complexity that we are not yet prepared to think effectively about them.

Suppose I had once borrowed your boat and, secretly, replaced each board with a similar but different one. Then, later, when I brought it back, did I return your boat to you? What kind of question is that? It's really not about boats at all, but about what people mean by "same." For "same" is never absolute but always a matter of degree. If I had merely changed one plank, we'd all agree that it's still your boat—but after *all* its parts are changed, we're not so sure of its identity. In any case, we do not doubt that the second boat will behave in much the same way —to the extent that all those substituted boards are suitably equivalent.

What has this to do with brains? Well, now suppose that we could replace each of your brain cells with a specially designed computer chip that performs the same functions, and then suppose that we interconnect these devices just as your brain cells are connected. If we put it in the same environment, this new machine would reproduce the same processes as those within your brain. *Would that new machine be the same as you?* Again, the real question is not what we mean by "you," but what we mean by "same." There isn't any reason to doubt that the substitute machine would think and feel the same kinds of thoughts and feelings that you do—since it embodies all the same processes and memories. Indeed, it would surely be disposed to declare, with all your own intensity, that it *is* you. Would that machine be right or wrong? As far as I can see, this, too, is merely a matter of words. A mind is a way in which each state gives rise to the state that follows it. If that new machine had a suitable body and were placed in a similar environment, its sequence of thoughts would be essentially the same as yours— since its mental states would be equivalent to yours.

> *Modifying or replacing the physical parts of a brain will not affect the mind it embodies, unless this alters the successions of states in that brain.*

You might object to this idea about duplicating minds on the grounds that it would never be practical to duplicate enough details. You could argue the same about that borrowed boat: no matter how carefully a carpenter were to copy every board, there would always remain some differences. This plank would be a little too stiff, that one would be a little too weak, and no two of them would bend in *exactly* the same way. The copied boat would never be precisely the same—even though you might need a microscope to see the differences. For similar reasons, it would be impractical to duplicate, with absolute fidelity, *all* the interactions in a brain. For example, our brain cells are all immersed in a liquid that conducts electricity, which means that every cell has at least a small effect on every other cell. If we tried to imitate your brain with a network of computer chips, many of those tiny interactions would be left out.

Can you then boast that your duplicated brain-machine would not have the same mind as yours because its computer chips don't work exactly like the brain cells they purport to replace? No; you'd get more than you bargained for if you argued that the new machine was not the same as you, merely because of microscopic differences. Consider that as you age, you're never the same as a moment ago. If such small differences matter that much, this would prove that you yourself are not the same as you.

28.8 OVERLAPPING MINDS

Consider the popular idea that a person is capable of two kinds of thinking at once—a "right brain" kind and a "left brain" kind—as though there were two different individuals inside each human brain. This raises some odd questions, since there are many other ways to draw imaginary boundaries through brains.

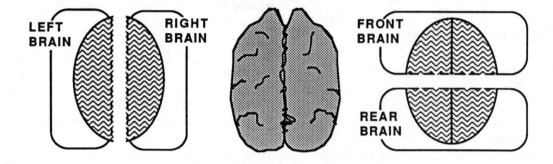

If you agree that each person has both a *left-brain mind* and *right-brain mind*, then you must also agree that each person also has a *front-brain mind* and a *back-brain mind*! Can a single large mind contain so many smaller ones, with overlapping boundaries? It makes sense to think of part of a structure as being a "thing" in its own right only when the relationships among parts of that structure have some significant type of coherency. Before you'd say that a certain arbitrary section of brain contains a mind, you'd want to have some evidence that what happens inside that boundary is something you would consider to be a mind.

The less another entity resembles you, the less it means for you to say that it, like you, must have a mind. Do our very smallest agencies have minds? No, because it would make no more sense to say this than to say that two trees form a forest or that two bricks form a wall. But there are indeed some agencies inside our brains that *do* have humanlike abilities to solve, by themselves, some types of problems that we regard as hard. For example, your agencies for locomotion, vision, and language may contain within their boundaries some processes that are quite as intricate as those "you" use for your own conscious thought. Possibly, some of those processes are actually more "conscious" than you are yourself, in the sense that they maintain and use even more complete records of their own internal activities. Yet what happens in those agencies is so sealed off that you have no direct experience of how "you" distinguish a cat from a dog, retrace "your" last few steps, or listen and talk without knowing how "you" do it.

All this suggests that it can make sense to think there exists, inside your brain, a society of different minds. Like members of a family, the different minds can work together to help each other, each still having its own mental experiences that the others never know about. Several such agencies could have many agents in common, yet still have no more sense of each other's interior activities than do people whose apartments share opposite sides of the same walls. Like tenants in a rooming house, the processes that share your brain need not share one another's mental lives.

If each of us contains several such mini-minds, could any special exercise help put them all "in closer touch"? Certainly there are ways to become selectively aware of processes that are usually not conscious at all. But becoming aware of everything that happens in brains would leave no room for thought. And the reports of those who claim to have developed such skills seem singularly uninformative. If anything, they demonstrate that it's even harder than we think to penetrate those unresisting barriers.

CHAPTER 29

THE REALMS OF THOUGHT

*"Elsewhere" is another view—possibly from philosophy—
or other "elsewheres" as well, since the views of man are
multiple. Each view has its own questions. Separate views
speak mostly past each other. Occasionally, of course, they
speak to the same issue and then comparison is possible, but
not often and not on demand.*

—ALLEN NEWELL

29.1   THE REALMS OF THOUGHT

Our view of the body and the mind as separate entities is only one example of our many ways to view the world as divided into different realms. Imagine that a committee were commissioned to write down everything about the universe in a perfectly organized book.

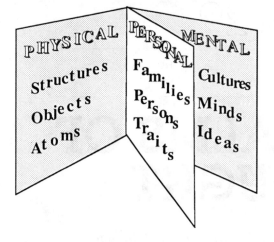

This suggests how people think about the world. The "pages" represent our physical, personal, and psychological conceptions, while the lines on each page correspond to the levels of detail that are distinguished in each realm.

Why do the gaps between the lines seem smaller than those that separate the pages? It is because we better understand what happens in between. We understand how walls relate to bricks because they represent closely related "levels of organization." Similarly, we understand the relation between houses and walls. But it would be hard to cross the gap between houses and bricks without having enough intermediate concepts such as that of a wall. It simply isn't practical to think of the place where someone lives as a network of relationships among a million boards and bricks.

It's much the same in other realms; we need to be able to describe things at many levels of detail. We all belong to families or companies, and sometimes we can think of each group as "nothing but" a network of agreements and relationships. But when we need a larger view, as when thinking about the politics of an entire country, we cannot think effectively without regarding entire families or companies as though they were single objects in a different realm. The same applies to how we think about our minds. Even if you knew all the details of each little agent in your brain, your higher-level processes would still need coarser summaries.

Why is it easier to understand how walls relate to bricks, or families to individuals, than to understand how thoughts relate to things? It's not because there's any single mystery. It is because the level gap between walls and bricks is *really* much smaller than that between minds and brain cells. Suppose we actually had that wonderful encyclopedia of "all possible knowledge," arranged according to the nearness of topics. There we'd find the essays on walls quite close to those about bricks. But the sections that cover the nature of thoughts would lie volumes away from the ones on the nature of things.

29.2 SEVERAL THOUGHTS AT ONCE

To see that we can think in several mental realms at once, consider the role of the word "give" in this simple sentence:

Mary gives Jack the kite.

We can see at least three distinct meanings here. First, we could represent the idea of the kite's motion through physical space by using a *Trans*-frame whose *Trajectory* begins at Mary's hand and ends at Jack's.

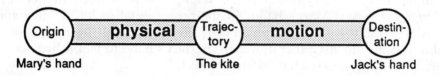

But quite beyond that realm of space, we also find a different significance in what Mary did —in another realm that we'll call "estates." This involves a different sense of "give," in which the object need not actually move at all! Instead, what happens is the transfer of its *ownership.*

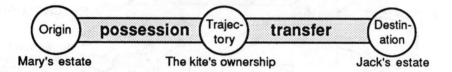

Each of us has an "estate"—the collection of possessions we control. And this "realm of estate" is more important than it might seem, because it lies between the realms of objects and ideas. In order to carry out our plans, it is not enough only to know what things or ideas are required and how to adapt them to our purposes. We must also be able to take possession of those objects or ideas, either by right or by might.

> *Possession plays essential roles in all our plans, because we can't use any materials, tools, or ideas until we gain control of them.*

We can also interpret Mary's act within a social realm, in which we understand that giving gifts involves yet other kinds of relationships. No sooner do you hear of Mary's gift than certain parts of your mind become concerned with why she was so generous and how this involved her affections and obligations.

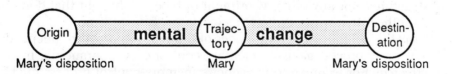

How can so many different thoughts proceed at the same time, without interfering with one another? I suspect that it is for the same reason that we have no trouble imagining an apple as both round and red at the same time: in that case, the processes for color and shape use agents that do not compete. Similarly, the different processes involved with ideas like "give" may operate in agencies so different that they rarely need to compete for the same resources.

29.3 PARANOMES

What enables us to comprehend "*Mary gives Jack the kite*" in so many ways at once? Different meanings don't conflict when they apply to separate realms—but that can't be quite what's happening here, since the physical, social, and mental realms are closely linked in many ways. So now I'll argue just the opposite, *that these meanings are so similar they don't conflict!* Here is my hypothesis about what holds together all these aspects of our thoughts:

> *Many of our higher level conceptual-frames are really parallel arrays of analogous frames, each active in a different realm.*

Consider all the different roles played by the *Actor* pronome of our sentence. In the physical realm, the *Origin* of *give* is Mary's hand. In the possessional realm of "give and take," that *Origin* is in Mary's estate—since Mary can only give Jack what she owns. Similarly, in the physical realm, it is the kite itself that moves from Mary's hand to Jack's; however, in the realm of estates, the kite's *ownership* is what "changes hands."

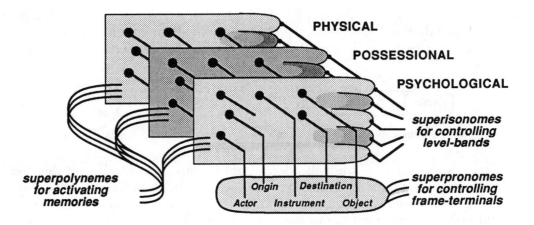

This suggests that certain pronomes can operate in several different realms at once. Let's call them "paranomes" to emphasize their parallel activities. When the language-agency activates some polynemes and paranomes, these agents run crosswise through the agencies of various realms to arouse several processes and frames at the same time; these correspond to different interpretations, in different realms, of the same phrase or sentence. Then, because each major agency contains its own memory-control system, the agencies within each realm can simultaneously apply their own methods for dealing with the corresponding aspect of the common topic of concern. In this way, a single language-phrase can at the same time evoke different processes involved with social dispositions, spatial images, poetical fancies, musical themes, mathematical structures—or any other assortment of types of thought that don't interfere too much with one another.

This is not to say that all these different modes of thought will proceed independently of one another. Whenever any process gains momentary control over a paranome, many other processes can be affected. For example, one agency's memory-control process might thus cause the agencies in several other realms simultaneously to "blink" on and off their *Origin* and *Destination* paranomes. This would force the agencies active in each of those realms to focus upon whichever types of differences they then discern; then, in between such episodes, each agency can apply its own way of thinking to the corresponding topic, difference, or relationship. By using these cross-connecting polynemes and paranomes, the activity in each realm can proceed sometimes independently, yet at other times influence and be affected by what happens in the other realms.

29.4 CROSS-REALM CORRESPONDENCES

We often describe the things we like as *"elevated," "lofty,"* or *"heavenly."* Why do we see such things in terms of altitude in space? We often speak of time itself in spatial terms, as though the future were "ahead" of us while the past remains behind. We think of problems as "obstacles" to go around and turn to using diagrams to represent things that don't have shapes at all. What enables us to turn so many skills to so many other purposes? These tendencies reflect the systematic "cross-realm correspondences" embodied in our families of polynemes and paranomes.

> *At each instant, several realms may be engaged in active processing. Each has separate processes but must compete for control of the ascending nemes that lead into the language-agency. Which polyneme will play the role of **Origin** in the next sentence-frame? Will it be Mary's physical arm or hand, or Mary's social role as party guest? It sometimes seems as though the language-agency can focus on only one realm at a time.*

This could be one reason why language scientists find it hard to classify the roles words play in sentence-frames. No sooner does a language-agency assign some polynemes and isonomes to a phrase than various mind divisions proceed to alter how they're used inside each different realm. Every shift of control from one realm to another affects which particular nemes will be next to influence the language-agency. This causes moment-to-moment changes in the apparent meaning of a phrase.

> *At one moment, control over language may reside in the realm of thought that is working most successfully; at the next moment, it may be the one experiencing the most difficulty. Each shift in attention affects how the various expressions will be interpreted, and this in turn can affect which realm will next take center stage.*

For example, the sentence *"Mary gives Jack the kite"* might start by arousing a listener's concern with Mary's social role as party guest. That would cause the pronomes of a social-frame to represent Mary's obligation to bring a present. But then the listener's possession realm might become concerned with Mary's ownership of that gift or with how she got control of it. This shift from social to possessional concern could then affect the processing of future sentences. For example, it will influence whether a phrase like *"Jack's kite"* is interpreted to refer to the kite that Jack happens to be holding or to a different kite that Jack happens to own.

Every mental realm accumulates its own abilities but also discovers, from time to time, how to exploit the skills of other realms. Thus the mind as a whole can learn to exploit the frames developed in the realm of space both for representing events in time and for thinking about social relationships. Perhaps our chaining skills are the best example of this; no matter which realm or realms they originate in, we eventually learn to apply them to any collection of entities, events, or ideas (in any realm whatever) that we can arrange into sequences. Then chains assume their myriad forms, such as spatial order, psychological causality, or social dominance.

29.5 THE PROBLEM OF UNITY

What makes our minds form many separate mental realms, instead of attempting, as scientists do, to see all aspects of the world in a unified way? Because, at least in everyday life, the latter simply isn't practical. Consider how different the rules are within the physical and social realms. If you want your furniture inside a different room, you normally would push it there. But when you want to move your party guests, it would be rude to push *them* there. Contrast the principles of physics and geometry with those we use in the social realm. In the physical realm, the rules seem very orderly:

> -- *A stationary object stays where it is unless another object pushes it.*
> --- *A moving object continues in its course until some external force makes it stop.*
> ---- *All unsupported objects start to fall.*
> ----- *No two things can occupy the same location.*
> ------ *Etc.*

These principles seem clear to us—but infants cannot appreciate them until they've built up ways to represent ingredients like "thing," "shape," "place," "move," and "near." It takes each child many years to develop these abilities.

Our comprehensions of social acts are based on different principles. When an ordinary object moves around, we usually see an obvious cause; most likely, another object pushed it. But when we see a person move, we rarely *see* the cause at all—because it's buried in a brain. In predicting how a person will react to an expression or gesture, we have little use for physical properties like color, shape, or place. Instead, we employ almost entirely different conceptions. To guess the outcome of a social interaction, we have to be able to represent each person's mental state—and to do that, we must develop concepts about traits, dispositions, motives, and plans. The concepts that serve so well for physical objects are of little help within the social realm—and *vice versa*.

When normal children start to talk, among the early aspects of their speech are words that distinguish animate things. Frequently, a child will use a single expression for all kinds of animals, and for everything else that can move by itself—for example, an automobile. According to our view of things, this surely is no accident.

To adults, the laws that govern the physical world seem simpler and more orderly than those that apply to human events. Does this mean that for infants, too, it should be easier first to master the physical world and later to proceed toward social and psychological understanding? No. Paradoxically, the social realm is initially the easier! Imagine that an infant wants a certain toy and that there's a sympathetic person near. The easiest thing is to *make a request*—that is, a gesture, smile, or cry—and this will probably achieve the goal. It would be far more difficult for the infant to coordinate all the complicated machinery for planning and executing the trajectory for propelling the object from where it is to where the infant wishes it to be. From the point of view of a physically helpless infant, the social realm is by far the simpler one.

29.6 AUTISTIC CHILDREN

Isn't it curious that infants find social goals easier to accomplish than physical goals, while adults find the social goals more difficult? One way to explain this is to say that *the presence of helpful people simplifies the infant's social world*—since because of them, simpler actions solve harder problems. Another explanation might be that the infant's social world is just as complicated as that of the adult, except *the presence of helpful people makes the infant's mind more powerful*—by making the agencies inside those other people's brains available for exploitation by the agencies in the infant's brain. Both explanations are the same, except for drawing different boundaries.

How do children start on the path toward distinguishing between psychological and physical relationships? In the appendix I'll suggest that our infant brains are genetically equipped with machinery for making it easy to learn social signals. But what if that machinery should somehow fail, so that by chance—or by neglect or accident—the realm-divisions never form? Then all those different kinds of thoughts would fuse together into one—and the child would face the impossible task of formulating principles that work in all domains. A child that tried to see the world without dividing it into realms would find no simple rules at all that work across so large a range.

This is why each child must learn different rules for the physical and psychological realms. But this means that the child must face not merely two formidable problems, but three. In addition to developing two different sets of concepts, the child must also develop agencies to *manage* those concepts by keeping them apart in different agencies, as we saw when we talked about Papert's principle.

This could explain some aspects of the disorders of the children psychiatrists call "autistic." These unhappy individuals do not establish effective communication with other people, although they may acquire some competence at dealing with physical things. No one knows the causes of those disorders. Some might begin when certain mental realms do not develop normally. Other kinds of problems could emerge *after* those divisions form, if their separateness were compromised by some too intense attempt to unify them. To be sure, that is what scientists do, but unlike those whom we regard as mentally ill, scientists also manage to maintain their ordinary views. Once a child is deprived of the normal ways to divide those realms—no matter what the cause of this—that hapless mind is doomed to fail.

29.7 LIKENESSES AND ANALOGIES

You are a full-spread, fair-set vine,
And can with tendrils love entwine,
Yet dried ere you distil your wine.
— ROBERT HERRICK

We always try to use old memories to recollect how we solved problems in the past. But nothing's ever twice the same, so recollections rarely match. Then we must force our memories to fit—so we can see those different things as similar. To do this, we can either modify a memory or change how we represent the present scene. For example, suppose you need a hammer but can only find a stone. One way to turn that stone to your purposes would be to make it fit your memory of a hammer's appearance—for example, by making your description of the stone include an imaginary boundary that divides it into two parts, to serve as handle and as head. Another way would be to make your hammer frame accept the entire stone as *a hammer without a handle.* Either scheme will make the memory match the description, but both will lead to conflicts in other agencies.

How hard it will be to make such a match depends both on which agents are now active in your mind and on the levels of their priorities—in short, upon the context already established. It will be easy for you to see two things as similar when you only need to change relatively weak attachments at the conceptual fringes of familiar things. But frequently the ease of comprehension will also depend upon how readily you can switch from one mental realm to another.

Consider what must happen in our minds when poets speak about their loves in romantic, floral terms. We all have learned a certain common way to represent a woman's beauty in terms of flowers that are prone, alas, to fade. For centuries this formula has been established in our language and literature; however, at first it must have seemed bizarre. We cannot possibly match our descriptions of women and flowers if we insist on interpreting such phrases and poems "literally"—that is to say, "illiterately"—entirely within the physical realm of the appearance, composition, and behavior of a typical flower.

To be sure, the colors, symmetries, and smells of flowers can certainly arouse the sorts of states we associate with things we've come to see as beautiful. But the more essential trick is in knowing how to turn entirely away from the physical realm and dwell instead upon the images and fantasies that flowers evoke in other spheres—such as the sense of a thing so sweet and innocent, so helpless and delicate, that it invites affection, nurture, and protection. Features like these must be made to fit the listener's private love ideal—only then can the metaphor match.

This, Herrick's bitter verse defeats. By holding us so tightly to the usual frames for human shapes, he steers us into fantasies of vegetables with hands and feet.

29.8 METAPHORS

Listen closely to anything anyone says, and soon you'll hear analogies. We speak of time in terms of space, as like a fluid that's *running out*; we talk of our friends in physical terms, as in *"Mary and John are very close."* All of our language is riddled and stitched with curious ways of portraying things as though they belonged to alien realms.

We sometimes call these "metaphors," our ways to transport thoughts between the various mental realms. Some metaphors seem utterly pedestrian, as when we speak of "taking steps" to cause or prevent some happening. Other metaphors seem more miraculous, when unexpected images lead to astonishing insights—as when a scientist solves a problem by conceiving of a fluid as made of tubes or of a wave as an array of overlapping, expanding spheres. When such conceptions play important roles in our most productive forms of thought, we find it natural to ask, *"What is a metaphor?"* But we rarely notice how frequently we use the same techniques in ordinary thought.

What, then, *is* a metaphor? It might be easy to agree on functional definitions like "A *metaphor is that which allows us to replace one kind of thought with another.*" But when we ask for a structural definition of "metaphor," we find no unity, only an endless variety of processes and strategies. Some are simple, as when we make an analogy by stripping away so many details that two different objects seem the same. But other forms of metaphor are as complex as can be. In the end there is little to gain by cloaking them all under the same name "metaphor," because there isn't any boundary between metaphorical thought and ordinary thought. *No* two things or mental states ever are identical, so *every* psychological process must employ one means or another to induce the illusion of sameness. Every thought is to some degree a metaphor.

Once scientists like Volta and Ampere discovered how to represent electricity in terms of the pressures and flows of fluids, they could transport much of what they already knew about fluids to the domain of electricity. Good metaphors are useful because they transport uniframes, intact, from one world into another. Such cross-realm correspondences can enable us to transport entire families of problems into other realms, in which we can apply to them some already well-developed skills. However, such correspondences are hard to find since most reformulations merely transform the uniframes of one realm into disorderly accumulations in the other realm.

From where do we obtain our most productive, systematic, cross-realm correspondences? Some must be virtually born into our brains through the wiring of our paranomes; other metaphors we discover by ourselves as individuals; but most of them are learned from other members of our cultural communities. Finally, from time to time, someone discovers a new reformulation that is both so fruitful and so easy to explain that it becomes part of the general culture. Naturally, we'd like to know how the greatest metaphorical discoveries were made. But because this is buried in the past, the best and rarest of those events may never be explained at all. Our greatest ideas, like our evolutionary genes, need form only once, by accident, and then can spread from brain to brain.

CHAPTER 30

MENTAL MODELS

The world has kept sentimentalities simply because they are the most practical things in the world. They alone make men do things. The world does not encourage a perfectly rational lover, simply because a perfectly rational lover would never get married. The world does not encourage a perfectly rational army, because a perfectly rational army would run away.

—GILBERT K. CHESTERTON

30.1 KNOWING

What does "knowing" really mean? Suppose Mary (or some other creature or machine) can answer certain questions about the world—without the need to do any actual experiments. Then we'd agree that Mary *knows* those things about the world. But what would it mean, to you or to me, to hear *Jack* say that *"Mary knows geometry"*? For all we know, Mary might believe that circles are square, and it happens that Jack agrees! Jack's statement tells us more about Jack than about Mary.

> When Jack says, "Mary knows geometry," this indicates to us that Jack would probably be satisfied by Mary's answers to the questions about geometry that he would be disposed to ask.

The meaning of *"Mary knows geometry"* depends on who is saying it. After all, no one knows *everything* about geometry; that statement would not mean the same to us as to a mathematician, whose concepts of geometry are different from those of ordinary persons. In the same way, the meanings of many other terms depend upon the speaker's role. Even an apparently unambiguous statement like *"This is a painting of a horse"* shares this character: you can be sure of little more than that it displays a representation that *in someone's view* resembles a horse in some respects.

Then why, when we talk about knowledge, don't we have to say who all those speakers and observers are? Because we make assumptions by default. When a stranger says that Mary knows geometry, we simply assume that the speaker would expect *any typical person* who knows Mary to agree that she knows geometry. Assumptions like that allow us to communicate; unless there is some reason to think otherwise, *we assume that all the things involved are "typical."* It does not bother us that a professional mathematician might not agree that Mary knows geometry—because a mathematician doesn't fit our stereotype of a "typical person."

You might maintain that none of this applies to you, since *you* know what you know about geometry. But there's still an observer on the scene, only now it is hiding inside your mind—namely, the portion of "you" that claims you know geometry. But the part of you that makes this claim has little in common with the other parts that actually *do* geometry for you; *those* agencies are probably incapable of speech and probably devoid of thoughts about your knowledge and beliefs.

Naturally, we'd all prefer to think of knowledge as more positive and less provisional or relative. But little good has ever come from trying to link what we believe to our ideals about absolute truths. We always yearn for certainty, but the only thing beyond dispute is that there's always room for doubt. And doubt is not an enemy that sets constraints on what we know; the real danger to mental growth is perfect faith, doubt's antidote.

30.2 KNOWING AND BELIEVING

We often speak as though we classify our thoughts into different types called *facts*, *opinions*, and *beliefs*.

> *"The red object is on the table."*
> *"I think the red block is on the table."*
> *"I believe that the red block is on the table."*

How do these statements differ from one another? Some philosophers have argued that "knowing" must mean *"true and justified belief."* However, no one has ever found a test to prove what's justified or true. For example, we all know that the sun rises in the morning. Once, long ago, some people thought this was due to godlike agents in the sky, and that the sun's trajectory was where Apollo steered his chariot. Today our scientists tell us that the sun doesn't really rise at all, because "sunrise" is simply what we each experience when the planet Earth's rotation moves us into the sun's unchanging light. This means we all "know" something that isn't true.

To comprehend what knowing is, we have to guard ourselves against that single-agent fallacy of thinking that the "I" in "*I believe*" is actually a single, stable thing. The truth is that a person's mind holds different views in different realms. Thus, one part of an astronomer's mind can apply the common view of sunrise to down-to-earth affairs, regarding the sun as like a lamp that wakes us up and lights our way. But at the same time, that same astronomer can apply the modern physical view to technical problems in astronomy. We each use many different views, and which we choose to use depends, from one moment to the next, upon the changing balance of power among our agencies.

Then if what we "believe" is so conditional, what makes us feel that our beliefs are much more definite than that? It is because whenever we commit ourselves to speak or act, we thereby have to force ourselves into clear-cut, action-oriented states of mind in which most of our questions are suppressed. As far as everyday life is concerned, decisiveness is indispensable; otherwise we'd have to act so cautiously that nothing would get done. And here lies much of what we express with words like "guess," "believe," and "know." In the course of making practical decisions (and thereby turning off most agencies), we use such words to summarize our various varieties of certainty.

The notion that only certain of a person's beliefs are "genuine" plays vital roles in all our moral and legal schemes. Whenever we censure or applaud what other people do, we're taught to be more concerned with what those other people "genuinely" expected or intended to happen than with what actually happened. This doctrine underlies how we distinguish thoughtlessness and forgetfulness from lies, deceit, and treachery. I do not mean that such distinctions are not important, only that they do not justify the simplistic assumption that, among all the mind's activities, certain special kinds of thoughts are essentially more "genuine" than others. All such distinctions seem less absolute when every deeper probe into beliefs reveals more ambiguities.

30.3 MENTAL MODELS

Does a book know what is written inside it? Clearly, no. Does a book *contain* knowledge? Clearly, yes. But how could anything contain knowledge, yet not know it? We've seen how saying that a person or machine possesses knowledge amounts to saying that *some observer could employ that person or machine to answer certain kinds of questions.* Here is another view of what it means to know.

> *"Jack knows about A"* means that there is a *"model"* **M** of **A** inside Jack's head.

But what does it mean to say that one thing is a model of another and how could one have a model in one's head? Again, we have to specify some standard or authority. Let's make Jack be the judge of that:

> *Jack considers* **M** *to be a good model of* **A** *to the extent that he finds* **M** *useful for answering questions about* **A**.

For example, suppose that **A** is a real automobile, and **M** is the kind of object we call a "toy" or "model" car. Then Jack will be able to use **M** to answer certain questions about **A**. However, we would think it strange to say that **M** is Jack's "knowledge" about **A**—because we reserve the word "knowledge" for something inside a head, and Jack can't keep a toy inside his head. But we never said that a model must be an ordinary physical object. Our definition allows a model to be *anything* that helps a person answer questions. Accordingly, a person could possess a "mental model," too—in the form of some machinery or subsociety of agents inside the brain. This provides us with a simple explanation of what we mean by knowledge: *Jack's knowledge about* **A** *is simply whichever mental models, processes, or agencies Jack's other agencies can use to answer questions about* **A**. Thus, a person's mental model of a car need not itself resemble an actual car in any obvious way. It need not itself be heavy, fast, or consume gasoline to be able to answer questions about a car like *"How heavy is it?"* or *"How fast can it go?"*

Our mental models also work in social realms to answer questions like *"Who owns that car?"* or *"Who permitted you to park it there?"* However, to understand questions like these, we have to ask what people mean by "who"—and the answer is that *we make mental models of people, too.* In order for Mary to "know" about Jack's dispositions, motives, and possessions, Mary has to build inside her head some structure to help answer those kinds of questions—and that structure will constitute her mental model of Jack. Just think of all the different things our person-models do for us! If Mary knows Jack well enough, she'll be able to reply not only to physical questions like *"How tall is Jack?"* but also to social inquiries such as *"Does he like me?"* and even to psychological queries like *"What are Jack's ideals?"* Quite possibly, Mary's model of Jack will be able to produce more accurate answers to such questions than Jack himself could produce. People's mental models of their friends are often better, in certain respects, than their mental models of themselves.

We all make models of ourselves and use them to predict which sorts of things we'll later be disposed to do. Naturally, our models of ourselves will often provide us with wrong answers because they are not faultless ways to see ourselves, but merely self-made answering machines.

30.4 WORLD MODELS

Now let's look at Mary's model of the world. (By "world" I mean the universe, not just the planet Earth.) This is simply *all* the structures in Mary's head that Mary's agencies can use to answer questions about things in the world.

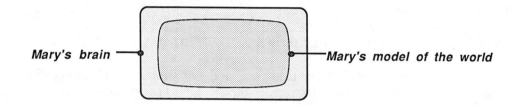

Mary's brain ———   ———Mary's model of the world

But what if we were to ask Mary a question not about any particular object, but one like *"What sort of thing is the world itself?"* That would put Mary in a curious predicament. She cannot answer this by using her world model, because each part of that model is designed only to answer questions about particular things. The trouble is that the world itself is not a particular thing inside the world.

One way to deal with this (and a method that, surely, many children use) is adding to the model of the world an additional "object"—that represents the world itself. Then, since any object must have properties, the child might then assign to this the features, say, of *an extremely large ball*. Naturally this will lead to trouble should that child persist in asking ordinary questions about this extraordinary object—such as *"What keeps the universe in place?"* or *"What's outside the universe?"*—for these then lead to strange and inconsistent images. Eventually we learn some ways to deal with this—for example, by learning which questions to suppress. But as in the case of a perfect point, we may always feel uncomfortable with the thought of a thing that is unimaginably large in size, but has no shape at all.

When you get right down to it, you can never really describe any *worldly* thing, either—that is, in any absolute sense. Whatever you purport to say about a thing, you're only expressing your own beliefs. Yet even that gloomy thought suggests an insight. Even if our models of the world cannot yield good answers about the world as a whole, and even though their other answers are frequently wrong, they can tell us something about ourselves. We can regard what we learn about our models of the world as constituting *our models of our models of the world*.

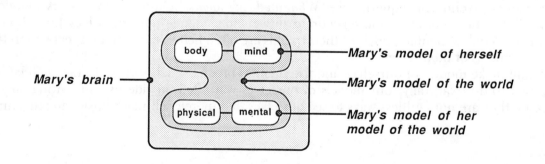

Mary's brain ———   ———Mary's model of herself
———Mary's model of the world
———Mary's model of her model of the world

30.5 KNOWING OURSELVES

Now let's ask Mary to describe herself—to tell us everything she can about her shape and weight and size and strength, her dispositions and her traits, her accomplishments and ambitions, wishes, fears, possessions, and so on. What might be the general character of what we'd hear? At first it would be hard to assemble any coherent sense of all those details. But gradually we'd notice that various groups of items were closely related, while others were rarely mentioned in connection with one another. Little by little, we would discern structure and organization in what Mary had said, and finally we'd start to see the outlines of at least two different mental realms.

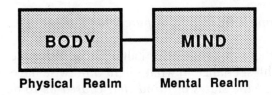

Now, what would happen if we asked Mary to speak not about specific things, but about the general subject of "What kind of entity am I?" Since she has no direct way to examine her entire self, she can only summarize what she can discover about her mental model of herself. In doing so, she'll probably find that almost everything she knows appears to lie in two domains, with relatively little in between. This means that Mary's *model of her model of herself will have an overall dumbbell shape*, one side of which represents her physical self, the other side of which represents her psychological self.

Do people go on to make *models of their models of their models of themselves?* If we kept on doing things like that, we'd get trapped in an infinite regress. What saves us is that we get confused and lose track of the distinctions between each model and the next—just as our language-agencies lose track when they hear about "*the rat that the cat that the dog worried killed.*" The same thing must happen whenever we ask ourselves questions like "*Did John know that I knew that he knew that I knew that he knew that?*" And the same thing happens whenever we try to probe into our own motivations by continually repeating, "What was my motive for wanting *that?*" Eventually, we simply stop and say, "*Because I simply wanted to.*" The same when we find things hard to decide: we can simply say, "*I just decide,*" and this can help us interrupt what otherwise might be an endless chain of reasoning.

30.6 FREEDOM OF WILL

We each believe that we possess an Ego, Self, or Final Center of Control, from which we choose what we shall do at every fork in the road of time. To be sure, we sometimes have the sense of being dragged along despite ourselves, by internal processes which, though they come from within our minds, nevertheless seem to work against our wishes. But on the whole we still feel that we can choose what we shall do. Whence comes this sense of being in control? According to the modern scientific view, there is simply no room at all for "freedom of the human will." Everything that happens in our universe is either completely determined by what's already happened in the past or else depends, in part, on random chance. Everything, including that which happens in our brains, depends on these and only on these:

A set of fixed, deterministic laws.   A purely random set of accidents.

There is no room on either side for any third alternative. Whatever actions we may "choose," they cannot make the slightest change in what might otherwise have been—*because those rigid, natural laws already caused the states of mind that caused us to decide that way.* And if that choice was in part made by chance—it still leaves nothing for us to decide.

Every action we perform stems from a host of processes inside our minds. We sometimes understand a few of them, but most lie far beyond our ken. But none of us enjoys the thought that what we do depends on processes we do not know; we prefer to attribute our choices to *volition, will,* or *self-control.* We like to give names to what we do not know, and instead of wondering how we work, we simply talk of being "free." Perhaps it would be more honest to say, *"My decision was determined by internal forces I do not understand."* But no one likes to feel controlled by something else.

Why *don't* we like to feel compelled? Because we're largely made of systems designed to learn to achieve their goals. But in order to achieve any long-range goals, effective difference-engines must also learn to resist whatever other processes attempt to make them *change* those goals. In childhood, everyone learns to recognize, dislike, and resist various forms of aggression and compulsion. Naturally we're horrified to hear about agents that hide in our minds and influence what we decide.

In any case, both alternatives are unacceptable to self-respecting minds. No one wants to submit to laws that come to us like the whims of tyrants who are too remote for any possible appeal. And it's equally tormenting to feel that we're a toy to mindless chance, caprice, or probability—for though these leave our fate unfixed, we'd still not play the slightest part in choosing what shall come to be. So, though it's futile to resist, we continue to regard both Cause and Chance as intrusions on our freedom of choice. There remains only one thing to do: we add another region to our model of our mind. We imagine a third alternative, one easier to tolerate; we imagine a thing called "freedom of will," which lies beyond both kinds of constraint.

30.7   THE MYTH OF THE THIRD ALTERNATIVE

To save our belief in the freedom of will from the fateful grasps of Cause and Chance, people simply postulate an empty, third alternative. We imagine that somewhere in each person's mind, there lies a Spirit, Will, or Soul, so well concealed that it can elude the reach of any law —or lawless accident.

I've drawn the box for Will so small because we're always taking things out of it—and scarcely ever putting things in! This is because whenever we find some scrap of order in the world, we have to attribute it to Cause—and whenever things seem to obey no laws at all, we attribute that to Chance. This means that the dominion controlled by Will can only hold what, up to now, we don't yet understand. In ancient times, that realm was huge, when every planet had its god, and every storm or animal did manifest some spirit's wish. But now for many centuries, we've had to watch that empire shrink.

Does this mean that we must embrace the modern scientific view and put aside the ancient myth of voluntary choice? No. We *can't* do that: too much of what we think and do revolves around those old beliefs. Consider how our social lives depend upon the notion of *responsibility* and how little that idea would mean without our belief that personal actions are voluntary. Without that belief, no praise or shame could accrue to actions that were caused by Cause, nor could we assign any credit or blame to deeds that came about by Chance. What could we make our children learn if neither they nor we perceived some fault or virtue anywhere? We also use the idea of freedom of will to justify our judgments about good and evil. A person can entertain a selfish impulse, yet turn it aside because it seems wrong, and that must happen when some self-ideal has intervened to overrule another goal. We can feel virtuous when we think that we ourselves have chosen to resist an evil temptation. But if we suspected that such choices were not made freely, but by the interference of some hidden agency, we might very well resent that interference. Then we might become impelled to try to wreck the precious value-schemes that underlie our personalities or become depressed about the futility of a predestination tempered only by uncertainty. Such thoughts must be suppressed.

No matter that the physical world provides no room for freedom of will: that concept is essential to our models of the mental realm. Too much of our psychology is based on it for us to ever give it up. We're virtually forced to maintain that belief, even though we know it's false —except, of course, when we're inspired to find the flaws in *all* our beliefs, whatever may be the consequence to cheerfulness and mental peace.

30.8 INTELLIGENCE AND RESOURCEFULNESS

How could anything as complex as a human mind work so well for so many years? We all appreciate those splendid feats of writing plays and symphonies. But we rarely recognize how wonderful it is that a person can traverse an entire lifetime without making a single really serious mistake—like putting a fork in one's eye or using a window instead of a door. How do we do such amazing feats as to imagine things we've never seen before, to overcome obstacles, to repair things that are broken, to speak to one another, to have new ideas? What magical trick makes us intelligent? *The trick is that there is no trick.* The power of intelligence stems from our vast diversity, not from any single, perfect principle. Our species has evolved many effective although imperfect methods, and each of us individually develops more on our own. Eventually, very few of our actions and decisions come to depend on any single mechanism. Instead, they emerge from conflicts and negotiations among societies of processes that constantly challenge one another. In this book we've seen many such dimensions of diversity:

> **The accumulation of myriad subagents.**
> *We learn many different ways to achieve each kind of goal.*
> **The many realms of ordinary thought.**
> *When one viewpoint fails to solve a problem, we can adopt other perspectives.*
> **The endowment of several "instinctive" protominds.**
> *We embody different kinds of organizations for achieving many kinds of goals.*
> **The hierarchies of administration grown in accord with Papert's principle.**
> *When simple methods fail, we can build new levels of organization.*
> **The evolutionary vestiges of animals that still remain inside our brains.**
> *We use machinery evolved from fish, amphibia, reptiles, and earlier mammals.*
> **The sequence of stages of the growing child's personality.**
> *We accumulate different personalities that we can apply to different situations.*
> **The complex, ever-growing heritage of language and culture.**
> *We can use methods and ideas developed by millions of our ancestors.*
> **The subordination of thought processes to censors and suppressors.**
> *We do not need perfect methods, since we can remember how imperfect methods fail.*

Each of these dimensions gives you toughness and versatility. They offer alternative ways to proceed when any system fails. If part of your society of mind proposes to do what other parts find unacceptable, your agencies can usually find another way. Sometimes you merely need to turn to another branch of the same accumulation. When that fails, you can ascend to a higher level and engage a larger change in strategy. Then, even if an entire agency should fail, your brain retains earlier versions of it. This means that every facet of your personality may have the option to "regress" to an earlier stage, which already has proved itself competent to deal with the usual problems of life. Finally, when even that won't work, you can usually switch to an entirely different family of agencies. Whenever anything goes wrong, there are always other realms of thought.

APPENDIX

1 HEREDITY AND ENVIRONMENT

Sometimes we ask why people are so similar. At other times, we wonder why they differ so much from one to the next. Often we try to classify our differences into those with which we're born and those we later learn—and then we find ourselves arguing about which virtues are inherited, and which we acquire from experience. Most arguments about "nature vs. nurture" are based on two mistakes. One is to talk about intelligence as though a person's quality of mind were like a quantity one could pour into a cup. The other mistake is to assume that there is a clear distinction between *what* we learn and *how* we learn—as though experience had no effect on how we learn.

Chance plays a major role in human differences, because each of us starts out by drawing lots from among our parents' genes. A gene is a unit of heredity—a specific chemical whose structure affects some aspects of the construction of the body and the brain. We inherit each of our genes from one parent or the other, more or less by chance, so as to receive about half of each parent's genes. Within the population as a whole, each particular kind of gene has variants that work in somewhat different ways, and there are so many possible combinations of these alternative genes that every child is born unique—except in the case of identical twins, who carry identical copies of genes. One thing that makes people both so different and so similar is this: *we are similar because those alternative genes are usually quite similar—and we are different because those genes are not identical.*

Every cell of the body contains identical copies of all of that person's genes. But not all genes are active at the same time—and this is why the cells in our different organs do different things. When a particular gene is "turned on" inside a cell, that cell manufactures copies of a particular chemical (called a *protein*), whose structure is determined by the structure of that gene. Proteins are used in many ways. Some of them are assembled into permanent structures, some are involved in manufacturing other chemicals, and certain proteins move around in the cell, to serve as messages that alter other processes. Since certain combinations of these messages can turn other genes on and off, the gene-constructed chemicals in cells can act like small societies of agencies.

Every cell has windows in its walls and special genes that control which chemicals can enter or leave through those windows. Then, certain of *those* chemicals can act as messages that change the states of specific genes in *other* cells. Thus groups of cells can also be societies. The effects of most of those messages between cells are temporary and reversible, but some of them can permanently change the character of other cells by altering the types of messages they can transmit and receive. In effect, this converts them into other "types" of cells. When new types of cells are produced this way, certain of them remain in place, but other types proceed to move and reproduce—to form new layers, strands, or clumps. Inside the brain, certain types of cells emit specific chemicals that drift out like scents; this causes certain other types of mobile cells that are sensitive to those particular chemicals to sniff out those scents and track them back to their sources—leaving tubelike trails behind. These traces of the travels of those migratory cells then form the nerve-bundles that interconnect various pairs of far-apart brain-agencies. With all this activity, the embryonic brain resembles a complex animal ecology—which even includes predators programmed to find and kill the many cells that happen to reach "wrong" destinations.

All human brains are similar in size and shape but differ in many smaller respects because of different alternative genes. Why does the human population support so many variant genes? One reason is that genes are sometimes altered by accidents. When this happens to a gene that lies within a reproductive cell—that is, inside an ovum or a sperm—the change will be inherited. We call this a "mutation." Most often, a mutant gene will simply fail to manufacture some vital chemical, and this will so badly impair the offspring that natural selection will quickly remove the mutated gene from the population. But occasionally a mutant gene will endow those offspring with some substantial advantage. Then natural selection will spread copies of that gene so widely among the population that its predecessor gene becomes extinct. Finally, a variant gene may provide an advantage only in certain circumstances; this type of mutation may spread only to a certain proportion of the population, and both the new and the old variants will continue to coexist indefinitely. The richness of this reservoir of alternative genes can determine how quickly a population adapts to changes in ecological conditions—and thus determines whether the entire species can escape extinction over longer periods of time.

Now let's return to what genes do. Not all genes turn on at once; some start early and some start late. In general, the sooner a gene starts working, the greater its effect on what comes afterward. Accordingly, it is the early-starting genes that most affect the basic, large-scale architecture of our bodies and our brains. A mutation in an early-working gene is likely to cause such a drastic alteration of an animal's basic architecture that the embryo will not survive to be born, grow up, and reproduce successfully. Accordingly, most mutations of early-working genes are swiftly weeded out of the population by natural selection. Mutations in later-starting genes tend to cause less drastic differences, hence are not so swiftly weeded out and can accumulate in the population—for example, as variations in the genes that affect the sizes of connections between various brain-agencies. Every different combination of such variant genes produces a person with a somewhat different brain.

The early-starting genes thus frame the large-scale outlines of the brain—and their uniformity explains why people are so similar on the broadest scale. These must be the genes responsible for what we call "human nature"—that is, the predispositions every normal person shares. Generally, the uniformity of early-starting genes is what makes all the members of each animal species seem so similar; indeed, it is partly why the earth is populated with distinct, recognizable species like lions, turtles, and people, rather than with an indistinct continuum of all conceivable animals. No human mother ever bears a cat, since that would require too many different early-starting genes.

2 THE GENESIS OF MENTAL REALMS

All normal children come to recognize the same sorts of physical objects. Is that because the concept of an object is innate in the human mind? Each of us becomes attached to certain other individuals. Does this mean that the concept of person, and the notion of love, are part of our inheritance? Every human child forms "realms of thought" that represent the physical, possessional, and psychological. But how could genes build concepts into minds when genes themselves are merely linked-together chemicals?

The problem is that thoughts proceed on levels so far removed from those of chemicals. This makes it hard for genes, which are merely chemicals, to represent such things as objects, persons, or ideas—at least in anything like the way that strings of words express our thoughts. Then how *do* genes encode ideas? The answer lies in the concept of "predestined learning" discussed in section 11.7. Although groups of genes cannot *directly* encode specific ideas, they can determine the architecture of agencies that are *destined to learn* particular kinds of processes. To illustrate this principle, we'll outline the architecture of an agency destined to learn to recognize a human individual.

When we first introduced the concept of a recognizer, we suggested a simple way to represent a physical object in terms of properties like color, texture, size, and shape—by combining evidence from several agencies, each containing sensors especially designed to react to certain particular properties. Now we'll take another step, by dividing each of those agencies into two sections that are similar in architecture, and which both receive sensory inputs from the eyes, ears, skin, and nose. The first system is destined, as before, to learn to represent physical objects in terms of simple properties. However, because the second system's inputs come from different types of agents, it is destined to learn to represent "social objects"—that is, people.

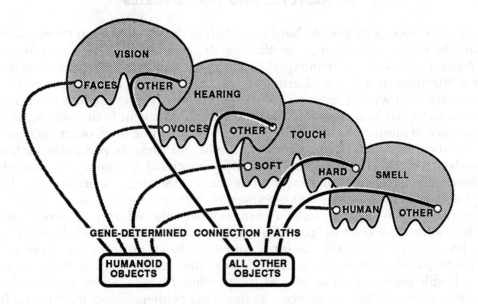

Our second "social object" agency takes all of its inputs from sensors that detect stimuli which usually signify the presence of a person—namely, human odors, voices, and faces. Because of this—and even though the genes that assemble this system know nothing of people —this system has no alternative but to learn to represent relations among the features of human individuals. Accordingly, this agency is destined to learn to recognize people!

The large-scale outline of this agency poses no engineering mystery—but we have to ask how genes could produce the sensory detectors that the system needs to do its job. There is ample evidence that the recognition of both voices and faces does indeed take place in special sections of the brain—since certain injuries of the brain leave their victims unable to distinguish voice sounds yet still able to recognize many other kinds of sounds, while other brain injuries destroy the ability to recognize faces but leave other visual functions intact. No one yet knows just how these recognition-systems work, but let's consider each in turn.

> ODOR RECOGNITION: *It is easy to build recognizers for particular odors because an odor is merely the presence of a certain combination of chemicals in the air, and a specific gene can make a cell sensitive to a particular chemical. So, to build agents for recognizing the odors of particular objects or people requires little more than connecting a variety of evidence-weighing agents to a variety of specific chemical detectors.*

VOICE RECOGNITION: *To distinguish the sounds of a human voice requires more machinery because vocal expressions are complicated sequences of events. Machines have been built that can make such distinctions.*

FACE RECOGNITION: *No one has yet been able to build vision machines that approach our human ability to distinguish faces from other objects—or even to distinguish dogs from cats. This remains a problem for research.*

In their first few days, human infants learn to distinguish people by their odors; then, over the next few weeks, they learn to recognize individuals by sound of voice; only after several months do they start to reliably distinguish the sights of faces. Most likely we learn to make each of these distinctions by several different methods, and it is probably no accident that these abilities develop in a sequence that corresponds to their increasing complexity.

3 GESTURES AND TRAJECTORIES

To recognize a voice or face seems hard enough; how does a child learn to recognize another person's mental state—of being angry or affectionate, for example. *One way is by distinguishing trajectories.* Just as we learn to interpret certain types of changes as representing the motions of objects in the physical realm, we learn to classify other types of changes as signifying mental events; these are what we call "gestures" and "expressions." For example, to identify a certain sound to be a particular language-word, some agencies inside your brain must recognize that a certain sequence of phonetic features has occurred. At the same time, other agencies interpret sequences of vocal sounds as having significance in other realms. In particular, certain types of vocal sounds are recognized as signifying specific emotional qualities. For example, people almost universally agree on which expressions seem most angry or imperative. In general, abruptly changing sounds evoke alarm—perhaps by inducing the sort of narrowing of interest that comes with pain; in any case, sudden changes in volume and pitch demand that we attend to them. In contrast, we react to "gentle" sounds in ways that people label "positive," as with affection, love, or reverence; the smoother time-trajectories do somehow serve to "calm us down," thus frequently inducing us to set aside our other interests. It's quite the same for sight and touch; hostile persons tend to jab and shout, while friendly people speak and wave with gestures and trajectories that we perceive as signifying gentleness and tenderness. Indeed, as shown in Manfred Clynes's book, *Sentics*, Doubleday, New York, 1978, people show similar emotional responses to certain types of time-trajectories regardless of the sensory domain. We consistently identify certain sudden, jerky types of action patterns as indicating anger—regardless of whether these are presented as visual motions, as envelopes of voice sounds, or as pushing, shoving tactile stimuli. In the same way, we consistently identify certain other, smoother action patterns to indicate affection. Clynes concludes that at least half a dozen distinct types of trajectories are universally associated with particular emotional states. What sort of brain-machinery could cause us to react in such similar ways to such different kinds of stimuli? I propose a three-part hypothesis. First, each of our sensory-agencies is equipped with special agents that detect certain types of time-trajectories. Second, the outputs of all the agents that detect similar "trajectory types" in different agencies are connected, through special connection bundles, to converge upon agents in some central "gesture-recognizing" agency. Finally, genetically established nerve-bundles run from each gesture-recognizing agent to a particular infantile "proto-specialist" of the sort described in section 16.3.

According to this hypothesis, each sensory-agency contains agents that are specialized to react to various types of temporal trajectories. For example, one kind might react only to stimuli that increase slowly and then decrease quickly; another kind might react only to signals that increase quickly and decay slowly. Inside the brain, although the agencies for hearing, sight, and touch lie far apart, the signals from agents that detect similar trajectories converge on a

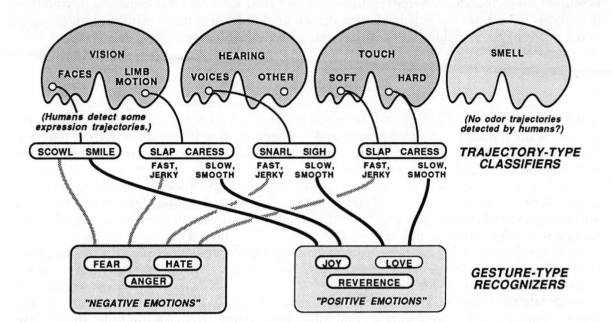

common agency composed of evidence-weighing agents. Notice that the architecture of this system is so similar to that of our "person-recognizing" agency that the two systems could form parallel layers; however, the destiny of each central "trajectory-type" agent is to learn to recognize, not a particular person, but a particular type of gesture or expression. For example, one such agent might learn to react in similar ways to a snarl, grimace, or shaken fist—and thus become an "anger-recognizing" agent whose function is "abstract" in the sense of being detached from any particular class of sensations.

To be sure, recognizing anger is not the same as comprehending or sympathizing with anger —nor does merely learning to make such a recognition, by itself, teach us to identify an "anger-type" trajectory of another person with our own personal experience of being angry. But if our genes equip us with connections from particular central trajectory-type agents to specific "proto-specialist" agencies, then each particular trajectory-type recognition would tend to activate a specific kind of emotional reaction.

Some of these connections could endow us with certain "empathies"—for example, to feel elated upon recognizing another person's joyous gestures. Other connections could make us become defensive at signs of aggression—or even aggressive at signs of weakness and withdrawal. There are innumerable examples, in animal behavior, of particular types of gestures evoking "instinctive" types of reactions, as when a sudden motion toward a bird provokes a fear-reaction flight. Surely our human genes provide us with a great deal of instinctive wiring. However, far more than any other kind of animal, we also have machinery that can bridge new agencies across the older ones, so that we can learn to bury ancient instincts under modern social disciplines.

We've seen how a gene-built agency could predispose us to use trajectory types to represent emotional and other sorts of states of mind. Once this is done, higher-level agencies could use the signals from trajectory-type agents to learn to recognize and represent more complex successions of mental states. In time, those representations could be assembled into models we could use for predicting and controlling our own mental processes. This illustrates how architectures framed by genes could serve our minds as stepping-stones toward learning how to think about ourselves.

As soon as you enter a certain room, you may experience the feeling that you can directly sense its history. Many people attribute such perceptions to imaginary influences with names like "intuitions," "spirits," "atmospheres," and "vibrations." Yet very likely all such perceptions come from inside the mind of those observers, as various mental agencies accomplish clever syntheses from clues derived from features and trajectories. In my view, believing in vibrations and ghosts diminishes our capabilities for mental growth by diverting attention from the mind and attributing those abilities to imaginary entities outside ourselves.

4 BRAIN CONNECTIONS

What possible sort of brain-machine could support a billion-agent society of mind? The human brain contains so many agencies and connections that it resembles a great nation of cities and towns, linked by vast networks of roads and highways. We are born with brain centers that are specialized for every sense and muscle group: for moving every eye and limb; for distinguishing the sounds of words, the features of faces, and all sorts of touches, tastes and smells. We're born with protospecialists involved with hunger, laughter, fear and anger, sleep, and sexual activity—and surely many other functions no one has discovered yet—each based upon a somewhat different architecture and mode of operation. Thousands of different genes must be involved in laying out these agencies and the nerve-bundles between them—and those brain-growth genes must generate at least three kinds of processes. Those genetic systems first must form the clumps and layers of brain cells that eventually become groups of agents; they must dictate the inner workings of those agencies; and, finally, they must determine the sizes and destinations of the nerve-bundles that interconnect those agencies—in order to constrain "who talks to whom" within each mind-society.

Now every population will include some variants among the genes that shape those highways in the brain, and this must influence their bearers' potential styles of thought. A person born with unusually sparse connections between the agencies for sight and speech might develop powerful machinery in both those realms but find it hard to make direct connections between them. On the surface, that might seem to constitute a disability. However, it might also lead to an advantage—if it served to force one's higher-level agencies to seek out indirect connections that lead to more articulate ways to represent reality. Similarly, one might suppose there would be advantages in having an uncommonly large capacity for short-term memory. Yet for all we know, our evolution has disfavored that because it tends to lead to less effective use of hard-learned long-term memories. Other differences in how we think could stem from variations in connection paths. An individual whose K-lines had more branches than usual might become inclined to assemble larger-than-usual accumulations in cases where a person whose memory-agents had fewer branches might be more disposed toward building uniframes. But the same genetic disposition can lead to different styles of thought: one person who is genetically disposed toward making uniframes might succumb to the chronic use of superficial stereotypes, while another person similarly endowed might compensate by building more deeply layered agencies that lead to more profound ideas. Although each particular variation will dispose each individual toward certain traits of personality, the final effect of any gene depends upon how it interacts with the structures built by other genes—and upon countless other accidents. This makes it almost meaningless to ask which particular genes lead to "good" forms of thought. It is better to think of a developing brain as a forest within which many different creatures grow, in conflict and in harmony.

Let's return to the architecture of machines that could hold societies of mind. How complicated this must be depends in part upon how many agents must be active at each moment. We can clarify the problem by considering two extremes. If only a few agents need to work at any time, then even an ordinary, serial, one-step-at-a-time computer could support billions of such agents—because each agent could be represented by a separate computer program. Then the computer itself could be quite simple, provided it has access to enough memory to hold all those little programs. On the other hand, no such arrangement would suffice to simulate societies of mind in which each of billions of agents constantly interact with all the others, all at once, because that would need more wires than any animal could carry in its head. I suspect that the human brain works somewhere in between; we do indeed have billions of nerve cells working at the same time, but relatively few of them have any need to communicate with more than a small proportion of the rest; this is simply because most agents are too specialized to deal with many types of messages. Accordingly, we'll propose an architecture that lies between those serial and parallel extremes—namely, a compromise in which a typical agent has comparatively few direct connections to other agents but can still influence a great many other agents through several indirect steps. For example, we can imagine a society in which each of a billion agents is connected to thirty other agents, selected at random. Then most pairs of agents should be able to communicate through merely half a dozen intermediates! This is because a typical agent can reach thirty others in one step, a thousand others in two steps, and a million others in only four steps. Thus a typical agent could reach any of the other billion agents in only six or seven steps!

However, randomly selected connections would not be very useful, because very few randomly selected pairs of agents would have any messages that might be useful to one another. When we actually examine the human brain, we find that connections between cells are not made either uniformly or randomly. Instead, within any typical small region, we see a great many direct connections between nearby cells but only a relatively small number of bundles of connections to other regions of cells that lie farther away. Here is an idealized representation of this arrangement:

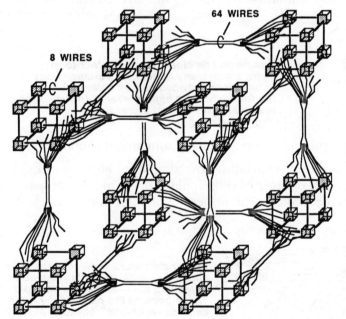

64 WIRES

8 WIRES

Here, 8 agents make a little cube, and 8 such cubes make a 64-agent supercube.

If we join 8 of these supercubes, we'll have 512 agents. And if we repeat this cube-on-cube pattern ten times, the resulting supercube will contain a billion agents!

But if we link each agent to 30 others instead of only 6, then each agent could communicate with a billion others in only 6 steps.

An embryonic brain might assemble such a structure by repeating a sequence of cell divisions and migrations perhaps half a dozen times. If only that were done, the resulting structure would be uselessly repetitive. However, in a real brain's growth, this underlying building plan is modified at every step by many other processes, and this produces many agencies that are similar in general form but different in specific details. Some of these gene-controlled interventions modify the properties of specific layers and clumps of cells, and this determines the internal workings of particular agencies. Other interventions affect the sizes and destinations of the nerve-bundles that interconnect particular pairs of agencies. Such highway-routing processes could be used, for example, to lead the nerves that emerge from the trajectory-type sensors in different agencies to the same central destination. This would be quite easy to arrange because the trajectory agents of similar types would tend to have similar genetic origins —and that would predispose them to be able to "smell" the same varieties of embryonic message chemicals and thus grow toward the same destination.

The same genetic argument can be applied to other aspects of a child's development—for example, to why all children seem to grow such similar Societies-of-More. When we discussed Jean Piaget's experiments, we left it as a mystery how children form the agencies called *History* and *Appearance*. What leads all those different minds to similar conceptions of comparisons? In section 10.7 we hinted that this might happen because similar agents like *Tall* and *Thin* originate in related sections of the brain. Consider that despite the fact that we do not know the brain-machinery for agents like *Tall* and *Thin*, we can be virtually sure that they are similar internally, because they both respond to the same sorts of spatial differences. Therefore, they almost surely have a common evolutionary origin and are constructed by the same or similar genes. Consequently, the embryonic brain cells that form these agents will tend to have similar "senses of smell" and are therefore likely to send out nerves that converge upon the same (or similar) agencies. From this point of view, the formation of a *Spatial* agency on which such properties converge need not be an unlikely chance event, but could be virtually predestined by inheritance.

Papert's principle requires many agencies to grow by inserting new layers of agents into older, already working systems. But this poses a problem because, once brain cells reach maturity, they no longer have much mobility. Consequently, adding new layers to old agencies must involve using brain cells in other locations. As far as we know, the only way this could be done is by using connections already available in the neighborhood of the original agency. Here's one way embryonic cells could provide frameworks for future multilayered mind-societies:

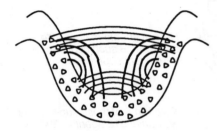

Agents that lie close to one another form clumps with many direct connections. Longer connections between nearby clumps form the foundation for higher-level agencies. This is repeated on several increasingly larger scales.

Any agency that is potentially capable of expanding to assimilate a lifetime of experience would need more space than any clump or layer of cells could provide in any compact neighborhood. This must be why the cerebral cortex—the newest and largest portion of the brain—evolved its convoluted form.

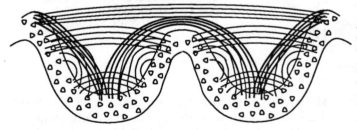

As our ancestors evolved, certain genes appeared, to cause certain agencies to bulge and fold—and bulge and fold again. These became the parts of the brain we call "convolutions" because of their appearance. They form early in life and may constrain how large each division of the mind is likely to grow.

If the connections in the cortex of the brain develop this way, through sequences of cell migrations, it could provide each local neighborhood with potential access to several other areas, through fanlike bundles and arrays of nerves. I have the impression the human cortex becomes thus folded upon itself perhaps five or six times, so that agents in each neighborhood have potential access to several other levels of convolution. This makes it possible for a typical agent to become connected to millions of other agents through only a few indirect connections. Presumably, only a small minority of cells ever actually acquire many such connections for their own exclusive use; however, such an arrangement makes any particular group of cells potentially capable of acquiring more significance—for example, by gaining control of a substantial bundle of connections that comes to represent some useful microneme. In its evolutionary course of making available so many potential connections, the human brain has actually gone so far that the major portion of its substance is no longer in its agencies but constitutes the enormous bundles of nerve fibers that potentially connect those agencies. The brain of Homo sapiens is mainly composed of cabling.

5 SURVIVAL INSTINCT

Many people seem to think that living things are born endowed with built-in instincts to survive. And certainly all animals do many things to stay alive. They build defenses against threats; they reproduce at any cost; they shy away from extremes of cold or heat, and from unfamiliarity. Now it usually is sensible, when one sees similarities, to seek some common cause for them. But I'll argue that it's usually wrong to seek a common *force*. There are many different reasons why animals do many things that help keep them alive—and, as we'll see, there is even a reason why there are so many different reasons. But to attribute this to any single, central force or to some basic, underlying survival instinct is as foolish as believing in special powers that attract corpses to cemeteries or broken cars to scrapyards.

No animal requires any central reason to survive, nor does the process of evolution itself require any reason to produce all those survival aids. On the contrary, evolution's versatility stems from its very lack of any fixed direction or constraint that might restrict its possibilities.

To understand why animals survive, one must see evolution as a sieve—that only passes through its mesh those animals who leave more offspring than the rest.

Many people also think that evolution favors life—although it is a painful fact that most mutated animals must die before they reproduce. But hindsight makes us tend to count only the survivors we see, while overlooking all the misfits that have disappeared; it is the same mistake that one might make from looking only at the sky—to then conclude that all the animals were birds. The animals we see today are precisely those whose ancestors accumulated a great many survival aids—and *that* is why so much of their behavior seems directed toward promoting their welfare—if only in the surroundings in which their ancestors evolved. It is an illusion that all those accumulated mechanisms have anything in common; actually, that seeming uniformity has no coherence of its own: it is nothing but the shadow of that evolutionary sieve. The myth of an underlying survival instinct explains nothing that cannot better be explained without it, and blinds us to the fact that each of those survival aids may exploit an entirely different mechanism.

I certainly don't mean to deny that people learn to love life and to fear death. But this is no simple matter of obeying some elemental instinct. It involves the development over many years of elaborate societies of concepts. Nor do I mean to say that people are born without any instincts at all and must learn everything from experience. On the contrary, we start with many built-in fragments of machinery, and these predestine us to learn to shy away from diverse forms of pain, discomfort, insecurity, and other forms of bodily and mental harm. But compared to those instinctive fears, the state of nonexistence we call death is a far more strange and difficult idea, of which no infant can conceive.

6 EVOLUTION AND INTENT

Could animals have evolved as they did, if "nature" had no sense of goal? A century ago, the world of biologists split in two: on one side stood the "evolutionists," who held that animals evolve through nothing more than accidents of chance. Their antagonists were called the "teleologists"; they disbelieved that such excellent animals could evolve without any purposeful guidance. The evolutionists turned out to be right, for now we can watch small animals and plants evolve before our very eyes and, at a correspondingly slower pace, see similar developments in creatures that have longer lives. In fact, we can actually observe how random accidents to genes lead to the selective survival of particular individuals in various environments—without the faintest reason to suspect that any goals must be involved. So why *do* so many people feel that evolution must have purposes? I suspect that this belief is based on combining a sound insight about problem solving with an unsound image of how evolution works. For example, common sense tells us that a person might never hit upon a design for a flying machine entirely by trial and error, without having any goals or purposes. This leads one to suppose that nature, too, must be subject to that same constraint. The error comes from thinking of "nature" as being concerned with such problems as finding a way to make animals fly.

The trouble is that this confuses *uses* with *purposes*. For example, suppose one asked how birds evolved, while thoughtlessly assuming that feathers and wings evolved exclusively for use in flight. One would be confronted with a formidable problem, since any organ as complex as a wing would require too many different genes to ever appear by random chance.

So long as one's mind is fixed on flight, one might feel that the only solution is to find some flight advantage in each and every earlier stage that merely produced a protofeather or protowing too small and weak for actual flight. This is why so many antievolutionists demand that evolutionary advocates must fill in every imagined "gap" along a direct path toward a specified goal. However, once we abandon that fixed idea, it is easier to see how various intermediate developments could have provided those animals with advantages quite unrelated to flying. For example, the early ancestors of birds could have accumulated genes to manufacture various sorts of feathered appendages that helped to wrap those protobirds in body cloaks *that kept them warm*. This sort of fortuitous "preparation" unrelated to any goal of flight would have made it much more likely that other accidents, perhaps millions of years later, might have brought a few such elements together to lend some genuine aerial advantage to an animal already prone to making leaps.

Incidentally, I do not mean to say that evolutionary processes must *by their nature* be devoid of purposes. We can actually conceive of how machinery *could* exist inside an animal, to purposefully direct some aspects of its evolution in much the way a farmer can promote the evolution of chickens that bear more meat or sheep that grow more wool. Indeed, the reproductive machinery inside our cells has already evolved so as to produce variations that are more likely to be useful than would otherwise occur by purely random chance; this idea is explained in a brilliant essay by Douglas Lenat, entitled "The role of Heuristics in learning by Discovery," in *Machine Learning: An Artificial Intelligence Approach*, edited by R. Z. Michalski, J. J. Carbonell, and T. M. Mitchell; Tioga Publishing Co., Palo Alto, Calif., 1983. It is even conceivable that our genetic systems might even contain some forms of difference-engine-like machinery that, over very long periods of time, generate variations in a somewhat purposeful manner. To be sure, this is mere speculation, since no such system has yet been discovered.

In any case, one aftermath of the controversy with teleologists was that many scientists in other realms became so afraid of making similar mistakes that the very concept of purpose became taboo throughout science. Even today, most scientists regard it as an abomination to use "anthropomorphic" or "intentional" language in connection with anything but persons or higher animals. This burdened the science of psychology with a double-barreled handicap. On one side, it made psychologists regard many of their most important problems as outside the scope of scientific explanation. On the other side, it deprived them of many useful technical ideas—because such concept-words as "want," "expect," and "recognize" are among the most effective ever found for describing what happens in human minds. It was not until the "cybernetic revolution" of the 1940s that scientists finally realized there is nothing inherently unscientific about the concept of goal itself and that attributing goals to evolution was bad not because it was impossible, but simply because it was wrong. Human minds do indeed use goal-machinery, and there is nothing wrong with recognizing this and bringing technical theories about intentions and goals into psychology.

7 INSULATION AND INTERACTION

*The hardest thing to understand is why we can understand
anything at all.*
—ALBERT EINSTEIN

What hope is there for any human mind to understand a human brain? No one could ever memorize the whole of all its small details. Our only hope is in formulating their principles. It wouldn't be much use, in any case, to know how each separate part works and how it interacts with the rest—because that simply isn't practical. Even if you knew all those details, if someone asked you to describe—in *general terms*—how brains work and how they change, you would have no way to reply.

We usually like to think in positive terms about how various parts of systems interact. But to do that, we must first have good ideas about which aspects of a system do *not* interact—since otherwise there would be too many possibilities to consider. In other words, we have to understand *insulations* before we can comprehend interactions. To put this in a stronger form: *No complicated society would actually work if it really depended on interactions among most of its parts.* This is because any such system would be disabled by virtually any distortion, injury, or environmental fluctuation. Nor could any such society evolve in the first place.

The science of biology was itself shaped by the discovery of insulations. Plants and animals were scarcely understood at all until it was found that they were made of separate cells. Then little more was learned so long as scientists thought of cells as bags of fluid within which countless chemicals could freely interact. Today we know that cells are more like factories, containing systems that are kept apart by sturdy walls, with doors that open only to those substances that bear the proper keys. Furthermore, even within these compartments, most pairs of chemicals cannot interact except by permission of particular genes. Without those insulations, so many chemicals would interact that all our cells would die.

For the purposes of this book, I have emphasized highly insulated systems—that is, mechanisms in which different functions are embodied in different agents. However, it is important to put this in perspective. For example, in chapter 19, we drew a sharp distinction between memorizers and recognizers; this made it easy to explain those ideas. However, in section 20.9 we mentioned very briefly the idea of a "distributed memory," in which both those functions are combined in the same network of agents. Now I do not want the reader to take the brevity of that discussion to suggest the subject is not important. On the contrary, I suspect that most of the human brain is actually composed of distributed learning-systems and that it is extremely important for us to understand how they can work. It is possible to combine even more functions; for example, John Hopfield has demonstrated a single distributed network that not only combines memory and recognition, but also "correctly yields an entire memory from any subpart of sufficient size"—in other words, an agency that "closes the ring," much as described in section 19.10. See Hopfield's article in the *Proceedings of the National Academy of Science*, 79, p. 2554, 1982, or the book *Parallel Distributed Processing* by D. E. Rumelhart and J. L. McLelland, M.I.T. Press, 1986.

The advantages of distributed systems are not alternatives to the advantages of insulated systems; the two are complementary. To say that the brain may be *composed* of distributed systems is not the same as saying that it *is* a distributed system—that is, a single network in which all functions are uniformly distributed. I do not believe any brain of that sort could work, because the interactions would be uncontrollable. To be sure, we have to explain how different ideas can become connected to one another—but we must also explain what keeps our separate memories intact. For example, we have praised the power of metaphors that allow us to combine ideas from different realms—but all that power would be lost if all our metaphors got mixed! Similarly, the architecture of a mind-society must encourage the formation and maintenance of distinct levels of management by preventing the formation of connections between agencies whose messages have no mutual significance. Some theorists have assumed that distributed systems are inherently both robust and versatile, but actually those attributes are likely to conflict. Systems with too many interactions of different types will tend to be fragile, while systems with too many interactions of similar types will be too redundant to adapt to novel situations and requirements. Finally, distributed systems tend to lack explicit, articulated representations, and this makes it difficult for any such agency to discover how any other such agency works. Thus, if distributed memory-systems are widely used within our brains, as I suspect they are, that could be yet another reason for the shallowness of human consciousness.

8 EVOLUTION OF HUMAN THOUGHT

What are the origins of human thought? Today, we're almost sure that our closest living relatives branched out according to the diagram below. It shows that none of the species that still exist are directly descended from any of the others, but that they all share common ancestors, now long extinct.

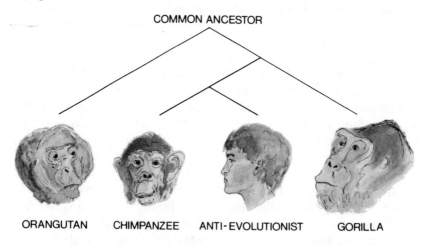

How different are we human beings from all the other animals? We recognize how similar those various brains and bodies are. But in view of our exceptional abilities to speak and think, we certainly seem to be unique. Could chimpanzees or gorillas ever learn to speak the way we do? Experience has shown that these wonderful animals can indeed learn to make connections among hundreds of different words and ideas, enabling them to produce speechlike strings of symbol-signs for expressing *Trans*-actions such as *"Put the candy into the box."* However, the same experiments appear to show that these animals find it much more difficult to construct language-strings in which the terminals of certain frames are filled with other filled-in frames. In other words, no one has succeeded in teaching these animals to use expressions that involve interruption clauses, such as *"Put the candy that is in the pail into the box."* To be sure, our inability to teach such things does not prove that these animals are inherently incapable of them. Still, no one can doubt that we have capabilities our ancestors did not possess. What sorts of brain developments could have given rise to our new and mighty forms of thought? Here are some possible candidates:

> *The capacity to attach new K-lines to old ones enabled us to build hierarchical memory-trees.*

> *The availability of more versatile temporary memories enabled us to pursue subgoals and to tolerate more complicated kinds of interruptions.*

> *The evolution of paranomes—that is, of isonomes that bridge across multiple realms—enabled us to examine the same problem from several viewpoints.*

> *The emergence of additional layers of agents allowed each child to grow through more stages of development.*

None of those advances by itself would seem to pose any special evolutionary difficulty. But what could have caused so many changes to have appeared so rapidly? Our ancestors diverged from their relatives, the gorillas and the chimpanzees, only a few million years ago, and our human brains have grown substantially in only the last few hundred thousand years. Little is known of what happened in that interval because we have found very few fossil remains of our ancestors. (This could be partly because their population was never very large but could also be because they had become too smart to permit themselves to be fossilized.) The evolutionary interval was so brief that most of our genes and brain structures remain nearly the same as those of the chimpanzees. Was it merely an increase in the brain's size and capacity that produced our new abilities? Consider that, by itself, an increase in the size of the brain might only cause the disadvantage of mental confusion and the inconvenience of a heavier head. However, if we first evolved significant advances in the ability to *manage* our memories, we could then take advantage of more memory. Similarly, inserting new layers of agents into old agencies might only lead to bad results—unless this were preceded by mechanisms for using such layers as "middle-level managers" without disrupting older functions. In other words, our evolution must have worked the other way: first came enhancements in abilities that made it feasible for us to manage larger agencies. Then, once we had the capability for *using* more machinery, natural selection could favor those who grew more massive brains.

POSTSCRIPT AND ACKNOWLEDGMENT

Never speak more clearly than you think.
—JEREMY BERNSTEIN

This book assumes that any brain, machine, or other thing that has a mind must be composed of smaller things that cannot think at all. The structure of the book itself reflects this view: each page explores a theory or idea that exploits what other pages do. Some readers might prefer a more usual form of story plot. I tried to do that several times, but it never seemed to work; each way I tried to line things up left too many thoughts that would not fit. A mind is too complex to fit the mold of narratives that start out *here* and end up *there*; a human intellect depends upon the connections in a tangled web—which simply wouldn't work at all if it were neatly straightened out.

Many psychologists dream of describing minds so economically that psychology would become as simple and precise as physics. But one must not confuse reality with dreams. It was not the ambitions of the physicists that made it possible to describe so much of the world in terms of so few and simple principles; that was because of the nature of our universe. But the operations of our minds do not depend on similarly few and simple laws, because our brains have accumulated many different mechanisms over aeons of evolution. This means that psychology can never be as simple as physics, and any simple theory of mind would be bound to miss most of the "big picture." The science of psychology will be handicapped until we develop an overview with room for a great many smaller theories.

To assemble the overview suggested in this book, I had to make literally hundreds of assumptions. Some scientists might object to this on the ground that successful sciences like physics and chemistry have found it more productive to develop theories that make the fewest assumptions, eliminating everything that does not seem absolutely essential. But until we have a more coherent framework for psychology, it will remain too early for the task of weeding out unproved hypotheses or for trying to show that one theory is better than another—since none of our present-day theories seem likely to survive very long in any case. Before we can have an image of the forest of psychology, we'll have to imagine more of its trees and restrain ourselves from simplifying them to death. Instead, we have to make ourselves complicated enough to deal with what is actually there.

It is scarcely a century since people started to think effectively about the natures of the brain-machines that manufacture thoughts. Before that, those who tried to speculate about this were handicapped on one side by their failure to do experiments, particularly with young children, and on the other side by their lack of concepts for describing complicated machinery. Now, for the first time, mankind has accumulated enough conceptual tools to begin comprehending machines with thousands of parts. However, we are only beginning to deal with machines that have millions of parts and we have barely started to acquire the concepts that we'll need to understand the billion-part machines that constitute our brains. New kinds of problems always arise when one encounters systems built on larger, less familiar scales.

Since most of the statements in this book are speculations, it would have been too tedious to mention this on every page. Instead, I did the opposite—by taking out all words like "possibly" and deleting every reference to scientific evidence. Accordingly, this book should be read less as a text of scientific scholarship and more as an adventure story for the imagination. Each idea should be seen not as a firm hypothesis about the mind, but as another implement to keep inside one's toolbox for making theories of the mind. Indeed, there is a sense in which that can be the only realistic way to think about psychology—since every particular person's mind develops as a huge machine that grows in a somewhat different way. *Are minds machines?* Of that, I've raised no doubt at all but have only asked, *what kind of machines?* And though most people still consider it degrading to be regarded as machines, I hope this book will make them entertain, instead, the thought of how wonderful it is to be machines with such marvelous powers.

Scientists like to credit those who first discovered each idea. But the central concept of this book, that the mind is a society of many smaller mechanisms, involved so many years of work to bring it to its present form that I can mention only a few of the people who had the most influence on it. In this research I shared the greatest privilege a human mind can have: to work on new ideas together with the foremost intellects of one's time. As a student at Harvard, I immersed myself in mathematics and psychology and attached myself to two great young scientists, the mathematician Andrew Gleason and the psychologist George A. Miller. This was the era of the scientific movement that was later called cybernetics, and I was especially entranced with the works of Nicholas Rashevsky and of Warren McCulloch, who were making the first theories of how assemblies of simple cell-machines could do such things as recognize objects and remember what they'd seen. By the time I started graduate school in mathematics at Princeton in 1950, I had a clear enough idea about how to make a multi-agent learning machine. George Miller obtained funds for building it; this was the *Snarc* machine of chapter 7. Constructed with the help of a fellow student, Dean Edmonds, it managed to learn in certain ways, but its limitations convinced me that a more versatile "thinking machine" would have to exploit many other principles.

My teachers in the golden age of mathematics at Princeton were not particularly interested in psychology, but the ways of thought are more important than the subject matter, and I learned new mental strategies from Albert Tucker, Ralph Fox, Solomon Lefshetz, John Tukey, Salomon Bochner, and John von Neumann. I learned even more from my own generation of students at Princeton: particularly from John Nash, Lloyd Shapley, Martin Shubik, and John McCarthy. In 1954 I returned to Harvard as a Junior Fellow of the Harvard Society of Fellows, with no obligation but to pursue whatever goal seemed most important. There seemed no way to get around the apparent limitations of low-level, distributed-connection learning machines, so I turned toward a new theory being pioneered by Ray Solomonoff, about generalizing from experience. I attached myself to Warren McCulloch and Oliver Selfridge, with whom I worked most closely of all until becoming a professor of mathematics at MIT. It was from them that I derived my image of how to make a laboratory work.

In 1959, John McCarthy came to MIT from Dartmouth, and we started the MIT Artificial Intelligence Project. We agreed that the most critical problem was of how minds do commonsense reasoning. McCarthy was more concerned with establishing logical and mathematical foundations for reasoning, while I was more involved with theories of how we actually reason using pattern recognition and analogy. This combination of theoretical and practical research attracted students of great ability, and our laboratory had an atmosphere that combined mathematical power with engineering adventure; this led not only to new theories of computation, but also to developing some of the very first automatic robots. In 1963, McCarthy left to start a new AI laboratory at Stanford, and now there were three principal centers of research in

Artificial Intelligence, including the one that Allen Newell and Herbert Simon had started earlier at Carnegie-Mellon University. A fourth center soon emerged at Stanford Research Laboratory, and we all worked closely together.

The money to support the people and equipment for this work came mainly from an office of the Advanced Research Projects Agency concerned with information processing technology. This office was directed, in effect, by the scientists themselves, initially by Dr. J.C.R. Licklider, who had been my teacher and friend when I was a student at Harvard. Licklider had already organized a research center at the Bolt, Beranek, and Newman company in Cambridge, Massachusetts, and McCarthy and I and several of our students had worked closely with that group for several years. Later, when Licklider returned to become a professor at MIT, the Information Processing Technology Office was taken over successively by Lawrence G. Roberts and Ivan Sutherland (who had been students of ours at MIT) and then by Robert Taylor and Robert Kahn—all of whom made important intellectual contributions. The actual details of all these research contracts were managed in the Office of Naval Research by Marvin Denicoff, whose vision of the future had a substantial influence on the entire field. My own research was supported by the ONR over an even longer period, since it had previously financed my graduate studies in topology at Princeton, and, subsequently, Denicoff's successor, Alan Meyrowitz, supported my research during the completion of this book.

Jerome Wiesner and Philip Morse of MIT obtained the resources for our first laboratory. Our development at MIT was encouraged by William Ted Martin, Norman Levinson, Witold Hurewicz, Norbert Weiner, Claude Shannon, Peter Elias, and Robert Fano. I was given the privilege of sharing with Shannon the endowed chair of Donner Professor of Science at MIT and enjoyed the support of many other people and organizations over the years: John Williams, Paul Armer, and Merril Flood enabled me to work with Newell, Shaw, and Simon at the Rand Corporation; Oliver Selfridge and Gerald Dinneen encouraged research at MIT's Lincoln Laboratory; Michel Gouilloud supported my work from the Schlumberger Corporation; Edward David provided support from Exxon; and Alan Kay supported many of our students with funding from (successively) the Xerox, Atari, and Apple corporations. For several years, the Thinking Machines Corporation has supported both this research and the development of a new type of computer called the Connection Machine—designed by my student Danny Hillis for embodying societies of mind.

Most of all, I want to acknowledge the contributions to this book of Seymour Papert, who came to MIT in 1963 after five years of studying child development with Jean Piaget in Geneva. Papert and I worked so well together that for a decade we supervised the laboratory jointly, each able at any time to leave the other to decide what should be done. Together we evolved new mathematical techniques, designed laboratory experiments, built computer hardware and software, and supervised the same students. Such a partnership could not have worked so well had we not both developed in similar intellectual directions before we met; we were both involved with the same areas of mathematics, with similar concerns about machinery, and with similar attitudes about psychology. One of our projects was to build a machine that could see well enough to use mechanical hands for solving real-world problems; this was the origin of *Builder* and the insights that emerged from it. In trying to make that robot see, we found that no single method ever worked well by itself. For example, the robot could rarely discern an object's shape by using vision alone; it also had to exploit other types of knowledge about which kinds of objects were *likely* to be seen. This experience impressed on us the idea that only a society of different kinds of processes could possibly suffice. Papert and I worked together not only on robotic machines, but in many other areas; for example, we spent several years developing a new mathematical theory for the then mysterious Perceptron type of learning machine. In the middle 1970s Papert and I tried together to write a book about societies of mind but abandoned the attempt when it became clear that the ideas were not mature enough. The results of that collaboration shaped many earlier sections of this book.

Eventually Papert and I both turned away from large-scale scientific enterprises toward somewhat different individual goals, and we imposed the directorship of our laboratory upon one of our most original and productive students, Patrick Winston—who first worked out the idea of making uniframes. Papert went on to develop a host of new theories about mental development and education; these led to the computer language LOGO and to many other concepts that started to enter the mainstream of educational thought over the next decade. I focused my concern on searching for better theories about the little world of how a child might learn to build with blocks. The parts of the puzzle that form this book began to fit together in my mind in the mid-1970s, around the concept of frame-array, and this eventually led to the theories about communication-lines, K-lines, and level-bands, and then, during the book's final stages, to the ideas about pronomes, polynemes, and cross-realm correspondences.

As for the manuscript itself, Bradley Marx read through every draft, comparing each version with earlier ones, helping to maintain clarity, stylistic coherency, and especially protecting good ideas from destructive revisionary impulses. This was hard because the early manuscript was more than twice its present length. Robin Lakoff suggested neutering the English; this seemed at first impossible but soon became quite natural. Theodore Sturgeon reviewed an early draft; I wish he had lived to see it now. Kenneth Haase, Betty Dexter, and Tom Beckman made innumerable suggestions and corrections. Successive drafts were reviewed by Danny Hillis, Steve Bagley, Marvin Denicoff, Charlotte Minsky, Michel Gouilloud, Justin Lieber, Philip Agre, David Wallace, Ben Kuipers, Peter de Jong, and Sona Vogel. Richard Feynman contributed a variety of insights about memory and parallel processing. David Yarmush helped to organize the book into sections, to smooth out the transitions, and to establish the gradient wherein the words begin with commonsense meanings and gradually become more technical. Bob Whittinghill made many suggestions about language as well as about psychology. Douglas Hofstadter evaluated the entire theory, forcing me to make several substantial changes. Michael Crichton made many technical suggestions and helped me to refine the early chapters.

Russell Noftsker and Tom Callahan made substantial engineering contributions to our work. Hosts of ideas came from students at MIT, notably Howard Austin, Manuel Blum, Danny Bobrow, Eugene Charniak, Henry Ernst, Tom Evans, Scott Fahlman, Ira Goldstein, William Gosper, Richard Greenblatt, Adolfo Guzman, Kenneth Haase, William Henneman, Carl Hewitt, Danny Hillis, Jack Holloway, Tom Knight, William Martin, Joel Moses, Bertram Raphael, Larry Roberts, James Slagle, Jerry Sussman, Ivan Sutherland, David Waltz, Terry Winograd, Patrick Winston, and many others. Countless other thoughts came from working at various times with Maryann Amacher, Gregory Benford, Terry Beyer, Woodrow Bledsoe, Mortimer Casson, Edward Feigenbaum, Edward Fredkin, Arnold Griffith, Louis Hodes, Berthold Horn, Joel Isaacson, Russell Kirsch, David Kirsh, Robert Lawlor, Justin Leiber, Douglas Lenat, Jerome Lettvin, David MacDonald, Curtis Marx, Hans Moravec, Stewart Nelson, Nils Nillsson, Donald Norman, Walter Pitts, Jerry Pournelle, Charles Rosen, Carl Sagan, Roger Schank, Robert Sheckley, Stephen Smoliar, Cynthia Solomon, Ray Solomonoff, Luc Steels, Warren Teitelman, and Graziella Tonfoni. I wish I could acknowledge the inspirations of all the friends of earlier years, particularly W. Ross Ashby, Thomas Etter, Ned Feder, Heinz von Foerster, Donald Hebb, John Hollander, Arnold Honig, Gordon Pask, Roland Silver, Jan Syrjala, Carroll Williams, Bertram Wolfe, David Yarmush—and of all the teachers of my youth, particularly Dudley Fitts, Ruth Gordon, Alexander Joseph, Edward Lepowsky, and Herbert Zim. My development was also strongly influenced first by the writing and later by the friendship of Arthur C. Clarke, Robert Heinlein, Frederick Pohl, and most of all by Isaac Asimov.

Of course, the deepest influence on my style of thought came from my parents, Henry Minsky and Fannie Reiser. My wife, Gloria Rudisch, our children Margaret, Henry, and Juliana (who drew the illustrations and sometimes changed the text to make them fit), and my sister Ruth all helped to shape this book. My sister Charlotte also lives between these lines, for even in our childhood, she was a powerful artist and critic, and her dreams became the meanings of my ordinary words.

GLOSSARY AND BIBLIOGRAPHY

Because I thought this theory of the mind might interest not only specialists but everyone who thinks, I favored ordinary words over the technical language of psychology. This was rarely any sacrifice because so many psychological terms already stood for obsolete ideas. But since I also wished to speak to specialists, I tried to hide more technical ideas between the lines; I hope this second level does not show. However there still were certain points at which no ordinary words seemed satisfactory, and I had to invent new terms or assign new meanings to old ones.

Accumulation (12.6) A type of learning based on collecting examples of an idea without attempting to describe what they have in common. Contrast with *Uniframe*.

Agency (1.6) Any assembly of parts considered in terms of what it can accomplish as a unit, without regard to what each of its parts does by itself.

Agent (1.4) Any part or process of the mind that by itself is simple enough to understand—even though the interactions among groups of such agents may produce phenomena that are much harder to understand.

Artificial Intelligence (7.4) The field of research concerned with making machines do things that people consider to require intelligence. There is no clear boundary between psychology and Artificial Intelligence because the brain itself is a kind of machine. For an introduction to this field, I recommend Patrick Winston's textbook *Artificial Intelligence*, Addison-Wesley, 1984. For more connections with psychology, see Roger Schank and Kenneth Colby (eds.), *Computer Models of Thought and Language*, Freeman, 1973. Some influential early ideas about brains and machines can be seen in Warren McCulloch's *Embodiments of Mind*, MIT Press, Cambridge, Mass., 1966. See *Intelligence*.

Attachment Learning (17.2) The specific theory, proposed in this book, that the presence of someone to whom we are emotionally attached has a special effect on how we learn, especially in infancy. Attachment learning tends to cause us to modify our goals—rather than merely improve our methods for achieving the goals we already have.

***B*-Brain** (6.4) Any part of the brain connected not to the outside world, but only to another part of the same brain. Like a manager, a *B*-brain can supervise an *A*-brain without understanding either how the *A*-brain works or the problems with which the *A*-brain is involved—for example, by recognizing patterns of activity that indicate the *A*-brain is confused, wasting time in repetitive activity, or focused on an unproductive level of detail.

Block-Arch (12.1) A scenario adapted from Patrick Winston's doctoral thesis, "Learning Structural Descriptions by Examples," in *The Psychology of Computer Vision*, P. H. Winston (ed.), McGraw-Hill, 1975. The study of the world of children's building-blocks may at first seem childishly simple, but it has been one of the most productive areas of research about Artificial Intelligence, child psychology, and modern robotics engineering.

Censor (27.2) An agent that inhibits or suppresses the operation of other agents. Censorlike agents are involved with how we learn from our mistakes. This idea played a prominent role in Freud's theories but has been virtually ignored by modern experimental psychologists—presumably because it is hard to study what people do *not* think. See Freud's 1905 book *Jokes and Their Relation to the Unconscious*. I suspect censorlike agents may constitute the larger part of human memory. The discussion of censors and jokes in chapter 27 is based on my essay "Jokes and Their Relation to the Cognitive Unconscious," published in *Cognitive Constraints on Communication, Representations and Processes*," L. Vaina and J.K.K. Hintikka (eds.), Reidel, 1981. See *Suppressors*.

Challenger, Professor (4.4) A rival of mine, disguised as the treacherous archaeologist in Arthur Conan Doyle's novel *The Lost World*, who resembles Sherlock Holmes's nemesis, the mathematician Moriarty, except for being somewhat more honorable.

Closing the Ring (19.10) A technique by which an agency can recall many details of a memory from being given only a few "cues."

Common Sense (1.5) The mental skills that most people share. Commonsense thinking is actually more complex than many of the intellectual accomplishments that attract more attention and respect, because the mental skills we call "expertise" often engage large amounts of knowledge but usually employ only a few types of representations. In contrast, common sense involves many different kinds of representations and thus requires a larger range of different skills.

Computer Science (6.8) A science still in its infancy. While other sciences study how particular types of objects interact, computer science studies how interactions work in general—that is, how societies of parts can accomplish what those parts cannot do separately. Although computer science began with the study of serial computers—that is, of machines that could do only one thing at a time—it has grown to the point of studying the sorts of interconnected networks of processes that must go on inside societies of mind. (For an introduction to the theory of single-process machines, see my book *Computation: Finite and Infinite Machines*, Prentice-Hall, 1967.)

Consciousness (6.1) In this book, the word is used mainly for the myth that human minds are "self-aware" in the sense of perceiving what happens inside themselves. I maintain that human consciousness can never represent what is occurring at the present moment, but only a little of the recent past—partly because each agency has a limited capacity to represent what happened recently and partly because it takes time for agencies to communicate with one another. Consciousness is peculiarly hard to describe because each attempt to examine temporary memories distorts the very records it is trying to inspect. The description of consciousness in section 6.1 was adapted from my epilogue to Vernor Vinge's novel *True Names*, Bluejay Books, New York, 1984.

Context (20.2) The effect upon one's state of mind of all the influences present at the time. At each moment, the context within which each agency works is determined by the activity of the nemes that reach that agency. See *Neme*.

Cross-Exclusion (16.4) An arrangement in which each of several agents is connected so as to inhibit all the others—so that only one of them can remain active at a time.

Cross-Realm Correspondence (29.4) A structure that has useful applications in two or more different *mental realms*. Such correspondences sometimes enable us to transfer knowledge and skill from one domain to another—without needing to accumulate experience in that other realm. This is the basis of certain important kinds of analogies and metaphors.

Creativity (7.10) The myth that the production of novel ideas, artistic or otherwise, comes from some distinctive form of thought. I recommend the discussion of this subject in the chapter "Variations on a Theme as the Crux of Creativity," in Douglas Hofstadter's *Metamagical Themas*, Basic Books, 1985.

Default Assumption (8.5, 12.12) The kind of assumption we make when we lack reasons to think otherwise. For example, we assume "by default" that an unfamiliar individual who belongs to a familiar class will think and act like a "typical" member of that class. Default assumptions are more than mere conveniences; they constitute our most productive way to make generalizations. Although such assumptions are frequently wrong, they usually do little harm because they are automatically displaced when more specific information becomes available. However, they can do incalculable harm when they are held too rigidly.

Demon (27.1) An agent that constantly watches for a certain condition and intervenes when it occurs. Our discussion of demons is partly based on Eugene Charniak's doctoral thesis, *"Toward a Model of Children's Story Comprehension,"* MIT, 1972.

Difference-Engine (7.8) An agency whose actions tend to make the present state of affairs more like some *goal* or "desired state" whose description is represented in that agency. This idea was developed by Allen Newell, C. J. Shaw, and Herbert A. Simon into an important theory about human problem solving. See G. Ernst and Allen Newell, *GPS, A Case Study in Generality and Problem Solving*, Academic Press, 1969.

Direction-Neme (24.6) An agent associated with a particular direction or region in space. I suspect that bundles of direction-nemes are used inside our brains for representing not only spatial locations and directions, but also for representing many nonspatial concepts. Direction-nemes resemble isonomes in spatial realms but more resemble polynemes in other realms. See *Interaction-Square* and *Frame-Array*.

Distributed Memory (20.9) A representation in which each fragment of information is stored, not by making a single, substantial change in one agent, but by making small changes in many different agents. Many theorists have been led to believe that the construction of distributed memory-systems must involve "nondigital" devices such as holograms; that this is not so was shown by P. J. Willshaw, O. P. Buneman, and H. C. Longuet-Higgins in "Non-Holographic Associative Memory," *Nature*, 222, 1969. See *Memorizers*.

Duplication Problem (23.2) The question of how a mind could compare two similar ideas without possessing two identical agencies for representing both of them at the same time. This problem was never recognized in older theories of psychology, and I suspect it will be the downfall of most "holistic" theories of higher-level thought. See *Time Blinking*.

Emotion (16.1) A term used for too many different purposes. There is a popular view that emotions are inherently more complex and harder to understand than other aspects of human thought. I maintain that infantile emotions are comparatively simple in character and that the complexity of adult emotions results from accumulating networks of mutual exploitations. In adults, these networks eventually become indescribably complicated, but no more so than the networks of our adult "intellectual" structures. Beyond a certain point, to distinguish between the emotional and intellectual structures of an adult is merely to describe the same structures from different points of view. See *Proto-specialist*.

Exploitation (4.5) The act of one agency making use of the activity of another agency, without understanding how it works. Exploitation is the most usual relationship among agencies because it is so difficult for them to understand one another.

Exception Principle (12.9) The concept that it may not pay to change a well-established skill in order to accommodate an exception. The more one builds upon a certain foundation, the greater the disruption upon changing it. A system's growth will tend to cease, past the point at which the damage caused by any change outweighs the immediate gain. See *Investment Principle*.

Frame (24.2) A representation based on a set of *terminals* to which other structures can be attached. Normally, each terminal is connected to a *default assumption*, which is easily displaced by more specific information. Other ideas about frames that are not discussed within this book were published in my chapter "A Framework for Representing Knowledge," in *Psychology of Computer Vision*, P. H. Winston (ed.), McGraw-Hill, 1975. See *Picture-Frame*, Trans-*frame*.

Frame-Array (25.2) A family of frames that share the same terminals. Information attached to any terminal of a frame-array automatically becomes available to all the frames of that array. This makes it easy to change perspective, not only in regard to a physical viewpoint, but in other mental realms as well. Frame-arrays are often controlled by bundles of *direction-nemes*.

Functional Autonomy (17.4) The idea that specific goals can lead to subgoals of broader character.

For example, in order to please another individual, a child might develop more general goals of acquiring knowledge, power, or wealth—yet the very same subgoals might serve equally well an initial wish to injure that other individual. The term "functional autonomy" derives from Gordon Allport, who was one of my professors at Harvard.

Functional Definition (12.4) Specifying something in terms of how it might be used, rather than in terms of its parts and their relationships. See *Structural Definition*.

Generate and Test (7.3) Solving a problem by trial and error—that is, by proposing solutions recklessly, then rejecting those that do not work.

Genius (7.10) An individual of prodigious mental accomplishment. Although even the most outstanding human prodigies rarely develop even twice as quickly as their peers, many people feel that their existence demands a special explanation. I suspect that the answer is to be found not in the superficial skills such people learn, but in the early accidents that lead them into *learning better ways to learn*.

Gestalt (2.3) The unexpected emergence, from a complex system, of a phenomenon that had not seemed inherent in that system's separate parts. Such "emergent" or "collective" phenomena show that "a whole is more than the sum of its parts." However, further research usually shows that such phenomena can be explained completely, once we also take into account the interactions among those parts—as well as the peculiarities and deficiencies in the observer's own perceptions and expectations. There do not seem to be any important principles common to the phenomena that have been considered, from time to time, to be "emergent"—beyond the contemporary inability to understand them. Thus, "holistic" views tend to become scientific handicaps when they undermine our determination to extend the boundaries of our comprehension. See *Interaction*.

Goal (7.8) The representation in a *difference-engine* of an imagined final state of affairs. This definition of goal may at first seem too impersonal because it does not explain either the elation that comes with achieving a human goal or the frustration that accompanies failure. However, we should not expect to explain such complicated phenomena of adult psychology directly in terms of simple principles, since they also depend on many other aspects of our mental architecture. Basing our concept of goal on the *difference-engine* idea helps us to avoid the *single-agent fallacy* by permitting us to speak about a goal without needing to refer to the person who entertains that goal; a person's many agencies may each have different goals—without that person being "aware" of them.

Grammar-Tactic (22.10) An operation involved with speech that corresponds to a step in a process of constructing a mental representation. Grammar-tactics are not the same as "grammar rules," although these have a close relation. The difference is that grammar rules are both superficial and subjective—in the sense that they purport to describe regularities in one person's behavior as observed by someone else—while grammar-tactics are objective in the sense that they are defined to be the underlying processes that actually produce speech. Although it may be more difficult to discover just what those processes do, it is better to speculate on how language is produced and used than merely to describe its observed, external forms.

Holism (2.3) See *Gestalt*.

Homunculus (5.2) Literally, a tiny person. In psychology, the unproductive and paradoxical idea that a person's behavior depends upon the behavior of another personlike entity located deeper inside that person.

Interaction-Square (14.9) The idea of representing the interaction between two processes by linking pairs of examples to *direction-nemes*. We can use this same technique not only for representing spatial relationships, but for causal, temporal, and many other kinds of interactions. This makes the interaction-square idea a powerful scheme for representing *cross-realm correspondences*.

Interaction (2.1) The effect of one part of a system upon another part. It is remarkable that in the history of science virtually all phenomena have eventually been explained in terms of interactions between parts *taken two at a time*. For example, Newton's law of gravity, which describes the mutual attraction of two particles, enables us to predict the motions of all the planets, stars, and galaxies—without any need to consider three or more objects at a time! One could conceive of a universe in which whenever three stars formed an equilateral triangle, one of them would instantly disappear—but virtually no three-part interactions have ever been observed in the physical world.

Interruption (15.9) A term used in this book to refer to any process that can be suspended while the agency involved can do some other job—yet later return to where it left off. The ability to do this requires some sort of temporary memory. See *Recursion Principle*.

Intelligence (7.1) A term frequently used to express the myth that some single entity or element is responsible for the quality of a person's ability to reason. I prefer to think of this word as representing not any particular power or phenomenon, but simply all the mental skills that, at any particular moment, we admire but don't yet understand.

Introspection (6.5) The myth that our minds possess the ability directly to perceive or apprehend their own operations.

Intuition (12.10) The myth that the mind possesses some immediate (and hence inexplicable) abilities to solve problems or perceive truths. This belief is based on naive views of how we get ideas. For example, we often experience a moment of excitement or exhilaration at the moment of completing a complex and prolonged but nonconscious analysis of a problem. The myth of intuition wrongly attributes the solution to what happened in that final moment. As for being able directly to apprehend what is true, we simply forget how frequently our "intuitions" turn out wrong.

Investment Principle (14.6) The tendency of any well-developed skill to retard the growth of similar skills because the latter work less well in their early forms—and hence are used so infrequently that they never reach maturity. Because of this, we tend to invest most of our time and effort on elaborating a comparatively few techniques, rather than on accumulating many different ones. This can lead, at the same time, both to the formation of a coherent and effective personal style and to a decline in flexibility that may be wrongly attributed to aging. See *Exception Principle*.

Isonome (22.1) A signal or pathway in the brain that has similar effects on several different agencies.

K-Line (8.1) The theory that certain kinds of memories are based on turning on sets of agents that reactivate one's previous *partial mental states*. This idea was first described in my essay "K-lines: A Theory of Memory," *Cognitive Science*, 4 (2), April 1980.

Learning (7.5) An omnibus word for all the processes that lead to long-term changes in our minds.

Level-Band (8.5) The idea that a typical mental process tends to operate, at each moment, only within a certain range or portion of the structure of each agency. This makes it possible for one process to work on small details without disrupting other processes that are concerned with large-scale plans.

Logical Thinking (18.1) The popular but unsound theory that much of human reasoning proceeds in accord with clear-cut rules that lead to foolproof conclusions. In my view, we employ logical reasoning only in special forms of adult thought, which are used mainly to summarize what has already been discovered. Most of our ordinary mental work —that is, our commonsense reasoning—is based more on "thinking by analogy"—that is, applying to our present circumstances our representations of seemingly similar previous experiences.

Memorizer (19.5) An agent that can reset an agency into some previously useful state. See *Recognizer* and *Distributed Memory*.

Memory (15.3) An omnibus term for a great many structures and processes that have ill-defined boundaries in both everyday and technical psychology; these include what we call "re-membering," "re-collecting," "re-minding," and "recognizing." This book suggests that what these share in common is their involvement with how we reproduce our former *partial mental states*.

Mental State (8.4) The condition of activity of a group of agents at a certain moment. In this book we have assumed that every agent, at any moment, is either completely aroused or completely quiescent; in other words, we ignore the possibility of different degrees of arousal. This kind of "two-state" or "digital" assumption is characteristic of computer science and, at first, may seem too simplistic. However, experience has shown that the so-called analog theories that are alleged to be more realistic quickly become so complicated that, in the end, the simpler two-state models actually lead to deeper understandings—at least about basic principles. See *Partial Mental State*.

Metaphor (29.8) The myth that there is a clear distinction between representations that are "realistic" and those that are merely suggestive. In their book *Metaphors We Live By*, University of Chicago Press, 1980, Mark Johnson and George Lakoff demonstrate that metaphor is no mere special device of literary expression but permeates virtually every aspect of human thought.

Micromemory (15.8) The smallest components of our short-term memory-systems.

Microneme (20.5) A neme involved with agents at a relatively low level. See *Neme*.

Model (30.3) Any structure that a person can use to simulate or anticipate the behavior of something else.

Neme (25.6) An agent whose output represents a fragment of an idea or state of mind. The "context" within which a typical agent works is largely determined by the activity of the nemes that reach it. I called nemes "C-lines" in "Plain Talk About Neurodevelopmental Epistemology," in *Proceedings of the Fifth International Joint Conference on Artificial Intelligence*, Cambridge, Mass., 1977; the description in section 20.5 is also based on the idea of "microfeature" developed by David L. Waltz and Jordan Pollack in "Massively Parallel Parsing," *Cognitive Science*, 9 (1).

Nome (25.6) An agent whose outputs affect an agency in some predetermined manner, such as a *pronome, isonome,* or *paranome*; an agent whose effect depends more on genetic architecture than on learning from experience. The suffix *-nome* was chosen to suggest an atom-like, unchanging quality.

Noncompromise Principle (3.2) The idea that when two agencies conflict it may be better to ignore them both and yield control to yet another, independent agency.

Papert's Principle (10.4) The hypothesis that many steps in mental growth are based less on the acquisition of new skills than on building new administrative systems for managing already established abilities.

Paranome (29.3) An agent that operates on agencies of several different *mental realms* at once, with similar effects on all of them.

Partial Mental State (8.4) A description of the state of activity of some particular group of mental agents. This technical but simple idea makes it easy to understand how one can entertain and combine several ideas at the same time. See *Mental State*.

Perceptron (19.7) A type of recognition machine that learns to weigh evidence. Invented by Frank Rosenblatt in the late 1950s, Perceptrons use singularly simple procedures for learning which weights to assign to various fragments of evidence. Seymour Papert and I analyzed this type of machine in the book *Perceptrons*, MIT Press, 1969, and showed that the simplest kinds of Perceptrons cannot do very much by themselves. However, they can do much more when arranged into societies so that some of them can then learn to recognize relations among the patterns recognized by the others. It seems quite likely that some types of brain cells use similar principles.

Picture-Frames (24.7) A type of *frame* whose terminals are controlled by *direction-nemes*. Picture-frames are particularly suited to representing certain kinds of spatial information.

Polyneme (19.5) An agent that arouses different activities, at the same time, in different agencies—as a result of learning from experience. Contrast with *Isonome*.

Pronome (21.1) A type of agent associated with a particular "role" or aspect of a representation—corresponding, for example, to the *Actor, Trajectory,* or *Cause* of some action. Pronome agents frequently control the attachments of terminals of frames to other frames; to do this, a pronome must possess some temporary memory.

Proto-specialist (16.3) One of the genetically constructed subsystems responsible for some of an animal's "instinctive" behavior. Large portions of our minds start out as almost separate proto-specialists, and we interpret their activity as manifesting different, primitive emotions. Later, as agencies become more interconnected and learn to exploit one another, these differences grow less distinct. This conception is based on the societylike theory proposed by Niko Tinbergen in *The Study of Instinct*, Oxford University Press, 1951.

Puzzle Principle (7.3) The idea that any problem can be solved by trial and error—provided one already has some way to recognize a solution when one is found. See *Generate and Test*.

Realm, Mental (29.1) A division of the mind that deals with some distinct variety of concern by using distinct mechanisms and representations.

Recognizer (19.6) An agent that becomes active in response to a particular pattern of input signals.

Recursion Principle (15.11) The idea that no society, however large, can overcome every limitation—unless it has some way to reuse the same agents, over and over again, for different purposes. See *Interruption*.

Reformulation (13.1) Replacing one representation of something by another, different type of representation.

Representation (21.6) A structure that can be used as a substitute for something else, for a certain purpose, as one can use a map as a substitute for an actual city. See *Functional Definition* and *Model*.

Re-duplication Theory of Speech (22.10) My conjecture about what happens when a speaker explains an idea to a listener. A difference-engine-like process tries to construct a second copy of the idea's representation *inside the speaker's mind*. Each mental operation used in the course of that duplication process activates a corresponding *grammar-tactic* in the language-agency, and these lead to a stream of speech. This will result in communication to the extent that suitably matched "inverse grammar-tactics" construct, inside the listener's mind, an equivalent representation.

Script (13.5) A sequence of actions produced so automatically that it can be performed without disturbing the activities of many other agencies. The action script in section 21.7 accomplishes this by eliminating all the higher-level managers like *Put* and *Get*. A script-based skill tends to be inflexible because it lacks bureaucracy; one gains speed by removing higher-level anchor points but loses access to alternatives when things go wrong; script-based experts run the risk of becoming inarticulate. The book by Roger Schank and Robert Abelson, *Scripts, Goals, Plans and Understanding*, Erlbaum Associates, 1977, speculates about the human use of scripts.

Self (4.1) In this book, when written "Self," the myth that each of us contains some special part that embodies the essence of the mind. When written as "self," the word has the ordinary sense of a person's individuality. See *Single-Agent Fallacy*.

Sensor (11.1) An agent whose inputs are sensitive to stimuli that come from the world outside the brain.

Single-Agent Fallacy (4.1) The idea that a person's thought, will, decisions, and actions originate in some single center of control, instead of emerging from the activity of complex societies of processes.

Simulation (2.4) A situation in which one system mimics the behavior of another. In principle, a modern computer can be used to simulate any other kind of machine. This is important for psychology, because in the past, there was usually no way for scientists to confirm their expectations about the consequences of a complicated theory or mechanism. The theories in this book have not yet been simulated, partly because they are not specified clearly enough and partly because the older computers lacked enough capacity and speed to simulate enough agents. Such machines have recently become available; for an example, see W. Daniel Hillis's doctoral thesis, *"The Connection Machine,"* MIT Press, Cambridge, Mass., 1985.

Simulus (16.8) An illusion that a certain thing is present, caused by a process that evokes, at higher levels of the mind, a state resembling the state of mind that would be caused by that thing's actual presence. (A new word.)

Society (1.1) In this book, an organization of parts of a mind. I reserved the term "community" for referring to organizations of people because I did not want to suggest that a human mind resembles a human community in any particular way.

Society of More (10.2) The agents used by a mind to make comparisons of quantities.

Stage of Development (16.2) An episode in the growth of a mind. Chapter 17 offers several reasons why complicated systems tend to grow in sequences of episodes, rather than through processes of steady change.

State of Mind (8.4) See *Mental State*.

Structural Definition (12.4) Describing something in terms of the relationships among its parts. Contrast with *Functional Definition*.

Suppressor (27.2) A censorlike agent that works by disrupting a mental state that has already occurred. Suppressors are easier to construct than censors, and require less memory, but are much less efficient.

Time Blinking (23.3) Finding differences between two mental states by activating them in rapid succession and noticing which agents change their states. I suspect it is by using this method that our brains avoid the *duplication problem* mentioned in section 23.2. Time blinking might be one of the synchronized activities of brain cells that gives rise to "brain waves."

Trajectory (21.6) Literally, the path or route of an action or activity. However, we use this word not only for a path in space, but, by analogy, for other realms of thought. See *Pronome*.

Trans-Frame (21.3) A particular type of frame that is centered around the trajectory between two situations, one for "before" and the other for "after." The theories in this book about *Trans*-frames owe much to Roger Schank. See his book, *Conceptual Information Processing*, North-Holland, 1975.

Unconscious (17.10) A term often used, in common-sense psychology, to refer to areas of thought that are actively barred or censored against *introspection*. In this book we take "conscious" to mean aspects of our mental activity of which we are aware. But since there are very few such processes, we must consider virtually everything done by the mind to be unconscious.

Uniframe (12.3) A description designed to represent whichever common aspects of a group of things can be used to distinguish them from other things.

Will, Freedom of (30.6) The myth that human volition is based upon some third alternative to either causality or chance.

ADDITIONAL REFERENCES

Several sections of this book were adapted from my earlier publications. The discussion of mathematics in section 18.8 is based partly on "Form and Content in Computer Science," *J. Assoc. Computing Machinery*, January 1972, and partly on my introduction to *LogoWorks* by Cynthia Solomon, Margaret Minsky, Brian Harvey (eds.), McGraw-Hill, 1985. Section 2.6 is based on "Why People Think Computers Can't," in *AI Magazine*, Fall 1982. Some of chapter 30 was adapted from my essay "Matter, Mind and Models" in my book *Semantic Information Processing*, MIT Press, 1968. Some of the ideas about definitions came from my book *Computation: Finite and Infinite Machines*, Prentice-Hall, 1967. The Hogarth quotations are from *The Analysis of Beauty*, 1753. The Lavoisier quotation is from *Elements of Chemistry*, 1783.

INDEX

Bold type indicates new words, and old words used with new meanings. The names of "agents" are printed in italics.

intention, 79–88, 130, 142–43, 186, 196, 224, 233, 318; *see also* goal
interaction, 25–28, 48, 56, 91, 129, 149, 199, 289, 319–20, 327, 329
-square, 149, 249, 329
interest, 45, 120, 241, 286
interruption, 33, 135, 139, 153, 157–161, 224, 231–33, 269, 271–72, 320, 329; *see also* memory, control of; recursion
intonation, 272
introjection, 181
introspection, 29, 58, 60, 178, 196–197, 276, 329
intuition, 128, 329
Investment Principle, 146, 168, 180, 193, 222, 329
isonome, 198, 227–28, 235, 241, 248, 259, 316, 329
IT, 221, 224

jealousy, 19
Johnson, Mark, 330
Johnson, Samuel, 19, 50, 62, 113, 122, 195
joke, 19, 183, 201, 273, 278–80
as frame-shift, 278
repetition of, 281
joy, 162
Joyce, James, 108

K-line, 60, 82, 83, 84–90, 92, 158, 198, 200, 205, 211–13, 215, 245–246, 251–52, 259, 314, 329
-tree, 91, 320
temporary, 226, 240
kite, 86–90, 259, 261, 274, 293–95
knowledge, 57, 74, 92, 120
about knowledge, 100, 103
generality of, 72, 87, 89, 165, 177, 301
nature of, 23, 154, 303
organization of, 90, 142, 193, 222, 292, 302
Kuhn, Thomas, 140

Lakoff, George, 330
language, 20, 57, 66–67, 84, 88, 109, 113, 116, 131, 159–60, 220, 231–236, 241, 298, 308, 320
agency, 197, 205, 207, 223, 233–234, 266–67, 271, 295
compared with vision, 209, 232, 269
comprehension of, 84, 86, 160, 200, 207, 217, 220, 233, 234, 261–265, 269, 271–72, 274
functions of, 79, 197–98, 204, 266, 270–72
learning, 174, 184, 197–98, 242, 267, 270–71
origins of, 242, 269–70
phrase, 197, 217, 232–34, 265, 267–268
psychology of, 196, 207, 212, 231–232, 294
"re-duplication" theory, 235–36, 294, 331

sentence, 186, 208, 210, 218, 232–233, 235, 246, 261–63, 266–69, 294–95
speech, 212, 242
word order, 266–67, 268
see also grammar; meaning
lateral thinking, 145
laughter, 279–80, 314
Lavoisier, Antoine, 173
layer, 92, 107, 112, 115, 180, 215
learning, 68, 77, 92, 120, 128, 137, 168, 200, 203, 214, 258, 277, 329
attachment-based, 175
from failure, 96, 175, 183, 191, 279
predestined, 115, 179, 242, 255, 271, 311, 313, 316
to learn, 75, 80, 97, 177, 229, 309
transfer of, 106, 229
Lefshetz, Solomon, 323
left brain, 116–17, 290
legal code, 33, 50, 65, 167, 302
Lenat, Douglas, 318
level, 87, 170
of aspiration, 91
of description, 26, 292, 319
of detail, 59, 88, 90–91, 112, 121, 224
of organization, 32, 77, 174, 212, 292
visual processing, 170, 209
level-band, 86, 87, 90–92, 203, 212–213, 230, 247–48, 329
Licklider, J. C. R., 324
life, nature of, 19, 28, 30, 42, 309, 317
lifting, 128
liking, 52, 75, 94, 97
literature, 247, 298
location, 114, 129, 221, 249–51, 265
locking-in, 166, 210–11, 245, 254
logic, 39, 72, 74, 96, 100, 116, 184–189, 192, 230, 248, 277–79, 329
Look-for, 220
Lorenz, Konrad, 93
love, 37, 162, 181–83, 193, 284, 298
lying, 277, 302

machine, 63, 109, 160, 163–64, 194, 283, 288, 323
limitations of, 19, 30, 41, 63, 72, 185–86, 205
magnitude, 191, 284–85
management, 77, 80, 116, 179–80, 321
middle-level, 102–3, 220
mapping, of space, 112
mathematics, 19, 39, 65, 72, 96, 97, 110, 127, 189, 192–93, 294, 296, 301; *see also* number
maturation, 180
McCarthy, John, 323
McCulloch, Warren S., 19, 184, 323, 326
meaning, 54, 64, 67, 88, 130–31, 192, 196, 204, 206, 208, 210, 234, 236, 241, 293–94, 301
meditation, 68, 145

Melnechuk, Theodore, 47, 132, 282
memorizer, 200, 215, 252, 319, 329
memory, 21, 35, 43, 54, 60, 75, 82–89, 120, 154–56, 168, 200, 252, 288, 321, 329
amount of, 276, 316
control of, 158–59, 197, 224, 227, 230, 235, 248
duration of, 153–54, 161, 182
long term, 46, 68, 153, 156, 158, 198, 226
micro-, 157–61, 330
of memories, 89, 90, 92, 154
photographic, 153, 156
rote, 64, 223
short term, 151–60, 166, 197–98, 224–26, 230, 233–34, 257, 279–280, 320
see also amnesia of infancy; interruption
mental state, 82, 84–85, 151–53, 192, 204, 266, 280, 287–89, 306, 330–31
partial, 85, 111, 213, 215, 217, 234, 239, 330
total, 85
metaphor, 114, 122, 126, 143, 191, 231, 268, 279, 295, 298–99, 320, 327, 330
of space or time, 295
Michalski, Ryszard, 318
Micro-memory: *see* memory
microneme, 211–12, 215, 274, 317, 330
Miller, George A., 323
mind, 72, 287, 289, 290
and matter, 19, 28, 55, 287–92
nature of, 28, 159, 287–88
overlapping, 290
mind's eye, 170
Minsky, Marvin, 327, 330
mistake, 97, 125, 135, 146, 275, 277, 279, 281, 308
MIT (Massachusetts Institute of Technology), 29, 82, 259, 274, 324
Mitchell, Tom, 318
model, 181–82, 303, 330
mental, 303
of model, 304
of model of model, 305
of personality, 176
of self, 181, 303, 305
of world, 110, 256, 304
money: *see* currency
mood, 171–72
Mooers, Calvin, 214, 327
moon, 233–34
morality, 181, 184, 265–66, 302
More: see **Society-of-More**
motion, 114, 255–56, 258, 312–13
motivation, 97, 163, 168, 177, 196, 232, 287, 296, 305
mourning, 176, 182, 281
mousetightness, 28
Move, 20, 22, 29, 90, 148, 220–21, 240
music, 64, 68, 84, 97, 156, 294
mutation, 146, 310, 317
mystical experience, 65, 257

Names, 61
narrative, 265, 272, 274
Nash, John, 323
nearness, 100–12, 129, 183–84, 296
neatness, 53, 65
neme, 212, 245, 259, 266–67, 295, 330
nerve cell, 19, 25, 30, 107, 110–12
Newell, Allen, 76, 78, 259, 291
Newton, Isaac, 20, 104
No-Loss, 179
nome, 330
Non-Compromise Principle, 33, 83, 85, 102, 166, 330
nonsense, 266, 277–78
noun, 88, 232, 266–67
novelist: *see* personality
novelty, 57, 250
novice, 145
number, 30, 59, 62, 65, 99, 103, 120, 135–36, 139, 183, 192–93, 285

object, 105, 134, 199, 232, 247, 254, 268, 287, 296, 304, 311
objectivity, 27, 88, 155
obstacle, 78, 79, 147–48, 177, 222, 224, 295
odor, recognition of, 176, 311–12
Oedipus Complex, 181
ONR, 324
operant conditioning, 75
opposite, 37, 101, 117, 172, 241
orangutan, 320
Origin, 218, 222–23, 226, 230
originality: *see* creativity
ownership, 219, 293, 295

Packer, 159, 161
pain, 37, 97, 286, 312
Papert, Seymour, 29, 102, 106, 180, 202, 324, 330
Papert's Principle, 102, 180, 297, 308, 316, 320, 330
paradox, 35, 50, 65, 183–84, 192–93, 277, 304–5; *see also* circularity
parallel: *see* computer
paranome, 294, 299, 320, 330
paraphrase, 236
parent, 68, 106, 171–72, 175, 177, 179, 181–82, 242
    foster, 182
partial: *see* **mental state**
party, 259, 261–64, 296
passion, 62, 163, 172
Pauli, Wolfgang, 193
Pavlov, Ivan, 75
peer pressure, 182
Perceptron, 202, 214–15, 330
    limitations of, 202
personality, 27, 43, 45, 46, 53, 68, 155, 174, 176, 308, 314
    accumulator, 125, 184, 244, 314
    coherence of, 46, 53, 164, 174, 176, 181, 290, 294
    novelist, 26, 145, 184
    reductionist, 26, 145, 184
    uniframer, 125, 184, 244, 314
perspective, 134, 254–58, 292, 308
philosophy, 50, 65, 167, 193, 288
phobia, 183, 203

phoneme, 242, 312
phrase: *see* language
physics, 20, 26, 60, 296, 299, 322
Piaget, Jean, 19, 99, 102, 104, 106, 114, 138, 178–79, 237, 258, 316
picture-frame: *see* frame
Pitts, Walter, 19
planning, 20, 29, 34, 37, 42, 45, 67, 75, 88, 276, 296
play, 21, 29, 32–33, 36. 72, 87, 131, 270
pleasure, 37, 68, 75, 94–97, 280, 284
plot, of story, 272, 322
Poe, Edgar Allan, 23, 186
poetry, 298
point, mathematical, 110, 112
Pollack, Jordan, 210, 330
polyneme, 198, 200, 204, 208, 210–212, 223–28, 235, 246, 259, 266, 271, 295, 330
Pope, Alexander, 51
practice, 106, 120, 137
predestined learning, 104, 115, 179, 242, 255, 271, 313, 316
prediction, 29, 52, 53, 59, 62, 75, 129, 172, 177, 224, 232, 238, 245, 280, 296
    of self, 303
preposition, 218, 268
pride, 97
Princeton University, 323
priority, 83, 100–101, 166, 171, 210, 286, 298
problem solving, 29, 71, 73–75, 77, 95, 131, 144–45, 153, 160–61, 163–65, 177, 186, 222, 230–31, 236, 276, 286, 298, 318
    pessimistic strategy, 148
prodigy, 156, 328
progress, 73, 92, 161, 174
Progress Principle, 74
prohibition, 275, 278
pronome, 217–28, 230, 233, 240, 246, 251, 259, 268–71, 330
pronoun, 198, 217–18, 220, 234, 262, 268, 274
pronunciation, 235, 242
property, 128–29, 199–205, 211, 268
protein, 146, 309
proto-specialist, 163–68, 171–72, 308, 312, 314, 330
prototype, 182
Proust, Marcel, 118, 148, 170, 247
psychology, 21, 26, 50, 51, 63, 84, 120, 164, 179, 187, 204, 239, 244, 318, 322, 326
    adult, 104, 135, 171–72, 242, 280–281, 296–97
    behavioral, 75, 275–76
    of crowd, 172
    of education, 193
    of language, 198
    popular, 283
puberty, 242
punishment, 77, 96, 97, 183
purpose, 42, 88, 130, 139; *see also* goal; intention
Put, 25, 67, 87
Puzzle Principle, 73, 174, 330

quality, of sensation, 109–13, 198
    vs. quantity, 186
quantity, 100, 105, 186, 283–85; *see also* conservation of quantity
question, 233–34, 267, 303–5

rage, 171–72
RAND Corporation, 324
randomness, 73, 76, 203, 214, 306–307, 331; *see also* uncertainty
Rashevsky, Nicholas, 323
rationality, 116, 163, 184, 191
re-duplication, 235–36, 331
reading, 54
reality, 58, 59, 64, 112, 155, 199, 204, 231, 247, 251, 257, 304, 307
    sense of, 89, 91, 110
realm of thought, 51, 126, 141, 287, 292–93, 295, 298, 302, 308, 330
    formation of, 297, 311
    physical, 293–94, 296
    possessional, 293–94
    psychological, 287, 293, 297
    social, 296–97
reasoning, 100, 124, 163, 182–84, 186–87, 189–90, 197, 218, 222, 228, 230, 278–80, 302
    by analogy, 238
    from stereotypes, 248
    see also chain
receiving-agent, 214
recognition, 22, 61, 75, 89, 114, 144, 201, 244–45, 252, 274
    of thoughts, 205
recognizer, 202–4, 208, 214–15, 229, 252, 311, 319, 330
recollection, 134, 251
recursion, 58
Recursion Principle, 161, 197, 231, 268, 320, 330; *see also* interruption
reductionism, 23, 25, 26; *see also* personality
reflection, 231
reformulation, 120, 123, 133, 141, 144–45, 154, 186, 189, 191, 210, 236, 299, 331
regression, 174, 178, 308
reinforcement, 75–76, 96; *see also* reward
rejection, 176, 181, 281
relationship, 182, 201, 240, 252
religion, 41, 45, 50, 65, 127, 167, 307
remembering: *see* memory
remorse, 94
representation, 222, 224, 232, 238, 244, 248, 272, 301, 320, 327, 331
    body-support, 22, 121, 134, 139, 141–43, 146, 191
    of action, 177, 226, 230, 237
    of common sense, 72, 229
    of differences, 239, 241
    of goal, 79, 88
    of space, 134, 138, 157, 250
    of state of mind, 196, 204, 230
repression: *see* suppression
responsibility, 302, 307
Reversible, 100, 103, 179–80
reward, 68, 75–77, 94, 96–97, 174–176, 181, 284–85